Taking Stock of Industrial Ecology

Roland Clift • Angela Druckman
Editors

Taking Stock of Industrial Ecology

 **Springer** Open

Editors
Roland Clift
Centre for Environmental Strategy
University of Surrey
Guildford, UK

Angela Druckman
Centre for Environmental Strategy
University of Surrey
Guildford, UK

ISBN 978-3-319-20570-0   ISBN 978-3-319-20571-7   (eBook)
DOI 10.1007/978-3-319-20571-7

Library of Congress Control Number: 2015957425

Springer Cham Heidelberg New York Dordrecht London

Springer International Publishing AG Switzerland is part of Springer Science+Business Media (www.springer.com)

Foreword

Industrial ecology has come of age: it has its own journal, its own society and, in this volume, a first full retrospective – 'taking stock' of its first quarter of a century as a scientific field. It is a remarkable achievement.

To speak of society as having an industrial ecology would barely have been understood, as little as three decades ago. Early in the 1990s, I submitted an article for a newspaper with the phrase industrial ecology in it, only to have the editor send it back to me, corrected to industrial 'economy'. At the time, we all knew exactly what the industrial economy was (or thought we did), but clearly industrial 'ecology' could only be a typo. When I explained that it was not, the editor deemed it best to remove the term, because 'no one would understand it'. Two decades later industrial ecology is a clearer concept than industrial economy. The former even has its own Wikipedia entry; the latter, strangely, does not.

Language is a curious commodity. Its malleability appears sometimes to be almost infinite. Meanings change and mutate over time, as intellectual territory is created and destroyed. We can respond to this linguistic contortionism in several distinct ways: at least two of them are wrong.

One of the wrong ways is to suppose that the meanings embedded in terms are not just fixed but rightly so. We can spend a lot of time and energy defending the territory that language creates: defile my meanings at your peril; they are part of my identity and protect my legitimacy in the world. This is a subtly disguised variant on G. E. Moore's (1903) naturalistic fallacy: what is, is what ought to be; and woe betide offenders. The best way to avoid such an error is not to be too attached to the precision of language.

Alternatively we can celebrate the loss of meaning in late, postmodern, advanced consumer capitalism, where nothing is any longer sacred, and definition counts for naught. Accepting this fluidity of meaning, it is all too easy to allow ourselves to float above rigour and define away contestation. In fact, a cynical variation on this theme is to deliberately employ such tactics to create your own territory. Academics throughout the ages have fallen foul of this. Let us put two unfamiliar words together and build a career from it. Come on; it is easy: epigenetic precognition, categorical hyper-glaciation, collateral proto-determinism. Anyone can play. Does it matter what it all means? Probably not if it gets my papers published.

Alan Sokal (1996) famously highlighted the problem in a paper submitted to a leading sociological journal to see if they would 'publish an article liberally salted with nonsense if (a) it sounded good and (b) it flattered the editors' ideological pre-conceptions'. Sadly, the journal failed the test; they published it. But because they did, the sociological lesson still resonates: not every unfamiliar coupling of familiar words can be expected to last the course, let alone contribute to knowledge. The best way to avoid this error is to become a little more attached to the precision of language.

Science must somehow chart a course between these two positions. How should we ensure that our linguistic efforts amount to more than academic birdsong? How can we develop intellectual territory which contributes meaningfully to understanding? It seems to me that successful scientific terminology has to have three specific characteristics. First, it must resonate with the cultural context into which it falls. Second, it must have integrity, allowing its proponents to convey a coherent and articulate vision. Finally, it must express humility, showing a preparedness to extend its boundaries, change its focus and, occasionally, when no longer needed, to expire gracefully in favour of new meanings, better understandings and clearer visions. Industrial ecology must at least partly have satisfied these conditions. For otherwise, there would be nothing after 25 years to take stock of, and, as this volume shows, there clearly is.

Industrial ecology emerged at a time when detailed understandings of the ecological impacts of human activity were painfully thin. Business knew too little about their supply chains. Citizens understood too little of their footprints. Climate scientists had barely begun the extraordinary collaborative endeavour to chart the impacts and progress of anthropogenic climate change. Accounting systems, so rigorously developed for profit and loss, were woefully lacking in the raw material basis of the modern economy. Industrial ecology, or something akin to it, was clearly missing, not just in our vocabulary but in our understanding. Its emergence resonated with a real need: to understand better the complex links between industrial systems, human society and the biosphere. Industrial ecology was resonant.

Industrial ecology also puts forward a vision. From the outset the language conveyed both an idea and an ideal. The idea was that industrial systems are also, in and of themselves, a part of the natural world and not apart from it. Human systems are irrevocably entwined in natural systems. Separation is impossible. Two simple questions form the basis of material flow analysis, one of industrial ecology's most important tools. Where does it come from? Where does it go to? These two questions are amongst the most powerful tools we have in our search to understand the material connection between the economy and the earth; and this connection lies at the heart of industrial ecology's vision.

But Frosch and Gallopoulos's seminal 1989 paper in *Scientific American* went further than this. There was, in that early vision, an unashamedly normative component. Since industrial systems are inextricably connected to natural ecosystems, should they not seek to be more *like* natural ecosystems? Our economies preside over a largely linear material throughput. Raw materials are extracted from the earth, pass through the industrial metabolism (an early linguistic variant on industrial ecology) and are dumped unceremoniously into the environment afterwards, polluting our atmosphere, our oceans, our rivers and our soils. Nature appears to be more conservative than this: more circumspect in its operations, more responsive to the scarcity of available resources. Natural ecosystems tend to reuse, recycle, upcycle and, otherwise, re-employ materials, either in the same or in another ecosystem, prodigiously. Might it not be a good idea, if industrial systems were to do the same?

At first sight, this normative ideal looks suspiciously like Moore's naturalistic fallacy all over again. It is not generally advisable to argue from what one observes in nature to what ought to be. But there is a subtle difference here. There were certainly plenty of rather too linear ecosystems that collapsed entirely, their integrity damaged beyond repair, their species sometimes lost forever. What nature shows us then is not what ought to be, but what emerged through evolution as a successful adaptation in the face of scarce resources. To aim to adopt a successful strategy is not the same as accepting the naturalistic proposition. Industrial ecology's normative ideal amounts to a necessary (but not sufficient) strategy for survival.

Finally then, we come to the question of humility. Each new discipline convenes around a set of personal histories. The contributors to this volume are all serious scientists, who have dedicated their lives' work to improving our understanding of human society and perhaps its chances of survival. None of them arrived in the world as fully formed industrial ecologists – if such a person could even be said to exist. Their careers were forged in the furnace of uncertainty. I was one of those scientists. My early career was firmly anchored in the ideas that became embodied in industrial ecology. Its journal, its conferences and its society became a part of an intellectual home for me, as it did for many represented in this book. We did not always use precisely the same language, but that did not seem to matter. We did not always agree, but science is not about agreement. It is about conjectures and refutations. It is about falsifiability. Being prepared to be wrong is as important as actually being right. A degree of humility is essential for this to work; and industrial ecology, as an intellectual home, has provided for that.

As my own work evolved, I became a little more separated from the core discipline than many of those represented here. I found myself increasingly interested, first in the social and later in the macroeconomic drivers of the industrial metabolism. To its credit, far from rejecting these interests, the discipline of industrial ecology not only recognized them but sought to include them in its broader remit. The journal published my article (Jackson 2005) on sustainable consumption, for example, without ever questioning its relevance to industrial ecology. And when my work for the UK Sustainable Development Commission on prosperity and growth was published, Reid Lifset, long-serving editor of the *Journal of Industrial Ecology*, specifically insisted that I should write something on it for the journal (Jackson 2009).

I do not suppose for a minute I am alone in this experience. On the contrary, I am convinced that each of the contributors to this volume could tell a similar story. As a community of intellect, industrial ecology has been inclusive, adaptive and open to change. This is clearly one of the reasons for the success of its scientific language and the enormous relevance of its ongoing work.

Ultimately, of course, there are limits to the extent that scientific language can shift the boundaries of its own meaning. The difficult course between precision and adaptability is navigated over and over again, as society's needs also change. Circumstances alter; culture reinvents itself. The challenge for science is to respond to those changes. History is almost as replete with defunct disciplines, which failed in that task, as ecology is of extinct species. In the long run, perhaps, the arbiter of success is not longevity, but usefulness. To have survived and thrived for over 25 years is of course a remarkable achievement. But what really counts is the light that industrial ecology has shed on some of the most pressing issues of our time. This volume is a fitting testament to that success.

Guildford, Surrey, UK   Tim Jackson

References

Frosch, R. A., & Gallopoulos, N. E. (1989). Strategies for manufacturing. *Scientific American, 261*(3), 144–153.

Jackson, T. (2005). Live better by consuming less. Where is the double dividend in sustainable consumption. *Journal of Industrial Ecology, 9*(1–2), 19–36.

Jackson, T. (2009). Beyond the growth economy. *Journal of Industrial Ecology, 14*, 487–490.

Moore, G. E. (1903). *Principia ethica*. Cambridge: Cambridge University Press.

Sokal, A. (1996). Transgressing the boundaries: Towards a transformative hermeneutics of quantum gravity. *Social Text, 46/47*, 217–252.

Contents

Introduction

The Industrial Ecology Paradigm

The earliest use of the term 'industrial ecology', if not the first application of the concept, is generally agreed to be in the seminal paper by Frosch and Gallopoulos (1989), as quoted in several chapters in this book:

> The traditional model of industrial activity… should be transformed into a more integrated model: an industrial ecosystem. In such a system the consumption of energy and materials is optimized, waste generation is minimised, and the effluents of one process…serve as the raw material for another.

The International Society for Industrial Ecology has adopted the definition of industrial ecology coined by White (1994). White's definition, while building on the ideas of Frosch and Gallopoulos, introduces the role of the consumer and stresses the importance of the wider socio-economic arena:

> the study of the flows of materials and energy in industrial and consumer activities, of the effects of these flows on the environment, and of the influences of economic, political, regulatory and social factors on the flow, use and transformation of resources (White 1994).

A later definition by Allenby (2006: 33) defines industrial ecology as

> a systems-based, multidisciplinary discourse that seeks to understand emergent behaviour of complex integrated human/natural systems.

This definition highlights that industrial ecology takes a whole systems approach and also that it involves many disciplines – not just the technical, economic and environmental fields but also fields such as sociology and philosophy, ethical philosophy in particular.

The term industrial ecology draws an analogy between industrial systems and natural ecosystems and is founded upon the suggestion that understanding and applying what can be learnt from natural systems will help us design more sustainable industrial systems (Graedel and Allenby 1995; Ehrenfeld 1997). Whether natural ecosystems are really a helpful model for industrial ecology and whether the term 'industrial metabolism' would be more appropriate have been much discussed.

But the key concern behind the rise of industrial ecology is the acceptance that the way human activities are using, and using up, the planet's resources cannot continue unchecked: we (i.e. human society and our economy) must change to become sustainable. Part of industrial ecology is concerned with analysing economic systems to identify where unsustainability originates, but, as in the original statement by Frosch and Gallopoulos, this necessarily leads to suggestions on how the system should be changed. These two sides of industrial ecology – analytical and prospective/design oriented (in the broadest sense) – are illustrated by the chapters in this book.

A basic premise of industrial ecology is that it is concerned with sustainability. This demands an articulation of what we mean by *sustainability* and *sustainable development*, given that the terms are 'contested' (and arguably have been diluted by loose usage). We start by pointing out that *development*, as in *sustainable development*, is not to be equated with economic 'development' in the sense of increasing *per capita* GDP or disposable income. The Brandt Commission (1980) pointed out firmly that:

> One must avoid the persistent confusion of growth with development, and we strongly emphasise that the prime objective of development is to lead to self-fulfilment and creative partnership in the use of a nation's productive forces and its full human potential.

There is a body of literature exploring this interpretation and the related question of how happiness and quality of life can best be promoted. Although this question is only briefly touched upon in the chapters in this book, the point clearly articulated by the Brandt Commission underlies many of the chapters.

Jackson (2010) has offered one of the most succinct definitions of sustainability:

> Sustainability is the art of living well, within the ecological limits of a finite planet.

Here 'living well' is to be interpreted in two senses. First, it means living prosperously – with a decent level of material comfort, security and dignity. Second, it has a moral sense – not living at the expense of the well-being of others – and thus feeling your life is good in ethical terms.

Recognition of ecological limits is fundamental: if it were possible to expand economic activity without limits, sustainability of development would not be a concern. A widespread interpretation of sustainability, which is implicit in many of the chapters in this book, recognises three dimensions or types of constraints. This is shown schematically in Fig. 1 in the form of a Venn diagram in which each lobe represents a possible operating space bounded by constraints, which may be 'hard' or 'soft'. 'Techno-economic efficiency' represents the ranges of activities available to us, limited by our technical skills and ingenuity, by the laws of thermodynamics and by the need to be efficient as defined by the prevailing economic system. An important point, picked up in several chapters in this book, is that that the laws of thermodynamics are 'hard-wired' into the universe, whereas the 'laws' of economics are human constructs and therefore mutable, for example by changes to the fiscal system. 'Environmental compatibility' represents the range of activities which can

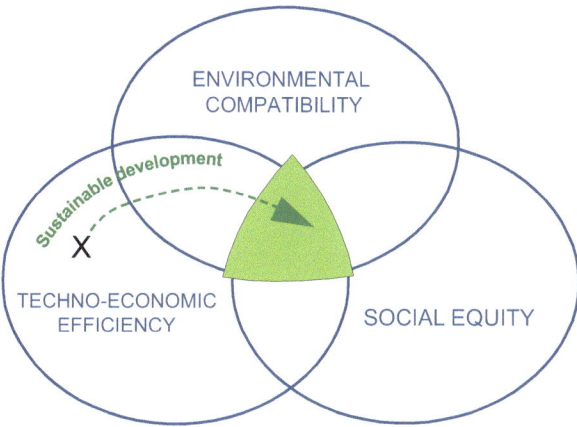

Fig. 1 Sustainability and sustainable development (adapted from Clift 1995 and Clift et al. 2013)

be pursued indefinitely within the resource and carrying capacity of the planet. A promising approach to defining and perhaps quantifying this space has been proposed by Rockström et al. (2009) who articulated the idea of 'a safe operating space for humanity' lying within 'Planetary Boundaries'. 'Social equity' represents the ethical imperative implicit in the original Brundtland articulation of the idea of sustainable development (WCED 1987) as development that:

> meets the needs of the present without compromising the ability of future generations to meet their needs.

This idea of sustainability has been summed up in some policy statements (e.g. DETR 1999) along the lines of

> the simple idea of ensuring a better quality of life for everyone, now and for generations to come.

A sustainable future must lie within all three of the operating spaces in Fig. 1. Thus 'sustainable' living is represented by the region at the centre of the figure. While the current human economy generally operates within the Techno-economic Efficiency lobe, as indicated by point X, it obviously does not lie within either of the other spaces; for example, Rockström et al. (2009) have identified domains in which planetary boundaries are already exceeded, while obvious disparities in quality of life between different populations and regions show that the global economy does not operate with the 'Social equity' lobe. 'Sustainable development' is then represented by a trajectory moving from present practice to the 'sustainable region'. Industrial ecology aspires to guide this trajectory.

Some of the discussion around sustainable living has been framed in terms of the so-called IPAT equation (Ehrlich and Holdren 1971), and several chapters in this book refer to it. The IPAT equation is actually an expression of a conceptual relationship rather than a formal equation

$$I = P \times A \times T$$

where I represents the impact of human activities on the environment, P is the human population, A is a measure of affluence (usually interpreted as average *per capita* GDP) and T is a measure of technological efficiency of consumption:

$$T = (Environmental\, impact) / (Unit\, of\, GDP).$$

The IPAT relationship in this simplistic form was originally deployed to frame discussions over the extent to which technological advances could offset growth in population and affluence; for example, 'clean technology' is conceived as an approach to reducing the T term (see Clift and Longley 1994). However, the recognition that well-being or quality of life is distinct from consumption as measured by GDP has led to a different discussion in industrial ecology and other circles: how can a good quality of life, the focus of the *Social equity* lobe in Fig. 1.1, be maintained within the constraints represented by the *Environmental compatibility* lobe? This means questioning whether A really needs to increase (see, e.g. Jackson 2009), as well as how far P will increase and whether reducing T can offset these increases.

Like White's (1994) definition of industrial ecology, the IPAT relationship is generally conceived in terms of consumption flows – GDP – but the debate in industrial ecology increasingly refers to the importance of social, environmental and manufactured stock. The distinction between *stocks* and *flows* is therefore important; many of the chapters in this book focus on it. The terminology comes from economics, but the concepts underlie many other disciplines, including most branches of engineering. A flow variable is one that has a time dimension or flows over time, like the flow in a stream. A stock variable measures a quantity at a particular instant, like the quantity of water in a lake. An individual item in the stock may have a limited life (or, in engineering terms, a limited 'residence time'). Income is a flow; wealth is a stock. A fleet of vehicles is a stock, but the fuel as it is burnt and the rate at which new vehicles enter service are both flows. A consumer making a purchase drives production, which is a flow; the object purchased then joins the stock of articles, such as appliances and clothing, in the consumer's home.

This Book

The motivation for this book, published to coincide with the 8th biennial conference of the International Society for Industrial Ecology in July 2015, was to review how industrial ecology has developed over the last 25 years; to provide some examples of where industrial ecology thinking has 'made a difference' in practice, strategy and policy; and to set out some of the current challenges in industrial ecology and how this new grouping of disciplines might develop in the future. The intention is to provide an introduction which will be valuable for students and other newcomers, but also a reference source for those already familiar with aspects of industrial ecology. Part I comprises a set of chapters which explore current thinking and practice in different aspects of industrial ecology and indicate how this thinking and application is likely to develop in the future. Part II gives a selection of areas where industrial ecology thinking has been applied, to provide valuable insights and hence to suggest practical actions and policies.

The book is not necessarily intended to be read from start to finish. Experienced industrial ecologists will be familiar with many of the concepts and should find the book useful to dip in and out of, according to their current needs and interests. We suggest that the reader who is new to industrial ecology should start with Chap. 1, in which Tom Graedel and Reid Lifset set out the history of the developments in North America which led to the Journal of Industrial Ecology (JIE) and the formation of the International Society for Industrial Ecology (ISIE). In Europe, the origins of the industrial ecology community are somewhat different: ISIE drew in a group of researchers and practitioners who had been involved primarily with Life Cycle Assessment (LCA) and Material Flow Accounting (MFA) but who recognised that they needed a new scientific and professional body with a broader scope. It is significant that the first conference of the International Society for Industrial Ecology, in 2001, was held in the Netherlands and organised by researchers at the University of Leiden best known for their work in LCA but who now thought of themselves as industrial ecologists.

Subsequent chapters in Part I discuss specific topics in industrial ecology. In Chap. 2, which exemplifies the 'whole-system' approach to stocks and flows, Stefan Pauliuk and Edgar Hertwich show how the basic principles of industrial ecology, such as taking a life cycle approach and stipulating mass balance consistency, are now being applied in dynamic, forward-looking (prospective) models to study the potential effect of sustainable development strategies at full scale. They explain how these newer models are related to more traditional techniques employed in industrial ecology such as LCA (see Guinée, Chap. 3), MFA (see Moriguchi and Hashimoto, Chap. 12) and Input-output Analysis (see Wiedmann, Chap. 8). Pauliuk and Hertwich also discuss the relationship between new prospective models and conventional Integrated Assessment Models (IAMs). IAMs are not the subject of a chapter of this book as, while they have, arguably, a more comprehensive scope, they lack the robust scientific basis fundamental to industrial ecology. However, Pauliuk and Hertwich see the integration of the core concepts of industrial ecology into IAMs as a promising approach to enable enhanced understanding of society's future metabolism.

From the first recognition of industrial ecology as a way of thinking, Life Cycle Assessment (LCA) has been one of its basic tools. LCA is concerned with identifying the impacts of a supply chain leading to delivery of a product or service. LCA started with industrial applications, only subsequently developing into a way of thinking which engaged the attention of the academic community. In Chap. 3, Jeroen Guinée explores one of the current challenges for LCA: how to develop the tool from its traditional role in identifying and quantifying environmental impacts to covering all three of the lobes in Fig. 1, i.e. to provide a tool to assess the sustainability of a product or service.

In Chap. 4, Chris Kennedy tracks the growing prominence of studies of cities in the industrial ecology literature. Kennedy conceptualises that urban metabolism studies are 'scale-delineated' components of wider socio-economic metabolism studies, and he sees two main challenges for future work. One is to address questions such as what are the limits to efficiency gains that can be achieved, for example, through closing material loops? This relates to the circular economy debate (see Chaps. 7 (Stahel & Clift) and 13 (Hill)). The other, related, challenge concerns industrial symbiosis, which is the subject of Chap. 5 (Chertow & Park) and Chap. 19 (Bailey & Gadd), and here Kennedy sees a need to explore how much potential there is for industrial symbiosis in cities, with a focus on the limitations of urban-scale agglomeration effects.

In Chap. 5, Marian Chertow and Jooyoung Park focus on industrial symbiosis, one of the areas within industrial ecology that derives most obviously from the analogy with natural ecosystems. They describe this as a subfield of industrial ecology which *engages traditionally separate industries and entities in a collaborative approach to resource sharing that benefits both the environment and the economy*. They start the chapter with a description of Kalundborg, the most famous and often cited example of industrial symbiosis, and the rest of their chapter charts the progress of industrial symbiosis worldwide. They conclude that while industrial symbiosis has achieved a great deal in the last quarter century, much remains to be done to understand the levels of material exchanges, institutional contexts, cultural changes and people-to-people collaborations that will enable industrial symbiosis to flourish worldwide.

Chapter 6, by Tim Baynes and Daniel Möller, again picks up the theme of stocks and flows in a whole-system perspective. Readers relatively new to industrial ecology are advised to read Chap. 4 (Urban metabolism) before embarking on Chap. 6. Increasing concentration of people in urban centres means that massive investment is needed in urban infrastructure (i.e. stock) in the coming decades, particularly in the 'global South'. This will require enormous flows of materials, and there are serious questions over how this can be accommodated within the Planetary Boundaries. Once the infrastructure is in place, society is 'locked in' to patterns of behaviour and economic activity. Baynes and Möller explore how the industrial ecology approach, drawing on the modelling approaches in Chap. 2, can lead to models of socio-economic metabolism which guide sustainable development by ensuring that the time perspective is long term.

The importance of the lifetime of stock is also one of the themes in Chap. 7 by Walter Stahel and Roland Clift. The idea of a 'circular economy', involving repeated rather than 'once through' use of materials, is currently enjoying widespread discussion; indeed, it has been one of the central ideas in industrial ecology from the outset. Stahel and Clift explore how focussing on the use of stocks gives insights which go beyond looking for circular flows, leading to the idea of a 'performance economy' which focuses on maintenance and exploitation of stock. The performance economy emphasises business models directed at providing services rather than selling material products and identifies reuse and remanufacturing as key parts of the economy. This represents a realisation of industrial ecology's aspiration of rethinking the economy and, in this case, exploring changes to the fiscal regime which would promote more resource-efficient economies.

In the next chapter, Chap. 8, Thomas Wiedmann considers the environmental and social impacts of the increase in global trade experienced during recent decades. He discusses how, for example, today's longer and increasingly complex global supply chains make it more difficult to connect the environmental impacts of production to the final consumer of the goods and services. In addition to discussing developments in analytical tools and data requirements, Wiedmann also discusses the broadening of indicators. He reviews how studies now extend beyond the traditional analysis of carbon emissions, to include other environmental indicators such as land and water use and, importantly, social indicators such as child labour, which extend the analysis into the 'Social equity' lobe of Fig. 1.

The chapter by Angela Druckman and Tim Jackson (9) follows neatly from Thomas Wiedmann's Chap. (8), as it takes the concept of 'footprinting', which Wiedmann applies to trade, and applies it to household consumption. Both approaches are based on consumption perspective accounting, which uses whole systems, life cycle thinking to attribute all impacts along supply chains to final consumers. Druckman and Jackson review the determinants of the carbon footprints of Western households, the composition of average household footprints, and also discuss the rebound effect. They conclude that while it is vital to address the systems of provision of food, energy, housing and transportation, structural income growth presents a key challenge and that we should seek solutions that provide low carbon lifestyles with high levels of well-being. Once again, 'development' must not to be equated with economic growth.

In Chap. 10 Marlyne Sahakian picks up on the importance of the structure of the economy and the well-being it provides by introducing the concept of the social and solidarity economy, fully embracing the 'Social equity' lobe in Fig. 1. As noted above, a high quality of life is the objective, not just economic activity as measured by GDP. The concept is people centric: it aims to place service to communities ahead of profit and embraces the notion of reciprocity. Examples of activities of the social and solidarity economy include social entrepreneurship, crowd funding, fair trade, community currencies and some forms of peer-to-peer sharing. Many of these activities are niche: they are often at the grassroots level operating at the margins of the dominant capitalist economy. Hence, to make progress, the issue of scale is, she says, the greatest challenge. Importantly for this book, Sahakian analyses the linkages between the social and solidarity economy and industrial ecology and finds many synergies. She also, however, notes a tension that is a challenge for the future, whereby industrial ecology generally prioritises the environment, whereas the social and solidarity economy prioritises people. In the terms of Fig. 1, a genuine balance must be found between the 'Environmental compatibility' and 'Social equity' lobes.

In Chap. 11, Megha Shenoy focuses attention on developing countries – the 'global South' – and so returns to the original focus of sustainable development. The holistic approach of industrial ecology can provide a valuable platform on which to develop strategies and policies for sustainable development. In turn, the development of industrial ecology thinking should be informed by experience in the developing world. Efforts in the global South have primarily focussed on cleaner production and eco-industrial parks (an application of the industrial symbiosis approach explored in Chap. 5), but awareness of the long-term socio-economic metabolism effects discussed in Chap. 6 is essential. Lack of essential data to inform policy is a common problem, and Shenoy suggests that industrial ecology thinking should be able to find simplified ways to collect and present data on economic and environmental performance.

Along with life cycle assessment, material flow analysis (MFA) is an essential tool in the industrial ecologist's kit. In Chap. 12, Yuichi Moriguchi and Seiji Hashimoto outline the development of MFA since the pioneering work of Robert Ayres and go on to show how MFA is already used to support waste management and recycling policy. They describe how in Japan, in particular, MFA has always been seen as one of the tools guiding policy for a circular economy. International efforts are needed to standardise compilation of MFA data. Input-output analysis, the approach mainly discussed in Chap. 8, is increasingly used as a way to estimate and represent material flows.

Following the general essays in Part I outlined above, Part II contains contributions on more specific areas and applications of industrial ecology.

The 'circular economy' is an increasingly popular concept with policy-makers and industrialists. Despite its flaws, discussed in Chap. 7, it is an approach which enables companies to implement some of the principles that are at the heart of industrial ecology. As such, the circular economy has been adopted as a model by countries as far apart as China (see Chaps. 5 and 12) and the nations of the European Union (EU). To demonstrate how EU policies are implemented in practice, in Chap. 13, Julie Hill discusses how the four administrations that make up the UK (England, Scotland, Wales and Northern Ireland) have implemented EU policies on the circular economy.

In Chap. 14, Paulo Ferrão, António Lorena and Paulo Ribeiro set out how one of the most direct applications of industrial ecology in public policy – Portugal's National Waste Management Plan – was developed in partnership between the Portuguese Environment Agency and the Instituto Superior Técnico of the University of Lisbon. The need for a life cycle approach to underpin waste management policies was recognised from the outset, leading to promotion of a circular economy to contribute to increasing resource efficiency. The benefits include reductions in the quantity of waste sent to landfill, increase in recycling of solid waste due to selective collection and more efficient recovery and treatment, leading to almost halving the GHG emissions associated with waste management. This is a benchmark example of what can be achieved by cooperation between government, academia and the private sector in applying industrial ecology thinking.

In Chap. 15, Sarah Sim, Henry King and Edward Price explain how scientific analysis is used to guide strategy in a major multinational company in the 'fast-moving consumer goods' (FMCG) sector. Unilever has a track record as a leader in applying life cycle thinking – LCA has been used for the design of many products, and the company's 'Sustainable Living Plan' includes extensive product footprinting. Sim and her colleagues foresee that the company will continue to use life cycle approaches but recognise the need to develop the framework further and to recognise and operate within absolute sustainability limits rather than working to achieve incremental improvements. This places their approach squarely within the framework of Fig. 1. They see the planetary boundaries approach initiated by Rockström et al. (2009) as the basis for conceptual developments and practical actions, to be developed by commercial organisations working with the academic industrial ecology community.

Chapter 16 by Kieren Mayers complements Hill's Chap. (13) by looking at the practical implications of product-based environmental legislation. Mayers discusses how producers address regulations concerning use of hazardous substances in their supply chains, energy efficiency of products during use and the management of products at end of life. In this chapter Mayers draws on his wide experience of working at Sony Computer Entertainment and introduces the reader to the reality of the challenges faced in the industry. He gives a frank assessment of the progress that legislation has made towards reducing the environmental impacts of products across their life cycles and stresses that producers are faced with administratively and logistically complex challenges.

Chapter 17, by Kirstie McIntyre and John A. Ortiz, returns to the theme of the circular economy discussed by Stahel and Clift in Chap. 7, and for which Hill set out the policy landscape with reference to the UK in Chap. 13. McIntyre and Ortiz discuss how a multinational corporation, Hewlett Packard, operationalise the circular economy in their global supply chains. They highlight the importance of placing customers centrally in the system and of viewing them as users rather than consumers. This links with the concept of the 'sharing' economy and the social and solidarity economy as discussed by Sahakian in Chap. 10.

In Chap. 18, Roland Geyer looks at the automobile through the lens of industrial ecology. Increasingly, decision-makers are looking for a new paradigm for the service of personal mobility. The literature is voluminous on how to make cars less unsustainable and whether the internal combustion engine should be supplanted by other technologies, including batteries and fuel cells. Lightweighting and end-of-life recycling are further approaches to reducing the impacts of automobiles. The basic question, which Geyer uses as his framework to assess their potential, is the extent to which possible new technologies would reduce rather than just shift the environmental impacts of personal transport.

Finally, in Chap. 19, Malcolm Bailey and Andrew Gadd review the practical application of industrial symbiosis (see Chertow, Chap. 5) in a major industrial cluster, at Humberside in North East England. The scope for industrial symbiosis applied to existing plants has been exploited, by both 'top-down' and 'bottom-up' initiatives, and the savings in generation of waste and greenhouse gases are impressive. Even larger potential savings have been identified, but they would require symbiotic interrelationships to be designed in, with installations co-located in eco-industrial parks. Bailey and Gadd raise some practical engineering problems that need to be addressed systematically, complementing the less technological issues raised by Chertow and Park in Chap. 5. This returns us to one of the themes in industrial ecology: the need for an economic and financial system which aligns economic performance and environmental performance.

The deadline imposed by having this book available at the International Society for Industrial Ecology's 2015 conference has meant that some topics we would have liked to include are not represented. Complex systems are obviously relevant to industrial ecology, but at least there are many good introductory texts on complex systems. As will be obvious through some of the chapters of the book (see, e.g. Chaps. 8, 9, 10 and 17), the social sciences are increasingly interested in and of interest to industrial ecology. Some of this work, e.g. Fischer-Kowalski and Haberl (2007), overlaps with another emerging field – ecological economics – but there is a growing interest in what the social sciences can contribute to and learn from industrial ecology. For this we refer the reader to Boons and Howard-Grenville (2009) and to Baumann et al. (forthcoming).

We hope that new readers and experienced industrial ecologists alike will find this a useful and interesting volume that takes stock of where industrial ecology has got to a quarter of a century after Frosch and Gallopoulos first introduced the term and that also provides insights into the challenges ahead.

Guildford, Surrey, UK   Roland Clift and Angela Druckman

Acknowledgement We would like to thank Linda Gessner for her help in editing and producing the book.

References

Allenby, B. (2006). The ontologies of industrial ecology? *Progress in Industrial Ecology – An International Journal, 3*(1/2), 28–40.

Baumann, H., Brunklaus, B., Lindqvist, M., Arvidsson, R., Nilsson-Linden, H., & Hildenbrand, H. (forthcoming). *Life cycle methods for management – How to analyze the social and organisational dimensions of product chains*. Cheltenham: Edward Elgar.

Boons, F., & Howard-Grenville, J. A. (Eds.). (2009). *The social embeddedness of industrial ecology*. Cheltenham: Edward Elgar.

Brandt, W. (1980). *North-south: A programme for survival. Report of the Brandt Commission* (p. 23). London: Pan Books.

Clift, R. (1995). The challenge for manufacturing. In J. McQuaid (Ed.), *Engineering for sustainable development* (pp. 82–87). London: Royal Academy of Engineering.

Clift, R., & Longley, A. J. (1994). Introduction to clean technology. In R. C. Kirkwood & A. J. Longley (Eds.), *Clean technology and the environment* (pp. 174–198). Glasgow: Blackie.

Clift, R., Sim, S., & Sinclair, P. (2013). Sustainable consumption and production: Quality, luxury and supply chain equity. In I. S. Jawahir, Y. Huang, & S. K. Sikdar (Eds.), *Treatise on sustainability science and engineering* (pp. 291–309). Dordrecht: Springer Netherlands.

DETR. (1999). *A better quality of life: A strategy for sustainable development for the United Kingdom*. London: (UK) Department of the Environment, Transport and the Regions.

Ehrenfeld, J. R. (1997). Industrial ecology: A framework for product and process design. *Journal of Cleaner Production, 5*, 87–95.

Ehrlich, P. R., & Holdren, J. P. (1971). Impact of population growth. *Science, 171*(3977), 1212–1217. doi:10.1126/science.171.3977.1212.

Fischer-Kowalski, M., & Haberl, H. (Eds.). (2007). *Socioecological transitions and global change – Trajectories of social metabolism and land use*. Cheltenham: Edward Elgar.

Frosch, R. A., & Gallopoulos, N. E. (1989). Strategies for manufacturing. *Scientific American, 261*(3), 144–153.

Graedel, T. E., & Allenby, B. R. (1995). *Industrial ecology*. Englewood Cliffs: Prentice Hall.

Jackson, T. (2009). *Prosperity without growth: Economics for a finite planet*. London: Earthscan/Routledge.

Jackson, T. (2010). Keeping out the giraffes. In A. Tickell (Ed.), *Long horizons* (p. 20). London: British Council.

Rockström, J., et al. (2009). A safe operating space for humanity. *Nature, 461*, 472–475.

WCED. (1987). *Our common future – Report of the World Commission on Environment and Development*. Oxford: Oxford University Press.

White, R. (1994). Preface. In B. Allenby & D. Richards (Eds.), *The greening of industrial ecosystems*. Washington, DC: National Academy Press.

Part I
State-of-the-Art and
Discussions of Research Issues

Chapter 1
Industrial Ecology's First Decade

T.E. Graedel and R.J. Lifset

Abstract Industrial ecology can be said to have begun with a 1989 seminal publication entitled "Strategies for Manufacturing." During the next decade, the field was initially defined and developed by researchers in industry and elsewhere who saw the opportunity for improving corporate and governmental performance related to the environment and sustainability. They introduced design for environment, industrial symbiosis, and resource use and loss assessments at national and global levels and enhanced the embryonic specialty of life-cycle assessment. In the same decade, industrial ecology became widely recognized as a scholarly specialty, with its own journals and conferences. This chapter reviews industrial ecology's emergence and evolution, largely from a North American perspective, with emphasis on the field's lesser-known first decade.

Keywords Emerging discipline • Evolution of industrial ecology • History of industrial ecology • International society for industrial ecology • Journal of industrial ecology

1 Origins of Industrial Ecology

The 1972 United Nations Conference on the Human Environment in Stockholm is often seen as milestone in the emergence of a global environmental movement. The declaration arising from that conference included twenty-six principles, including several that resonate with what we now know as industrial ecology:

Principle 2: The natural resources of Earth… must be safeguarded for the benefit of present and future generations.

Principle 5: The nonrenewable resources of Earth must be employed in such a way as to guard against the danger of their future exhaustion.

T.E. Graedel (✉) • R.J. Lifset
Center for Industrial Ecology, Yale University, 195 Prospect St., New Haven, CT 06511, USA
e-mail: thomas.graedel@yale.edu

© The Author(s) 2016
R. Clift, A. Druckman (eds.), *Taking Stock of Industrial Ecology*,
DOI 10.1007/978-3-319-20571-7_1

Principle 6: The discharge of toxic substances… in such quantities or concentrations as to exceed the capacity of the environment to render them harmless must be halted.

Slightly preceding the Stockholm conference, however, was the establishment of the US Environmental Protection Agency (EPA) in 1970. Over the next several decades, the EPA developed air and water pollution control activities seeking to achieve the sorts of goals articulated in Stockholm's Principle 6, creating regulatory oversight of emissions and the implications of those emissions on human health. The preservation of resources entered the picture with the widely read book *Limits to Growth* (Meadows et al. 1972), but this issue, with the exception of oil, received only passing attention in the 1970s, both by governments and corporations.

Erkman (1997) has demonstrated that several intellectual threads that eventually became part of industrial ecology were under development in the 1970s and 1980s: the concept of industry as an ecosystem, the quantification of material and energy flows, and the relationships of technology to the general economy. Japan, in particular, moved during this time toward using advanced technology to limit its demands for materials and energy (e.g., Watanabe 1972; MITI 1988). This approach embedded industrial ecology thinking in industry to a greater degree than existed elsewhere at that time, a distinction that to some degree remains true today.

Almost 30 years after the 1972 Stockholm conference, Robert Frosch and Nicholas Gallopoulos of the General Motors Research Laboratory published a paper with the modest title "Strategies for Manufacturing." In this paper, Frosch and Gallopoulos (1989) discussed the environmental impacts of manufacturing, speculated that resource depletion and waste accumulation would be challenges in the coming years, and provided an innovative approach to address these issues:

> The traditional model of industrial activity … should be transformed into a more integrated model: an industrial ecosystem. In such a system the consumption of energy and materials is optimized, waste generation is minimized, and the effluents of one process … serve as the raw material for another.

With these words, Frosch and Gallopoulos inaugurated the field of industrial ecology.

2 Constructing the Field of Industrial Ecology

Frosch and Gallopoulos were employees of General Motors, rather than university researchers, and were advocating an environmental ethic that went beyond complying with existing regulations. Their call to action was soon recognized as acknowledging what some corporations were already doing, or upon which they were soon to embark. Frosch and Gallopoulos were not the only ones thinking along these lines but were among the first to put a public face on these efforts. An admittedly incomplete list of some of the most active corporations during the seminal period 1988–1996 includes Volvo (1991; Horkeby 1997), 3M (Holusha 1991), BMW (Holusha

1991), Xerox (Murray 1993; Azar et al. 1995), Procter & Gamble (Pittinger et al. 1993), Pitney Bowes (Ryberg 1993), AT&T (Allenby 1994), Motorola (Hoffman 1995, 1997), IBM (Bendz 1993; Kirby and Pitts 1994), Hewlett-Packard (Bast 1994), Philips (Boks et al. 1996; Stevels 2001, 2009), and Bosch (Klausner et al. 1998). The mantle was also taken up by industrial associations, particularly the electronics industry (Microelectronics and Computer Technology Corporation 1994; Sony (Scheidt and Stadlbauer 1996); and NEC (Suga et al. 1996)).

It is fair to say that while some corporate initiatives were inspired by altruism to a significant degree, those who were involved had other motives as well: simplification of assembly and disassembly of products (Lundgren et al. 1994), reuse of material resources (Porada 1994), and recovery and recycling of components (Azar et al. 1995; Nagel 1997), among others. One of the most dramatic corporate initiatives in the early years of industrial ecology was that of Volvo, which worked with Swedish academic and governmental organizations to produce one of the first workable versions of life-cycle impact assessment (Steen and Ryding 1992) and then used the results to influence the design of Volvo products (Horkeby 1997).

Most of the early corporate and governmental initiatives related to industrial ecology were uncoordinated and ad hoc. This situation began to change with a conference held in 1991 at the US National Academy of Sciences (Patel 1992). Attendees at that meeting began the process of identifying what topics should be included in an industrial ecology framework (material cycles, energy efficiency, input–output analysis, etc.). A 1992 conference in Colorado (Socolow et al. 1994) expanded that framework to incorporate human impacts on natural cycles, IE in manufacturing, and IE in policy-making. In subsequent years, the field has proceeded in fairly straightforward fashion from those foundations. That conference also provided the name for the field. As Socolow (1994) describes in the introduction to the book that came from the conference (Socolow et al. 1994), the choice was between "industrial ecology" and "industrial metabolism" (Ayres 1989). As conference chair, Socolow chose the former as being the more encompassing of the two options and one that brought such ecological topics as food chains and resource reuse into the discussion. Perhaps largely because conference attendees included a number of those who went on to research and write about this area of study, Socolow's choice stuck. Nonetheless, metabolism has remained an important concept and analogy in the field (e.g., Octave and Thomas 2009; Gierlinger and Krausmann 2011), providing a rich and growing framework for much of the material flow analysis that is central to industrial ecology.

Governments joined the industrial ecology effort soon after its identification by corporations (e.g., MITI 1988; Office of Technology Assessment 1992; U.S. Environmental Protection Agency 1995). Nonetheless, it is noteworthy that, unlike many other fields of study, the origins of much of industrial ecology lay not in academia but in industry. IE is today regarded as an academic specialty, but it continues to rest on the foundation developed and practiced by industry and, to some extent, by governments.

Industrial ecology was thus becoming a recognized field in the mid-1990s, but what exactly *was* industrial ecology? An early definition by the president of the US

National Academy of Engineering attempted to encapsulate what the concept was all about:

> Industrial ecology is the study of the flows of materials and energy in industrial and consumer activities, of the effects of those flows on the environment, and of the influences of economic, political, regulatory, and social factors on the flow, use, and transformation of resources. (White 1994)

This definition remains, 20 years later, as a reasonably good synopsis of the field. However, an alternative and more expansive definition was provided a year later:

> Industrial ecology is the means by which humanity can deliberately and rationally approach and maintain sustainability, given continued economic, cultural, and technological evolution. The concept requires that an industrial system be viewed not in isolation from its surrounding systems, but in concert with them. It is a systems view in which one seeks to optimize the total materials cycle from virgin material, to finished material, to component, to product, to obsolete product, and to ultimate disposal. Factors to be optimized include resources, energy, and capital. (Graedel and Allenby 1995: 9)

This second definition extends the field outward from a solely industrial focus to a more societal one and introduces the issue of sustainability. In the twenty-first century, this enhanced concept has strongly influenced the way industrial ecology is practiced. In fact, a recent "sound bite" definition of industrial ecology, "Industrial ecology is the science behind sustainability," (Makov 2014) almost bypasses the industrial focus in the interest of a planetary focus.

Regardless of which definition a particular individual may prefer, a few key words appear to indicate the scope and focus of the field: industry, environment, resources, life cycle, loop closing, metabolism, systems, and sustainability.

3   Building the Tools of the Trade, 1990–2000

3.1   Life-Cycle Assessment

Life-cycle assessment (LCA) is the methodology that seeks to identify the environmental impacts of a product or process at each stage of its life cycle. Analytical efforts to quantify emissions and resource loss on a life-cycle basis date from the 1970s (e.g., Bousted 1972; Hunt and Welch 1972), but LCA's rapid growth and its close relationship with industrial ecology began about 1990, especially in Sweden (Steen and Ryding, 1992), and it first became codified in a 1993 handbook (Heijungs et al. 1992). Klöpffer (2006) has reviewed the key role of the Society for Environmental Toxicology and Chemistry (SETAC) in the early development of LCA. In Europe, a spur for the development of a standard methodology came from the adoption of LCA as the basis for product labeling (Clift et al. 1994). While the need for further development of the methodology was widely recognized (e.g., Field et al. 1993), adoption of LCA as an industrial ecology tool became increasingly widespread, both in industry and government (Harsch et al. 1996; Matsuno et al. 1998; Itsubo et al. 2000).

A lively community of methodology developers and practitioners emerged, and LCA benefited from database development, standards setting, and creation of software. However, some potential users (especially in industry) found the methodology too complex and contested to be workable on a routine basis. This led many to work with the LCA consulting industry that sprang up to respond to a demonstrated need and the increasing availability of LCA software. An alternative approach was to "streamline" LCA (e.g., Graedel et al. 1995; Weitz et al. 1995; Hoffman 1997; Christiansen 1997), and streamlined LCA (SLCA) has since been used in various forms throughout industry. As a consequence of these initiatives, LCA and SLCA activities in industry are much more significant than might be inferred by an outside observer.

In 2002, a new LCA guide addressed in detail many of the issues that had caused concern in the past (Guinée 2002). However, unresolved problems remained, as pointed out by Reap et al. (2008a, b). LCA remains today in the interesting position of being viewed as still in development as an academic tool but widely employed in industry. It will doubtlessly continue to undergo further development, as it continues to provide important perspectives on industrial product and process design activities.

3.2   Design for Environment

The recovery and reuse of a variety of "industrial resources" was rather common early in the twentieth century (Desrochers 2000) but became more challenging as materials, components, and products became increasingly complex and as resources appeared abundant. However, in the late 1980s, a number of corporations began to rethink their product design processes, especially as those processes related to recycling or resource loss (Henstock 1988). The result was methods that looked beyond product performance, appearance, and price to attributes such as efficient manufacturing, fewer parts suppliers, and less inventory (Watson et al. 1990). From that perspective, it was an easy step to consider environmental factors such as minimizing energy requirements, decreasing discards from manufacturing, choosing more sustainable materials, and the like (e.g., Hamilton and Michael 1992; Kirby and Pitts 1994; Azar et al. 1995; Sheng et al. 1995). Among several related books, the 1996 volume *Design for Environment* (Graedel and Allenby 1996) stimulated interest among industrial design groups throughout the world (e.g., Klausner et al. 1998; Stevels 2001). Aspects of disassembly, remanufacture, and recycling, widely discussed in the 1990s, have continued to be emphasized (Cândido et al. 2011; Go et al. 2011; Hatcher et al. 2011; Ryan 2014).

Design for environment is becoming increasingly embedded in both the educational and industrial aspects of product design. Perhaps the best evidence for this is the broad acceptance of the 2009 book *Materials and the Environment: Eco-Informed Materials Choice*, by Cambridge University engineering professor Michael Ashby (Ashby 2009). This volume is widely used in undergraduate education and in the industrial design sector, an achievement that is perhaps one of the more significant (if not the most visible) contributions of the industrial ecology field thus far.

3.3 Material Flow Analysis

Material flow analysis (MFA) is the methodology for quantifying the stocks, flows, inputs, and losses of a resource. It is sometimes used for mixed materials (e.g., construction minerals) but more commonly is directed to a specific resource such as a particular metal or plastic. For specific resource applications, the methodology is sometimes termed substance flow analysis (SFA). Early MFA research was conducted by Robert Ayres when he was at Carnegie Mellon University in Pittsburgh, PA. In 1968, he and Alan Kneese contributed to a US Congress report arguing that economic theory was at odds with the first law of thermodynamics: materials could not be "consumed" physically. Rather, emissions and wastes from economic activity could only be reduced by lowering the physical input into the economy. This material balance approach was truly revolutionary for the environmental and economic thinking of that time; it predated the book by Georgescu–Roegen (1971) which is widely regarded as one of the seminal works in ecological economics. The material balance approach provided the theoretical base for what today has become material flow accounting (MFA) as well as part of a number of nations' public statistics. Ayres's initial MFA application was for emissions from metal processing activities in the New Jersey–New York area (Ayres and Rod 1986), followed by a comprehensive study of chlorine (Ayres 1997, 1998, Ayres and Ayres 1997, 1999). In the same general time period, the MFA approach was also developed in Switzerland by Baccini and Brunner (1991), who produced an important book on the topic.

The distinction between bulk MFA and SFA was described by Bringezu and Moriguchi (2002), who categorized analyses from the perspective of substances, materials, products, firms, and geographical regions, although MFA studies tended to dominate early efforts.

The first metal-specific SFA was directed at zinc in the United States over the period of 1850–1990 (Jolly 1993); it showed that about three-quarters of potential zinc losses to the environment were due to dissipative uses and landfill disposal. Other early MFA studies included those for cobalt in the United States (Shedd 1993), vanadium in the United States (Hilliard 1994), and cadmium in the Netherlands (van der Voet et al. 1994). In another early effort, Socolow and Thomas (1997) produced a MFA study for lead in the United States that called for the integration of risk analysis and highlighted the importance of recycling and technological transformation. A seminal dynamic study (i.e., a time-dependent SFA) was completed for aluminum in Germany by Melo (1999).

By 2000–2010, MFAs had been completed for most metals and in several countries (Chen and Graedel 2012) and for some polymers (Kleijn et al. 2000; Diamond et al. 2010; Kuczenski and Guyer 2010). Data challenges continue to constrain the accuracy of these studies, and resource flows and stocks are highly dynamic, but the methodology is firm, and the results thereby produced have proven directly relevant to corporate and public policy (e.g., Pauliuk et al. 2012).

3.4 Socioeconomic Metabolism

Industrial ecology is often viewed as a natural science in that it tends to be directed at the quantification of such things as use of resources, emissions, recycling rates, and the like. These concerns are, of course, a consequence of human action. This realization inspired in the early to mid-1990s the specialty of socioeconomic metabolism, in which material input, processing, energy use, and loss are quantified and viewed from a socio-technical perspective (see Chap. 6). The ultimate task of this area of study is to relate resource transitions to societal change and to prospects for and measurement of sustainability (Fischer-Kowakski and Haberl 1998; Moriguchi 2001). A principal manifestation of this approach is the studies of economy-wide material flows at the level of various societal units, often on a national level.

Use of MFA of national economies began to surge in the 1990s, beginning with those independently developed for Austria (Steurer 1992), Germany (Schütz and Bringezu 1993), and Japan (Japanese Environment Agency 1992). The material flow balance approach was extended to consider transnational resource extractions induced by domestic demand and to indicate the total material use of an economy, including the so-called ecological rucksacks (Bringezu 1993; Bringezu and Schütz 1995). Bringezu (1993) related the ecological rucksack idea to national material flow balances, accounting for the ecological rucksacks of domestic production of raw materials (e.g., unused extraction) and the ecological rucksacks of imports and exports. This method provided the basis for the first international comparisons through the *Resource Flows* report (Adriaanse et al. 1997).

National material accounts (NMAs) began with a collaborative study among researchers from Germany, Japan, the Netherlands, and the United States (Adriaanse et al. 1997). An important contribution of this study was the identification and quantification of ecological rucksacks, or "hidden flows" – flows such as mineral wastes and agricultural debris. A subsequent effort (Matthews et al. 2000) added Austria to the group of countries that were represented and emphasized waste and hazardous material outputs.

National MFAs aggregate a variety of material inputs and outputs, generating thereby a number of indicators for the material use of national economies that have become internationally standardized, among them "domestic material input (DMI)," "domestic material consumption (DMC)," and "total material requirement (TMR)" (Fischer-Kowalski et al. 2011).

NMAs have now been completed for many countries and have become a required output of statistical offices in European countries. Bringezu et al. (2003, 2004) and Weisz et al. (2006) have compared the results of NMAs for a number of countries and found (among many other features) that the domestic material input per capita had not been decoupled on an absolute basis from gross domestic product per capita. At present, there exist NMAs for almost all countries of the world, documenting annual material extraction and use as well as trade for the past several decades (Schaffartzik et al. 2014).

3.5 Input–Output Analysis

Input–output tables (IOT) in economics quantify the transactions that occur between different industrial sectors in an economy. They are expressed as flows from one sector to another measured in either monetary or mixed units. After some early thoughts on how economics and industrial practice might be linked (Leontief 1970; Ayres 1978; Forsund 1985), IOA was proposed as relevant to industrial ecology in 1992 (Duchin 1992). The extension of IO tables to include specific data about industrial/environmental problems followed fairly soon thereafter in the form of "environmental IOTs" (EIOTs) or "physical IOTs" (PIOTs) (Lave et al. 1995; Kondo et al. 1998; Lenzen 2001; Nakamura and Kondo 2002).

An increasing number of national statistical offices produce input–output tables on a regular basis, an activity that provides substantial information that industrial ecology can draw upon. Enhancements to the earlier EIOT and PIOT methodologies have now rendered input–output analysis increasingly relevant to industrial ecology and increasingly practiced with the field (Suh 2009). Several environmentally extended input–output databases covering the global economy are now available (see Chap. 8). Empirical studies using IO databases are used to analyze problems of concern in industrial ecology; examples are Nakamura and Kondo (2009) on waste management (see Chap. 12) and Lopez–Morales and Duchin (2011) on water management. In addition, a number of environmentally extended multiregional input–output models allow the attribution of globally extracted natural resources to individual countries and economic sectors worldwide (Wiedmann et al. 2013).

3.6 UrbanMetabolism

In principle, urban metabolism might not necessarily be regarded as a distinct branch of the field, because it merely applies industrial ecology tools in a specific spatial location. In practice, however, cities are centers of population, of resource use, and of waste generation, and the data available for such systems is often richer than elsewhere. As a consequence, urban metabolism has become a subspecialty of industrial ecology and one that is increasingly widely practiced (see Chap. 4).

The concept of urban metabolism is attributed to a 1965 paper by Wolman. One of the earliest studies of a quantified urban metabolism is Newcombe et al.'s analysis of resource flows in Hong Kong (1978). This exceptionally detailed study, still a model for today's efforts, quantified flows of human and animal food, glass, plastics, sewage, sulfur dioxide emissions, and much more. This effort was repeated a quarter-century later (Warren-Rhodes and Koenig 2001), demonstrating strong per capita increases in food, water, and material consumption. Researchers in Europe took up the challenge, with examples in Switzerland (Brunner et al. 1994; Baccini 1996) and Sweden (Bergbäck et al. 2001) at about the same time period. Newman (1999) discussed the concept in some detail and applied it to Sydney. Urban industrial ecology is now rather common, a recent example being a metabolic analysis of six Chinese cities (Zhang et al. 2009).

3.7 Industrial Symbiosis

Industrial symbiosis is the organization of industrial organisms and their processes so that "the effluents of one process… serve as the raw material for another process" (Frosch and Gallopoulos 1989). Such arrangements occur because they make good business sense, often because of the proximity of facilities discarding resources to those reusing them, as in Kalundborg, Denmark (Anonymous 1990; Ehrenfeld and Gertler 1997). Increasingly, these systems are recognized as having significant environmental benefits as well (Klee 1999).

Many additional examples of industrial symbiosis have been described (e.g., Schwarz and Steininger 1997; Van Beers et al. 2007), and in the first few years of the twenty-first century, industrial symbiosis methodology became better codified (Chertow 2000). Industrial symbiosis is today recognized as a path to improved operational performance as well as to improved environmental performance in situations where resource exchanges can be efficiently achieved (see Chap. 5).

4 Becoming a Scholarly Field

4.1 Conferences

The first regular conferences in the field of industrial ecology were the IEEE Symposium on Electronics and the Environment that began in 1993 (they have been held each year since, now under the name International Symposium on Sustainable Systems and Technology). At about the same time, the AT&T Foundation began to award university grants for research in industrial ecology (Alexander 1994) and hosted invited meetings of largely industrial and academic participants each year from 1994 to 1997 (Laudise and Taylor-Smith 1999).

Having attended Gordon Research Conferences (GRC) on corrosion science in the late 1980s, one of the authors (T.E.G.) conceived the idea of organizing a GRC on industrial ecology. The 1996 proposal was successful, and the first GRC/IE was held in New London, New Hampshire in 1998. For the first few (biennial) conferences, it was a challenge to attract more than about 80–90 participants, and the conference was on probation during those years. Since that time, the value of a relatively small week-long meeting with invited speakers has become widely appreciated. The GRC/IE now routinely "sells out" at about 140–150 attendees, however, and has been held in the United Kingdom, Switzerland, and Italy as well as in the United States.

The IEEE and AT&T meetings gradually became less central to the field as the Gordon Research Conference on Industrial Ecology began and as the International Society for Industrial Ecology (ISIE) began its own biennial conferences; the first was held in the Netherlands in 2001. Regional ISIE conferences are now held as well, and the industrial ecology field has reached the point where colleagues from around the world meet each other at regular intervals on one continent or another around the world.

4.2 Scholarly Journals

A key attribute of a scholarly specialty is the professional journals in which the research of the field is published. In industrial ecology, the most widely known of these is the *Journal of Industrial Ecology* (*JIE*), whose first issue was published in 1997, predating the formation of ISIE by several years.

The genesis of the *JIE* was a meeting of international leaders in the emerging field in 1995. Convened by the Yale School of Forestry & Environmental Studies in collaboration with programs at MIT and UCLA and with funding from the AT&T Foundation, the meeting participants expressed substantial support for a peer-reviewed journal. MIT Press was chosen as the publisher and Yale University agreed to own the journal. The *JIE* aimed to reach both academics and professionals and helped link the North American industrial ecology community to researchers in Europe and Japan that were active in cleaner production and life-cycle assessment and to other researchers working in material flow analysis.

The *JIE* evolved as the market for academic journals changed. What began as a print-only quarterly journal progressed by adding electronic publication; supporting information on the Web, online early release of articles, article-level open access, and expansion of content, the last by transitioning to bimonthly publication (2008) and by increasing the size of the printed page (2012). In 2015 the journal shifted on online-only. Translations of abstracts of all journal articles into Chinese began in 2001 with support of the Henry Luce Foundation. The *JIE* changed publishers to Wiley–Blackwell in 2008.

Edited by Reid Lifset since its inception, it has published path-breaking issues on bio-based products largely before biofuels, and bio-based products became the intense focus of LCA and GHG emissions research, on e-commerce, bringing rigor to the discussion of whether bits replacing atoms has a desirable environment impact, and nanotechnology, drawing attention to the environmental impacts of nano-manufacturing which had been ignored in many parts of the world.

Other journals with substantial industrial ecology content are *Resources, Conservation, and Recycling* (first issue 1988), and the *Journal of Cleaner Production* (first issue 1993). Industrial ecology research and assessments have appeared as well in many additional journals, including *Ecological Economics* (e.g., Bringezu et al. 2004), *Environmental Science & Technology* (e.g., Graedel 2000), *Proceedings of the National Academy of Sciences of the U.S.* (e.g., Rauch 2009), *Nature* (e.g., Lenzen et al. 2012), and *Science* (Reck and Graedel 2012). It is clear that there is no longer a shortage of places for industrial ecology papers to be published or a lack of editors willing to accept them if they are of satisfactory quality.

4.3 The International Society for Industrial Ecology

By the late 1990s, it was apparent that the field of industrial ecology needed to organize itself into a professional society. The issue was discussed in detail in a meeting hosted by Jesse Ausubel at the New York Academy of Sciences in 2000. Once again, support was provided by the AT&T Foundation. Based on the consensus of the international group, the International Society for Industrial Ecology was launched a year later with the *Journal of Industrial Ecology* as its official journal. The Yale School of Forestry & Environmental Studies agreed to serve as the temporary international secretariat for the ISIE. As the society and the industrial ecology community grew, the ISIE created sections focused on industrial symbiosis/eco-industrial development, socioeconomic metabolism, life-cycle sustainability assessment, organizing sustainable consumption and production, sustainable urban systems, and environmentally extended input–output analysis. More intangibly, but perhaps more importantly, the ISIE provides continuity, an opportunity for interaction and, along with the *JIE*, academic legitimation for those seeking to work in industrial ecology.

4.4 Courses and Textbooks

As with any scholarly field, training the next generation is important, and doing so requires pedagogical materials. Aside from scholarly journal articles, several textbooks have been produced for use in industrial ecology-related programs in various academic fields. They include *Industrial Ecology* (Graedel and Allenby 1995 [Japanese translation, 1996]); 2nd edition, 2003 [Korean translation, 2004; Chinese translation, 2005; Russian translation, 2006]), *Green Engineering: Environmentally Conscious Design of Chemical Processes* (Allen and Shonnard 2002), *Applied Industrial Ecology: A New Platform for Planning Sustainable Societies* (Erkman and Ramaswamy 2003), *Industrial Ecology: konzeptionelle Grundlagen, zentrale Handlungsfelder, Kernwerkzeuge und erfolgreiche Praxisbeispiele* (Isenmann and von Hauff 2007), *Industrial Ecology Management: Nachhaltige Entwicklung durch*

Unternehmensverbünde (Von Hauff et al. 2012); *Crossing "Environmental Mountain" – Study on Industrial Ecology* (Lu 2008), *Ecologia Industrial* (Ferrao 2009), *Environmental Engineering: Fundamentals,Sustainability, Design* (Mihelcic & Zimmerman, 2009, 2nd edition, 2014), and *Industrial Ecology and Sustainable Engineering* (Graedel and Allenby 2010).

The first formal course in industrial ecology was apparently taught at the Norwegian University of Science and Technology in 1993 (Marstrander et al. 1999). A few years later, the Delft University of Technology and Philips Consumer Electronics jointly developed modules for teaching eco-design in industry and in universities (Stevels 2001). From that beginning, a number of universities in Europe and North America began offering industrial ecology courses in the latter 1990s and the early 2000s (Cockerill 2013). This area of study is growing rapidly: a recent survey identified industrial ecology courses and/or programs at 190 universities and colleges in 46 countries (Finlayson et al. 2014). As with many young fields, the scope, level, and content of these courses have considerable diversity, but it is clear that education in industrial ecology is widespread, growing, and evolving.

5   Epilogue

This review and recollection emphasizes the first decade or so of the industrial ecology field, largely because the authors had the good fortune to be involved in most of the developmental activities that occurred during that period and because many aspects of those activities are not very widely known. It is a great pleasure to us that the industrial ecology field has grown enormously in size and influence since its founding and has become "the science behind sustainability." It will be interesting and rewarding to watch industrial ecology's further development in the years to come.

Authors' Note Because we are not historians, this contribution inevitably emphasizes our own interactions and experiences at the close of the twentieth century. It also has a North American bias, partly because some of the most significant parts of industrial ecology's early days occurred there and partly because that is where we live. We are, however, grateful for the comments and suggestions from Faye Duchin, Marina Fischer-Kowalski, and Stefan Bringezu which have significantly improved an earlier version of this work. For what we perceive will be of maximum utility to most readers, the references are largely restricted to English.

References

Adriaanse, A., Bringezu, S., Hammond, A., Moriguchi, Y., Rogich, D., & Schütz, H. (1997). *Resource flows: The material basis of industrial economies* (65 pp). Washington, DC: World Resources Institute.

Alexander, A. S. (1994). Creating strategic philanthropic partnerships: The AT&T foundation's new programme in industrial ecology. *Industry and Higher Education, 8,* 103–106.

Allen, D. T., & Shonnard, D. H. (2002). *Green engineering: Environmentally conscious design of chemical processes* (552 pp). Upper Saddle River: Prentice Hall.

Allenby, B. R. (1994). Integrating environment and technology: Design for environment. In B. R. Allenby & D. J. Richards (Eds.), *The greening of industrial ecosystems* (pp. 137–148). Washington, DC: National Academy Press.

Anonymous. (1990, November 14). Group of businesses trade waste, water, and surplus energy on industrial estate. *Financial Times*, p. 15.

Ashby, M. F. (2009). *Materials and the environment: Eco-informed materials choice* (385 pp). Amsterdam: Butterworth-Heinemann.

Ayres, R. U. (1978). *Resources, environment, and economics.* New York: Wiley Interscience.

Ayres, R. U. (1989). Industrial metabolism. In J. H. Ausubel & H. E. Sladovich (Eds.), *Technology and environment* (pp. 23–49). Washington, DC: National Academy Press.

Ayres, R. U. (1997). The life cycle of chlorine, part 1, chlorine production and the chlorine-mercury connection. *Journal of Industrial Ecology, 1*(1), 81–94.

Ayres, R. U. (1998). The life-cycle of chlorine, part III: Accounting for final use. *Journal of Industrial Ecology, 2*(1), 93–115.

Ayres, R. U., & Ayres, L. W. (1997). The life-cycle of chlorine, part II: Conversion processes and use in the European chemical industry. *Journal of Industrial Ecology, 1*(2), 65–89.

Ayres, R. U., & Ayres, L. W. (1999). The life-cycle of chlorine, part IV: Accounting for persistent cyclic organo-chlorines. *Journal of Industrial Ecology, 3*(2–3), 121–159.

Ayres, R. U., & Rod, S. R. (1986). Patterns of pollution in the Hudson-Raritan basin. *Environment, 28*(11), 14–20 and 39–43.

Azar, J., Berko-Boateng, V., Calkins, P., deJong, E., George, J., & Hilbert, H. (1995). Agent of change: Xerox design for environment program. In *Proceedings of the IEEE international symposium on electronics and the environment* (pp. 51–61). Piscataway, NJ: IEEE.

Baccini, P. (1996). Understanding regional metabolism for a sustainable development of urban systems. *Environmental Science and Pollution Research, 3*(2), 108–111.

Baccini, P., & Brunner, P. H. (1991). *Metabolism of the anthroposphere* (157 pp). Berlin: Springer.

Bast, C. (1994). Hewlett-Packard's approach to creating a life cycle program. In *Proceedings of the IEEE international symposium on electronics and the environment* (pp. 31–36). Piscataway, NJ: IEEE.

Bendz, D. J. (1993, September). Green products for green profits. *IEEE Spectrum*, 63–66. Piscataway, NJ: IEEE.

Bergbäck, B., Johansson, K., & Mohlander, U. (2001). Urban metal flows – A case study of Stockholm. *Water, Air, and Soil Pollution Focus, 1,* 3–24.

Boks, C. B., Brouwers, W. C. J., Kroll, E., & Stevels, A. L. N. (1996). Disassembly modeling: Two applications to a Philips 21" television set. In: *Proceedings of the IEEE international symposium on electronics and the environment* (pp. 224–229).

Bousted, I. (1972). *The milk bottle.* Milton Keyes: Open University Press.

Bringezu, S. (1993). Towards increasing resource productivity: How to measure the total material consumption of regional and national economies? *Fresenius Environmental Bulletin, 2,* 437–442.

Bringezu, S., & Moriguchi, Y. (2002). Material flow analysis. In R. U. Ayres & L. M. Ayres (Eds.), *A handbook of industrial ecology.* Cheltenham: Edward Elgar.

Bringezu, S., & Schütz, H. (1995). How to measure the ecological sustainability of an economy? A contribution of materials flow accounting for the example of Germany (in German). In S. Bringezu & S. Berlin (Eds.), *Neue Ansätze der Umweltstatistik.* Basel: Birkhäuser Verlag.

Bringezu, S., Schütz, H., & Moll, S. (2003). Rationale for and interpretation of economy-wide materials flow analysis and derived indicators. *Journal of Industrial Ecology, 7*(2), 43–64.

Bringezu, S., Schütz, H., Steger, S., & Baudisch, J. (2004). International comparison of resource use and its relation to economic growth. *Ecological Economics, 51*, 97–124.

Brunner, P. H., Daxbeck, H., & Baccini, P. (1994). Industrial metabolism at the regional and local level: A case study on a Swiss region. In R. U. Ayres & U. E. Simonis (Eds.), *Industrial metabolism: Restructuring for sustainable development* (pp. 163–193). Tokyo: United Nations University Press.

Cândido, L., Kindlein, W., Demori, R., Carli, L., Mauler, R., & Oliveira, R. (2011). The recycling cycle of materials as a design project tool. *Journal of Cleaner Production, 19*, 1438–1445.

Chen, W., & Graedel, T. E. (2012). Anthropogenic cycles of the elements: A critical review. *Environmental Science & Technology, 46*, 8574–8586.

Chertow, M. R. (2000). Industrial ecology: Literature and taxonomy. *Annual Review of Energy and Environment, 25*, 313–337.

Christiansen, K (Ed.). (1997). *Simplifying LCA: Just a cut? Final report of the SETAC-Europe LCA screening and streamlining working group*. ISBN 90-5607-006-1. Brussels: SETAC-Europe.

Clift, R., Udo de Haes, H. A., Bensahel, J. F., Fussler, C. R., Griesshammer, R., & Jensen, A. A. (1994). *Guidelines for the application of life cycle assessment in the EU ecolabelling programme*. Brussels: DGXI of the Commission of the European Communities.

Cockerill, K. (2013). A failure reveals success: A comparative analysis of environmental education, education for sustainable development, and industrial ecology education. *Journal of Industrial Ecology, 17*, 633–641.

Desrochers, P. (2000). Market processes and the closing of 'industrial loops'. *Journal of Industrial Ecology, 4*(1), 29–43.

Diamond, M., Melymuk, L., Csisyer, S. A., & Robson, M. (2010). Estimation of PCB studies, emissions, and urban fate: Will our policies reduce concentration and exposure? *Environmental Science & Technology, 44*, 2777–2783.

Duchin, F. (1992). Industrial input–output analysis: Implications for industrial ecology. *Proceedings of the National Academy of Sciences of the United States of America, 89*, 851–855.

Ehrenfeld, J., & Gertler, N. (1997). Industrial ecology in practice: The evolution of interdependence at Kalundborg. *Journal of Industrial Ecology, 1*(1), 67–79.

Erkman, S. (1997). Industrial ecology: An historical view. *Journal of Cleaner Production, 5*, 1–10.

Erkman, S., & Ramaswamy, R. (2003). *Applied industrial ecology: A new platform for planning sustainable societies*. Bangalore: Aicra Publishers.

Ferrao, P. (2009). *Ecologia industrial*. Lisbon: Instituto Superior Tecnico.

Field, F. R., III, Isaacs, J. A., & Clark, J. P. (1993). Life cycle analysis and its role in product and process development. *International Journal of Environmentally Conscious Design & Manufacturing, 2*(2), 13–20.

Finlayson, A., Markewitz, K., & Frayret, J.-M. (2014). Postsecondary education in industrial ecology across the world. *Journal of Industrial Ecology, 18*, 931–941.

Fischer-Kowalski, M., & Haberl, H. (1998). Sustainable development: Socio-economic metabolism and colonization of nature. *International Social Science Journal, 50*, 573–587.

Fischer-Kowalski, M., Krausmann, F., Giljum, S., Lutter, S., Mayer, A., Bringezu, S., Moriguchi, Y., Schütz, H., Schandl, H., & Weisz, H. (2011). Methodology and indicators of economy wide material flow accounting. State of the art and reliability across sources. *Journal of Industrial Ecology, 15*(6), 855–876.

Forsund, F. R. (1985). Input–output models, national economic models, and the environment. In A. V. Kneese & J. L. Sweeney (Eds.), *Handbook of natural resource and energy economics* (Vol. 1, pp. 325–341). Dordrecht: Elsevier.

Frosch, R. A., & Gallopoulos, N. E. (1989). Strategies for manufacturing. *Scientific American, 261*(3), 144–152.

Georgescu-Roegen, N. (1971). *The Entropy Law and the Economic Process*. Cambridge, MA: Harvard University Press.

Gierlinger, S., & Krausmann, F. (2011). The physical economy of the United States of America: Extraction, trade, and consumption of materials from 1870 to 2005. *Journal of Industrial Ecology, 16*, 365–377.

Go, T. F., Wahab, D. A., Rahman, M. N. A., Ramli, R., & Azhari, C. H. (2011). Disassemblability of end-of-life vehicle: A critical review of evaluation methods. *Journal of Cleaner Production, 19*, 1536–1546.

Graedel, T. E. (2000). The evolution of industrial ecology. *Environmental Science & Technology, 34*, 28A–31A.

Graedel, T. E., & Allenby, B. R. (1995). *Industrial ecology* (412 pp). Englewood Cliffs: Prentice Hall.

Graedel, T. E., & Allenby, B. R. (1996). *Design for environment* (175 pp). Upper Saddle River: Prentice Hall.

Graedel, T. E., & Allenby, B. R. (2003). *Industrial ecology* (2nd ed., 363 pp). Englewood Cliffs: Prentice Hall.

Graedel, T. E., & Allenby, B. R. (2010). *Industrial ecology and sustainable engineering* (403 pp). Englewood Cliffs: Prentice Hall.

Graedel, T. E., Allenby, B. R., & Comrie, P. R. (1995). Matrix approaches to abridged life cycle assessment. *Environmental Science & Technology, 29*, 134A–139A.

Guinée, J. (Ed.). (2002). *Handbook on life cycle assessment – Operational guide to the ISO standards*. Dordrecht: Kluwer Academic Publishers.

Hamilton, A., & Michael, J. A. (1992). The NCR 7731: Ecologically responsible design, business & technology. *Innovation* (special issue), 21–23.

Harsch, M., Schukert, M., Eyerer, P., & Saur, K. (1996). Life-cycle assessment. *Advanced Materials & Processes, 6*, 43–46.

Hatcher, G. D., Ijomah, W. L., & Windmill, J. F. C. (2011). Design for remanufacture: A literature review and future research needs. *Journal of Cleaner Production, 19*, 2004–2014.

Heijungs, R., Guinée, J., Huppes, G., Lankreijer, R. M., Udo de Haas, H. A., Wegener Sleeswijk, A., Ansems, A. M. M., Eggels, P. G., van Duin, R., & de Goede, H. P. (1992). *Environmental life cycle assessment of products: Guide and backgrounds*. Leiden: CML, Leiden University.

Henstock, M. E. (1988). *Design for recyclability* (135 pp). London: The Institute of Metals.

Hilliard, H. E. (1994). *The materials flow of vanadium in the United States* (20 pp). Washington, DC: U.S. Department of the Interior.

Hoffman, W. F. III. (1995). A tiered approach to design for environment. In *Proceedings of the IEE conference on clean electronic products and technology*, London: Institution of Electrical Engineers, Edinburgh.

Hoffman, W. F. (1997). Recent advances in design for environment at Motorola. *Journal of Industrial Ecology, 1*(1), 131–140.

Holusha, J. (1991, May 28). Making disposal easier, by design. *New York Times*, pp. D1, D3.

Horkeby, I. (1997). Environmental prioritization. In D. J. Richards (Ed.), *The industrial green game*. Washington, DC: National Academy Press.

Hunt, R., & Welch, R. (1972). *Resource and environmental profile analysis of plastics and non-plastics containers*. New York: The Society of the Plastics Industry.

Isenmann, R., & Von Hauff, M. (2007). *Industrial ecology: konzeptionelle Grundlagen, zentrale Handlungsfelder, Kernwerkzeuge und erfolgreiche Praxisbeispiele*. Dordrecht: Elsevier.

Itsubo, N., Inaba, A., Matsuno, Y., Yasui, I., & Yamamoto, R. (2000). Current status of weighting methodologies in Japan. *International Journal of Life Cycle Assessment, 5*(1), 5–11.

Japanese Environment Agency. (1992). *Quality of the environment in Japan 1992*. Tokyo: Ministry of the Environment Japan.

Jolly, J. H. (1993). Materials flow of zinc in the United States, 1850–1990. *Resources, Conservation and Recycling, 9*, 1–30.

Kirby, J. R., & Pitts, D. (1994). Resource recovery strategies for end-of-life business machines. In *Proceedings of the IEEE international symposium on electronics and the environment* (pp. 167–170). Piscataway, NJ: IEEE.

Klausner, M., Grimm, W. M., & Hendrickson, C. (1998). Reuse of electric motors in consumer products: Design and analysis of an electronic data log. *Journal of Industrial Ecology, 2*(2), 89–102.

Klee, R. (1999). Zero waste system in paradise. *Biocycle, 40*(2), 66–67.

Kleijn, R., Huele, R., & van der Voet, E. (2000). Dynamic substance flow analysis: The delaying mechanism of stocks, with the case of PVC in Sweden. *Ecological Economics, 32*, 241–254.

Klöpffer, W. (2006). The role of SETAC in the development of LCA. *The International Journal of Life Cycle Assessment, 11*, 116–122.

Kondo, Y., Moriguchi, Y., & Shimizu, H. (1998). CO_2 emissions in Japan: Influences of imports and exports. *Applied Energy, 59*, 163–174.

Kuczenski, B., & Guyer, R. (2010). Material flow analysis of polyethylene terephthalate in the US, 1996–2007. *Resources, Conservation and Recycling, 54*, 1161–1169.

Laudise, R. A., & Taylor-Smith, R. E. (1999). Lucent industrial ecology faculty fellowship program: Accomplishments, lessons, and prospects. *Journal of Industrial Ecology, 2*(4), 15–27.

Lave, L. B., Cobas-Flores, E., Hendrickson, C. T., & McMichael, F. C. (1995). Using input–output analysis to estimate economy-wide discharges. *Environmental Science & Technology, 29*, 420A–426A.

Lenzen, M. (2001). Errors in conventional and input–output-based life-cycle inventories. *Journal of Industrial Ecology, 4*(4), 127–148.

Lenzen, M., Moran, D., Kanemoto, K., Foran, B., Lobefaro, L., & Geschke, A. (2012). International trade drives biodiversity threats in developing nations. *Nature, 486*, 109–112.

Leontief, W. (1970). Environmental repercussions and economic structure – Input–output approach. *Review of Economics and Statistics, 52*, 262–271.

Lopez-Morales, C., & Duchin, F. (2011). Policies and technologies for a sustainable use of water in Mexico: A scenario analysis. *Economic Systems Research, 23*, 387–407.

Lu, Z. (2008). *Crossing "environmental mountain": Study on industrial ecology*. Beijing: Science Press.

Lundgren, J., Franzén, B., & Storåkers, J. (1994). Design for environment can be profitable – A case study of the paper feeder in the IBM printer 4234. In *Proceedings of the IEEE international symposium on electronics and the environment* (pp. 128–133).

Makov, T. (2014). Private communication.

Marstrander, R., Brattebø, H., Røine, K., & Støren, S. (1999). Teaching industrial ecology to graduate students: Experiences at the Norwegian University of Science and Technology. *Journal of Industrial Ecology, 3*(4), 117–130.

Matsuno, Y., Inaba, A., & Betx, M. (1998). Valuation of electricity grid mixes in Japan with application of life-cycle impact assessment methodology. In *Proceedings of the third international conference on ecobalance* (pp. 97–100). Tokyo: Society of Non-Traditional Technology.

Matthews, E., Amann, C., Bringezu, S., Fischer-Kowalski, M., Hüttler, W., Kleijn, R., Moriguchi, Y., Ottke, C., Rodenburg, E., Rogich, D., Schandl, H., Schütz, H., van der Voet, E., & Weisz, H. (2000). *The weight of nations: Material outflows from industrial economies* (125 pp). Washington, DC: World Resources Institute.

Meadows, D. H., Meadows, D. L., Randers, J., & Behrens, W. H., III. (1972). *Limits to growth*. New York: New American Library.

Melo, M. T. (1999). Statistical analysis of metal scrap generation: The case of aluminium in Germany. *Resources, Conservation and Recycling, 26*, 91–113.

Microelectronics and Computer Technology Corporation. (1994). *Electronics industry environmental roadmap*. Austin: Microelectronics and Computer Technology Corporation.

Mihelcic, J., & Zimmerman, J. B. (2009). *Environmental engineering: Fundamentals sustainability, design*. Hoboken: Wiley.

Mihelcic, J., & Zimmerman, J. B. (2014). *Environmental engineering: Fundamentals, sustainability, design* (2nd ed.). Hoboken: Wiley.

Ministry of International Trade and Industry (MITI) (1988, September). *Trends and future tasks in industrial technology. Developing innovative technologies to support the 21st century, summary of a white paper on industrial technology* (29 pp). Tokyo.

Moriguchi, Y. (2001). Rapid socio-economic transition and material flows in Japan. *Population and Environment, 23*, 105–115.

Murray, F. E. S. (1993). *Xerox: Design for the environment*. Harvard Business School case study 9-794-022, Cambridge, MA: Harvard University.

Nagel, C. (1997). Single-use cameras within a multi-use concept – Ecological (non)sense!? In *Proceedings of the IEEE international symposium on electronics and the environment* (pp. 69–72). Piscataway, NJ: IEEE.

Nakamura, S., & Kondo, Y. (2002). Input–output analysis of waste management. *Journal of Industrial Ecology, 6*(1), 39–63.

Nakamura, S., & Kondo, Y. (2009). Waste input–output analysis, LCA and LCC. In S. Suh (Ed.), *Handbook of input–output economics for industrial ecology* (pp. 561–572). Dordrecht: Springer.

Newcombe, K., Kalma, J. D., & Aston, A. R. (1978). The metabolism of a city: The case of Hong Kong. *Ambio, 7*(1), 3–15.

Newman, P. W. G. (1999). Sustainability and cities: Extending the metabolism model. *Landscape and Urban Planning, 44*, 219–226.

Octave, S., & Thomas, D. (2009). Biorefinery: Toward an industrial metabolism. *Biochimie, 91*, 659–664.

Office of Technology Assessment. (1992). *Green products by design: Choices for a cleaner environment* (117 pp). Washington, DC: Congress of the United States.

Patel, C. K. N. (Organizer) (1992). Papers from a colloquium entitled "industrial ecology". In *Proceedings of the National Academy of Sciences of the U.S.*, 89, 793–884.

Pauliuk, S., Wang, T., & Müller, D. B. (2012). Moving toward the circular economy: The role of stocks in the Chinese steel cycle. *Environmental Science & Technology, 46*, 148–154.

Pittinger, C. A., Sellers, J. S., Janzen, D. C., Koch, D. G., Rothgeb, T. M., & Hunnicut, M. L. (1993). Environmental life-cycle inventory of detergent – Grade surfactant sourcing and production. *Journal of the American Oil Chemists' Society, 70*, 1–15.

Porada, T. (1994). Materials recovery: Asset alchemy. In *Proceedings of the IEEE international symposium on electronics and the environment* (pp. 171–173). Piscataway, NJ: IEEE.

Rauch, J. N. (2009). Global mapping of Al, Cu, Fe, and Zn in-use stocks and in-ground resources. *Proceedings of the National Academy of Sciences of the United States of America, 106*, 18920–18925.

Reap, J., Roman, F., Duncan, S., & Bras, B. (2008a). A survey of unresolved problems in life-cycle assessment. Part 1: Goal and scope and inventory analysis. *International Journal of Life Cycle Assessment, 13*, 290–300.

Reap, J., Roman, F., Duncan, S., & Bras, B. (2008b). A survey of unresolved problems in life-cycle assessment. Part 2: Impact assessment and interpretation. *International Journal of Life Cycle Assessment, 13*, 374–388.

Reck, B. R., & Graedel, T. E. (2012). Challenges in metal recycling. *Science, 337*, 690–695.

Ryan, M. J. (2014). Design for system retirement. *Journal of Cleaner Production, 70*, 203–210.

Ryberg, B. A. (1993). Design for environmental quality: Reap the benefits of closing the design loop. In *Proceedings of the IEEE international symposium on electronics and the environment* (pp. 37–42). Piscataway, NJ: IEEE.

Schaffartzik, A., Mayer, A., Gingrich, S., Eisenmenger, N., Loy, C., & Krausmann, F. (2014). The global metabolic transition: Regional patterns and trends of global material flows, 1950–2010. *Global Environmental Change, 26*, 87–97.

Scheidt, L.-G., & Stadlbauer, H. (1996). CARE "VISION 2000" – The environmental research platform of the electronics industry in Europe. In *Proceedings of the IEEE international symposium on electronics and the environment* (pp. 264–268).

Schütz, H., & Bringezu, S. (1993). Major material flows in Germany. *Fresenius Environmental Bulletin, 2*, 443–448.

Schwarz, E. J., & Steininger, K. W. (1997). Implementing nature's lesson: The industrial recycling network enhancing regional development. *Journal of Cleaner Production, 5*(1–2), 47–56.

Shedd, K. B. (1993). *The materials flow of cobalt in the United States* (26 pp). Washington, DC: U.S. Department of the Interior.

Sheng, P., Willis III B., & Shiovitz, A. (1995). Influence of computer chassis design on metal fabrication waste streams. In *Proceedings of the IEEE international symposium on electronics and the environment* (pp. 171–183). Piscataway, NJ: IEEE.

Socolow, R. H. (1994). Preface. In R. Socolow, C. Andrews, F. Berkhout, & V. Thomas (Eds.), *Industrial ecology and global change*. Cambridge: Cambridge University Press.

Socolow, R., & Thomas, V. (1997). The industrial ecology of lead and electric vehicles. *Journal of Industrial Ecology, 1*(1), 13–36.

Socolow, R., Andrews, C., Berkhout, F., & Thomas, V. (Eds.). (1994). *Industrial ecology and global change* (500 pp). Cambridge: Cambridge University Press.

Steen, B., & Ryding, S.-O. (1992) The EPS Enviro-Accounting Method. An Application of Environmental Accounting Principles for Evaluation and Valuation of Environmental Impact in Product Design, Report B 1080, Stockholm: Swedish Environmental Research Institute.

Steurer, A. (1992). *Stoffstrombilanz Östereich, Schriftenreihe Soziale Ökologie* (Band 26). Vienna: Institut für Interdisziplinäre Forschung und Fortbildung der Universitaten Innsbruck, Klagenfurt und Wien.

Stevels, A. (2001). Teaching modules on eco-design for competitive advantage. *Journal of Sustainable Product Design, 1*, 273–282.

Stevels, A. L. N. (2009). *Adventures in ecodesign of electronic products, 1993–2007*. Design for sustainability program publication no. 17. Delft: University of Delft, the Netherlands.

Suga, T., Saneshige, K., & Fujimoto, J. (1996). Quantitative disassembly evaluation. In *Proceedings of the IEEE international symposium on electronics and the environment* (pp. 19–24). Piscataway, NJ: IEEE.

Suh, S. (Ed.). (2009). *Handbook of input–output economics for industrial ecology*. Dordrecht: Springer.

U.S. Environmental Protection Agency. (1995). *Design for the environment: Building partnerships for environmental improvement, EPA/600/K-93/002*. Washington, DC: U.S.E.P.A..

Van Beers, D., Corder, G., Bossilkov, A., & van Berkel, R. (2007). Industrial symbiosis in the Australian mining industry: The case of Kwinana and Gladstone. *Journal of Industrial Ecology, 11*(1), 55–72.

Van der Voet, E., van Egmond, L., Kleijn, R., & Huppes, G. (1994). Cadmium in the European community: A policy-oriented analysis. *Waste Management & Research, 12*, 507–526.

Volvo Car Corporation. (1991, September 23). *Environmentally compatible product development, environmental report No. 27*. Göteborg.

Von Hauff, M., Isenmann, R., & Müller-Christ, G. (2012). *Industrial ecology management: Nachhaltige Entwicklung durch Unternehmensverbünde*. Wiesbaden: Springer Gabler.

Warren-Rhodes, K., & Koenig, A. (2001). Escalating trends in the urban metabolism of Hong Kong: 1971–1997. *Ambio, 30*, 429–438.

Watanabe, C. (1972). *Industry-ecology: Introduction of ecology into industrial policy* (12 pp). Tokyo: Ministry of International Trade and Industry (MITI).

Watson, R. G., Theis, E. M., & Janek, R. S. (1990). Mechanical equipment design for simplicity. *AT&T Technical Journal, 69*(3), 14–27.

Weisz, H., Krausmann, F., Amann, C., Eisenmenger, N., Erb, K.-H., Hubacek, K., & Fischer-Kowalski, M. (2006). The physical economy of the European Union: Cross-country comparison and determinants of material consumption. *Ecological Economics, 58*, 676–698.

Weitz, K. A., Malkin, M., & Baskir, J. N. (1995). *Streamlining life-cycle assessment conference and workshop*. Research Triangle Park: Research Triangle Institute.

White, R. M. (1994). Preface. In B. R. Allenby & D. J. Richards (Eds.), *The greening of industrial ecosystems*. Washington, DC: National Academy Press.

Wiedmann, T., Schandl, H., Lenzen, M., Moran, D., Suh, S., West, J., & Kanemoto, K. (2013). The material footprint of nations. *Proceedings of the National Academy of Sciences of the United States of America*. doi 10.1073/pnas.1220362110.

Wolman, A. (1965). The metabolism of cities. *Scientific American, 213*(3), 179–190.

Zhang, Y., Zhao, Y. W., Yang, Z. F., Chen, B., & Chen, G. Q. (2009). Measurement and evaluation of the metabolic capacity of an urban ecosystem. *Communications on Nonlinear Science and Numerical Simulation, 14*, 1758–1765.

Chapter 2
Prospective Models of Society's Future Metabolism: What Industrial Ecology Has to Contribute

Stefan Pauliuk and Edgar G. Hertwich

Abstract Scientific assessment of sustainable development strategies provides decision-makers with quantitative information about the strategies' potential effect. This assessment is often done by forward-looking or prospective computer models of society's metabolism and the natural environment. Computer models in industrial ecology (IE) have advanced rapidly over the recent years, and now, a new family of prospective models is available to study the potential effect of sustainable development strategies at full scale.

We outline general principles of prospective modeling and describe the current development status of two prospective model types: extended dynamic material flow analysis and THEMIS (Technology-Hybridized Environmental-Economic Model with Integrated Scenarios). These models combine the high level of technological detail known from life-cycle assessment (LCA) and material flow analysis (MFA) with the comprehensiveness of, respectively, dynamic stock models and input/output analysis (I/O). These models are dynamic; they build future scenarios with a time horizon until 2050 and beyond. They were applied to study the potential effect of a wide spectrum of sustainable development strategies, including renewable energy supply, home weatherization, material efficiency, and light-weighting.

We point out future applications and options for model development and discuss the relation between prospective IE models and the related concept consequential LCA (CLCA).

The prospective models for industrial ecology can answer questions that were previously in the exclusive domain of integrated assessment models (IAMs). A debate about the relation between the two model families is necessary.

S. Pauliuk (✉) • E.G. Hertwich
Industrial Ecology Programme and Department for Energy and Process Engineering,
Norwegian University of Science and Technology (NTNU),
Høgskoleringen 1, 7491 Trondheim, Norway
e-mail: stefan.pauliuk@ntnu.no

© The Author(s) 2016
R. Clift, A. Druckman (eds.), *Taking Stock of Industrial Ecology*,
DOI 10.1007/978-3-319-20571-7_2

We find that IAMs have a more comprehensive scope than the prospective IE models, but they often do not obey central IE principles such as the life cycle approach and mass balance consistency. Integrating core IE principles into IAMs would increase the scientific quality and policy relevance of the scenarios of society's future metabolism generated by IAMs, while placing industrial ecology concepts more prominently at the same time. We provide a sketch of what this integration could look like.

Keywords Socioeconomic metabolism • Socio-metabolic transition • Sustainable development strategy • Prospective model • Industrial ecology • Integrated assessment model • Policy assessment • Dynamic material flow analysis • Integrated hybrid LCA

1 Introduction

1.1 The Great Transformation Ahead

Human interference with global biogeochemical cycles has grown to a level that will trigger epochal changes, including climatic change and state shifts in the Earth's biosphere. These changes have substantial impact on humanity; they force humans to adapt or to be proactive and mitigate negative impacts on the environment. The spectrum of options for future action is wide. It includes technology development and deployment, economic instruments including taxes and subsidies, regulation and standards, and changes in consumer choices and lifestyle. Both adaptation and mitigation will lead to a transformation of the biophysical basis of our society, which includes agriculture, industry, infrastructure, building stocks and vehicle fleets, and consumer products, and of the way we build, maintain, and operate this basis.

The coming transformation is the continuation of a historic sequence of *socio-metabolic transitions* of mankind, first from the hunter/gatherer to the agrarian and later from the agrarian to the industrialized society (Fischer-Kowalski and Haberl 2007; Fischer-Kowalski et al. 2014; Krausmann and Fischer-Kowalski 2013). The coming transformation represents a special global challenge, however, because it is likely to happen under environmental conditions that are significantly different from those that enabled the previous two transitions (Barnosky et al. 2012).

1.2 Scientific Response: The Interdisciplinary Systems Approach and Prospective Models

Environmental literacy is a core feature of higher species. It is the "capability [...] to appropriately read, utilize, and adapt to environmental information, resources, and system dynamics" (Scholz and Binder 2011). Today, human environmental literacy is higher than ever before. Humanity uses scientific methods to study the biophysical basis of society, to anticipate future challenges associated with the transformation of that basis, and to offer quantitative and objective assistance to decision-makers.

The scientific approach to studying the transformation faces two major challenges: (1) The transformation affects many different aspects of society and the environment; it ignores traditional boundaries between scientific disciplines. (2) The transformation is a complex process and spans many different scales, which are interconnected: spatial (local biotopes, cities, regions, countries, the globe), organizational (households, companies, sectors, nations, global community), and temporal (from immediate consequences to long-term effects several centuries from now).

The necessity to study different scales follows from the nature of the problem: It is essential to consider the global scale for three reasons: the changes in the environment are global, our global economy is causing these changes, and relocation of production activities happens on a global scale. Smaller scales need to be studied as well because these scales represent the typical scope of decision-making; they form the arena where interventions take place.

To cope with these two challenges, scientists use an interdisciplinary systems approach, where the biophysical basis of human society is seen as a complex self-reproducing (autopoietic) system controlled by human agents (Binder et al. 2013; Fischer-Kowalski and Weisz 1999). The systems approach to studying the biophysical basis of human society is called socioeconomic metabolism (SEM) (Fischer-Kowalski and Haberl 1998); it forms the basis for scientific assessments from the angle of different disciplines (Pauliuk and Hertwich 2015). A major application of the systems approach is to quantify possible future impacts of specific transformation strategies, such as deployment of renewable energy supply or carbon taxation, on different spatial, temporal, and organizational scales. This forward-looking analysis is called *prospective assessment* of transformation strategies. It requires prospective models of socioeconomic metabolism that can capture its future development. These models are being developed in several scientific fields, including integrated assessment model (IAM), econometrics, and industrial ecology (IE).

1.3 Goal and Scope

Unlike IAMs and prospective econometric models, prospective models in industrial ecology were developed very recently, and so far, the community of researchers involved has been rather small. A general overview of prospective modeling in industrial ecology is not available, a gap that we try to fill in this chapter. Our review includes a discussion of general principles of prospective modeling, and it shows how the recently developed prospective IE models relate to the established IE method material flow analysis (MFA), life-cycle assessment (LCA), and input/output analysis (I/O), as well as IAMs.

The remainder of this chapter is structured as follows: First, we describe general principles of prospective models of society's metabolism. Then, we describe the state of the art of prospective models in industrial ecology (IE) and explain the relation of these models to consequential LCA, which is an IE concept that also has a prospective aspect. Finally, we discuss future applications and options for further development of prospective IE models, with special focus on the relation to integrated assessment models (IAMs).

2 Principles of Prospective Models of Socioeconomic Metabolism

2.1 Overview and General Principles

Prospective models of society's metabolism require certain features to be fit for purpose. They need to follow an interdisciplinary systems approach, as explained above, and in Table 2.1, we list salient features of the systems approach and mention briefly how they are commonly implemented.

(1) *Dynamic models* have an explicit time dimension and contain mechanisms to generate the future state of the system from its past and from additional exogenous information. Dynamic models link different time scales with each other, which is necessary to study how changes on short time scales affect the long-term dynamics of the system. For example, large-scale substitution of materials today will alter the recycling system in the future. Combining dynamic models of SEM with other models with an explicit time line, like climate models, allows us to study the interaction between socioeconomic metabolism and the environment over time. Dynamic models allow for flexible handling of time discounting, e.g., an artificial time horizon to calculate the global warming potential is not needed in models where time is explicit. Finally, dynamic models enable researchers to study changes that happen gradually over time, like the introduction of new tech-

Table 2.1 Salient features of prospective models of socioeconomic metabolism and common ways of implementing them

Feature	Common implementation
Ability to capture different spatial, organizational, and temporal scales	(1) Dynamic models
	(2) Assessment at full scale
Capability to produce results that are relevant for different scientific disciplines	(3a) Multilayer modeling
	(3b) Satellite accounts
Ability to determine future consequences of decisions in the past	(1) Dynamic models
Ability to deal with the indeterminacy ("uncertainty") of future development	(4) Scenario modeling with exogenous model parameters
Adherence to generally accepted scientific principles such as mass and energy conservation or economic balances	(3c) Balancing constraints for processes and regions

The numbers refer to the paragraphs below, where more detailed explanation is provided

nologies and the transformation of in-use stocks that leads to new recycling opportunities, resource depletion, and declining ore grades.

(2) *Assessment at full scale:* The ultimate goal of the coming transformation is to rescale human activity to a level that can be sustained by nature in the long run and that allows for future human development at the same time. Identifying the appropriate scale of human activity requires us to study socioeconomic metabolism on the global level, which was not necessary to understand the previous socioeconomic transitions.

Socioeconomic metabolism is a nonlinear system, which means that the impact of upscaling small modifications to the system is in general not proportional to the scaling factor. The upscaling of certain sustainable development strategies is subject to local and global constraints for, e.g., land, water, or mineral resources. Moreover, large-scale implementation of certain strategies feeds back into the system and causes structural change. Examples include changing recycling systems, technology learning, or rebound and spillover effects (Hertwich 2005). The total system-wide impact of the strategies' potential effect can therefore only be reliably assessed if the latter are studied at full scale, so that constraints and feedbacks can be included in the assessment.

(3a–c) *Multilayer modeling*, *satellite accounts*, and *balancing constraints* allow scientists from different disciplines, like industrial ecologists and economists, to use a consistent framework to describe society's metabolism and to address a variety of research questions. In multilayer modeling, the physical and economic properties of objects are quantified in consistent parallel frameworks (Pauliuk et al. 2015; Schmidt et al. 2012). Satellite accounts, like emissions to nature or labor requirements, contain additional information about how society's metabolism is connected to the environment and to human agents. They form the interface between models of SEM and those from other scientific disciplines, like climate models or environmental impact assessment. Balancing constraints for the physical and monetary layers, like industry or market balances, is the most fundamental way to check the validity of a prospective system description. Prospective models should always respect these fundamental balances.

(4) *Scenariomodeling with exogenous parameters* acknowledges the indeterminacy of future development and reduces system complexity to a manageable level. Only with scenario modeling one can build scientifically credible prospective models of complex indeterminate systems like socioeconomic metabolism. This central aspect of prospective modeling needs some more elaboration.

Socioeconomic metabolism is a non-isolated and non-deterministic complex system. It is not isolated, because it exchanges energy and matter with the inner of the Earth and with space. SEM is non-deterministic, because it is controlled by human agents that use their environmental literacy to intervene and divert the system from its current trajectory in a non-predictable manner. For such a system, there is no deterministic model that can predict its future development. Instead, scientists use *prospective models* of socio-economic metabolism to compute future trajectories of the system that are considered *possible but not necessarily likely* continuations of historic development. Such possible future trajectories are called *scenarios*. Prospective models use a trick to compute future scenarios for an indeterminate system: First, a number of *exogenous* parameters, assumptions, and model drivers are defined, and then these are fed into a dynamic model of socioeconomic metabolism that is deterministic relative to the exogenous parameters. Parameters like fertility or efficiency improvement rates, model drivers like GDP or population trajectories, and assumptions like "ceteris paribus" or "business as usual" describe the possible future development of certain indicators and system properties on the macro-scale. In a second step, the prospective model applies the exogenous assumptions to the system description and generates a detailed scenario for society's future metabolism. Specification of exogenous parameters not only eliminates indeterminacy from the model, it also reduces the complexity of the system description by fixing those system variables that one else would have to determine by modeling poorly understood feedback mechanisms or those where sufficient empirical data are not available. Scenario analysis is therefore an important way to handle our ignorance of human-environment systems in a productive and transparent way.

2.2 Credible, Possible, and Likely Scenarios

To decide what is a possible future and what is a credible scenario, scientists have established criteria that prospective models and their results need to fulfill. Exogenous assumptions need to be plausible and consistent. Often, they follow a certain scheme or idea that is called *story line*. Criteria for prospective models include process balancing constraints such as monetary, mass, or energy balances; the assumption that certain parameters, like efficiency improvement rates, do not leave empirically determined ranges; assumptions on human behavior; or the ability of the model to correctly determine the actual development from a given starting point in the past, using macro-indicators such as GDP as driver.

Criteria for plausibility, consistency, and properties of exogenous assumptions and prospective models differ across modeling fields, mainly because of different academic traditions. This scientific inconsistency has repeatedly led to criticism across modeling disciplines, like our criticism of integrated assessment models from an industrial ecology perspective presented below.

Are scenarios, or *possible* futures, also *likely* outcomes of future development? This often-raised question about the predictive capability of scenarios needs clarification. Strictly speaking, there cannot be a connection between possibility and likelihood in an indeterminate system, and a scenario can never be a prediction of the likely future outcome. This dogma, however, contradicts our intuition and the way the term scenario is often used. Especially when it covers only a short time span into the future, a scenario for the future development of SEM can appear to have predictive character (Börjeson et al. 2006). We assert that the apparent short-term determinacy of the indeterminate socio-metabolic system is a result of the slow turnover of in-use stocks, such as buildings, infrastructure, and products of different kinds, which adds considerable inertia to the system. The large amount of social and biophysical resources required to transform in-use stocks limits the speed at which the system can deviate from its present state (Pauliuk and Müller 2014). Hence, the spectrum of likely future states of SEM is the narrower the shorter the time horizon. Still, this predictability of an indeterminate system differs from the "absolute" predictions for truly deterministic systems because in indeterminate systems, unforeseen events such as the discovery of new technologies, sudden political changes, or natural catastrophes can substantially alter the trajectory even in the short run. To accommodate for the indeterminacy of SEM in the near future, prospective short-term models of society's metabolism are complemented by risk assessment.

The coming transformation of society's metabolism will require us to rebuild a substantial fraction of society's in-use stocks. The more complete the transformation, the less the future state of SEM is determined by present in-use stocks. Scenarios that cover time scales during which the coming transition may take place are therefore only little constrained by the inertia given by present stocks. Consequently, these scenarios have no predictive but explorative (What can happen?) or normative (How can a specific target be reached?) character (Börjeson et al. 2006). Prospective models in industrial ecology are used to study the transition ahead, and hence, the scenarios they generate are explorative or normative but not predictive.

3 Prospective Modeling in Industrial Ecology: State of the Art

3.1 Prospective Modeling with Established IE Methods

Industrial ecology methods and models, which allow us to study complex industrial systems, have been at the forefront of the interdisciplinary systems approach for more than three decades. Traditional industrial ecology methods include EE-I/O, LCA, MFA, urban metabolism, and industrial symbiosis. They cover a wide spectrum of spatial, temporal, and organizational scales, from static snapshots of the supply chain of local companies to studies of the evolution of aggregated material and energy flow accounts through the last centuries. They offer to decision-makers quantitative information about supply chains, environmental impacts embodied in trade, material and energy stocks and flows, and options for system-wide improvement.

Industrial ecology methods have reached high levels of sophistication and are broadly applied in companies and academia alike, but their use for prospective assessment of transformation strategies has remained a niche application. Prospective scenario exercises for I/O tables have repeatedly been conducted over the last decades (Cantono et al. 2008; De Koning et al. 2015; de Lange 1980; Idenburg and Wilting 2000; Leontief and Duchin 1986; Levine et al. 2007), but this modeling approach has not entered mainstream research on society's future metabolism. The reason may be twofold: (1) constructing I/O tables for future years requires many assumptions to be made and (2) using I/O tables in monetary units to measure interindustry flows, as in the studies above, makes it difficult to include physical process descriptions for specific technologies. Beyond IE, I/O tables form the core of computable general equilibrium (CGE) and prospective econometric models like the E3ME model (Burfisher 2011; Cambridge Econometrics 2014).

Most LCA studies are retrospective and attributional; they use historic data to model the life cycle of product systems and provide timeless indicators for environmental product performance. Prospective LCA (Lundie et al. 2004; Spielmann et al. 2005) and consequential LCA (CLCA) (Earles and Halog 2011; Finnveden et al. 2009; Whitefoot et al. 2011) add a forward-looking perspective to LCA. They typically assess transformation strategies on the small scale.

Prospective MFA studies mostly cover metals and building materials but do not include other layers or satellite accounts (Elshkaki and Graedel 2013; Hatayama et al. 2010; D. B. Müller 2006; Northey et al. 2014; Pauliuk et al. 2012; Sartori et al. 2008; Gallardo et al. 2014).

From the methods above, only MFA has been used to analyze preindustrial societies' socio-metabolic transitions (Krausmann 2011; Schaffartzik et al. 2014; Sieferle et al. 2006). These studies quantified trends in the total energy and material turnover of different socio-metabolic regimes, but they did not assess specific transformation strategies to shift from one regime to another.

3.2 New Approaches to Prospective Modeling in Industrial Ecology

The state of development of the above methods to conduct prospective studies of the next socio-metabolic transition is not satisfactory. The history of IE exhibits several examples for problems that were overcome by combining different IE methods into new frameworks. Examples include hybrid LCA (Suh et al. 2004) and WIO-MFA (Nakamura et al. 2007).

To come closer to the ultimate goal of studying a wide spectrum of transformation strategies at full scale in a common prospective modeling framework, the established IE methods have been combined in novel ways. As a result, a new family of prospective industrial ecology models is available, and we briefly present two of its members, extended dynamic MFA and THEMIS (Technology-Hybridized Environmental-Economic Model with Integrated Scenarios), and their application so far.

In-use stocks of buildings, infrastructure, or products are central in understanding the transition from the present to different possible future states (Pauliuk and Müller 2014). In-use stocks therefore need to be part of prospective models of socioeconomic metabolism. They are commonly represented as dynamic stock or population balance models, which are time series of stocks that are broken down into age cohorts and specific product types or technologies. The items in each age cohort and technology class can have specific material composition, energy efficiency, and other parameters necessary to determine the requirements and emissions of each item in the stock during its useful life.

Next to in-use stocks, prospective models of SEM contain descriptions of the industries to build up, maintain, and dispose of these stocks and markets that distribute products or product mixes across users. In-use stocks, industries, and markets are arranged into a general system description of socioeconomic metabolism (Fig. 2.1). The universal system structure of the socioeconomic metabolism in

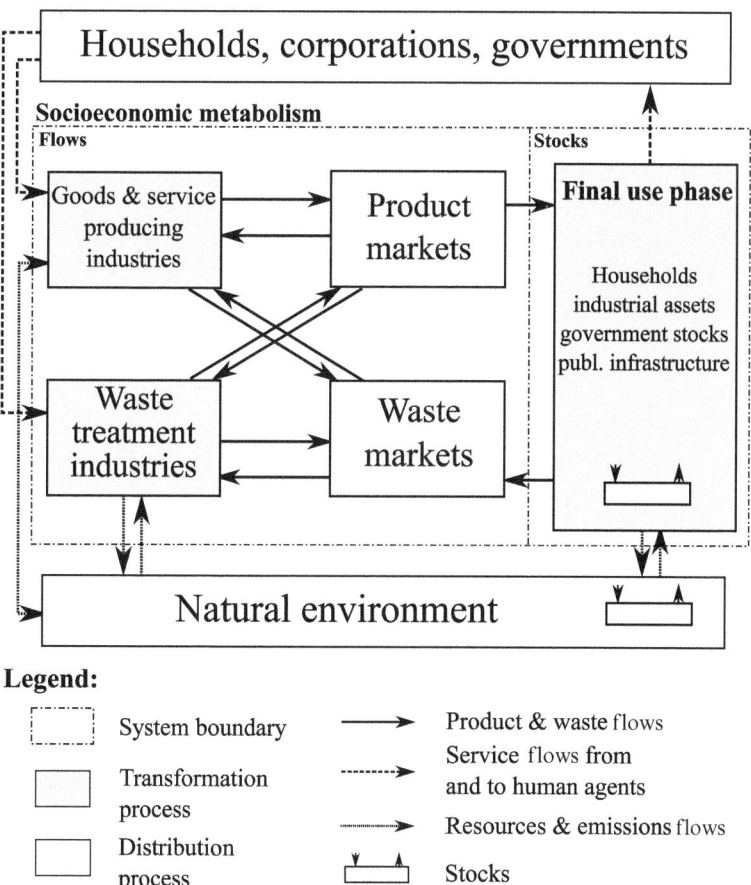

Fig. 2.1 The general structure of the system definitions of the prospective industrial ecology models (Adapted from Pauliuk et al. (2015))

Fig. 2.1 serves as blueprint for the structure of the system definitions of the different prospective models, including CGEs and IAMs (Pauliuk et al. 2015).

3.2.1 Prospective Modeling Using Extended Dynamic MFA

MFA models contain both flows and stocks; they are hence a natural starting point for dynamic and subsequently prospective modeling (Baccini and Bader 1996; Kleijn et al. 2000; D. B. Müller et al. 2004; van der Voet et al. 2002). MFA studies focus on a few materials or product groups at a time, and it was clear early on that prospective modeling with such a limited scope requires exogenous assumptions on the future development of material demand and technological change. In stock-driven modeling, the size of in-use stocks and the lifetime distribution of the different age cohorts are given exogenously, and deconvolution is applied to determine material demand and scrap supply (D. B. Müller 2006). The ability of dynamic stock models to determine future scrap supply from historic material consumption has enabled prospective modeling of mass-balanced recycling systems (Busch et al. 2014; Hatayama et al. 2009, 2010; Igarashi et al. 2007; E. Müller et al. 2014; Murakami et al. 2010; Tanikawa et al. 2002). These models often distinguish between different quality levels of secondary material and contain rules for the substitution of secondary for primary metal that are similar to system expansion in LCA or the by-product technology assumption in I/O (Daigo et al. 2014; Hashimoto et al. 2007; Løvik et al. 2014; Pauliuk et al. 2012, 2013a). Transformation strategies often affect products and the materials contained therein are not directly addressed. To understand the role of materials in different transformation strategies, there hence was a need to include product life cycles into dynamic MFA models, which led to the development of multilayer MFA and the combination of MFA with process-based LCA and life-cycle impact assessment (Milford et al. 2013; Pauliuk et al. 2013b; Sandberg and Brattebø 2012; Pauliuk 2013).

State-of-the-art extended dynamic MFA models comprise these different trends and provide large-scale and long-term dynamic assessments of specific transformation strategies, such as material efficiency (Milford et al. 2013) or passenger vehicle light-weighting (Modaresi et al. 2014). Starting from scenario assumptions on stock size and technology choice, these models apply stock-driven modeling to determine the levels of material production and energy supply that are required to build, operate, and dispose of the product stocks. They contain material-balanced process models of the industrial system and use satellite accounts to track resource consumption, energy supply, and emissions to the environment. A special feature of dynamic extended MFA is the high level of detail of the material cycles in the system, the distinction between open-loop recycling ("downcycling") and proper recycling, and their capability to quantify how changes in material production and recycling systems impact the overall effect of a certain transformation strategy.

3.2.2 Prospective Modeling Using the THEMIS Model

Large-scale deployment of more efficient and renewable energy technology can substantially reduce the environmental footprint of the global economy. It also leads to large changes in the carbon footprint of energy-intensive products and services such as materials or transportation. For example, the environmental superiority of electrically propelled passenger vehicles compared to gasoline-driven ones depends to a large extent on the carbon intensity of the electricity supply (Hawkins et al. 2013). Prospective LCAs of future technologies need to account for these different framing conditions, for example, by conducting a scenario analysis with different mixes of energy carriers and conversion technologies. Possible future mixes are commonly determined by integrated assessment models, such as the TIMES/MARKAL model family (Loulou et al. 2005), which stands behind the Energy Technology Perspectives of the International Energy Agency (OECD/IEA 2010). The technology mixes determined by such models can be used to build future scenarios for LCA databases, so that the market mix for certain products like electricity resembles the mix in the IAM scenarios. A thus modified LCA database can be used to conduct prospective attributional LCAs of future consumption.

The THEMIS model (Technology-Hybridized Environmental-Economic Model with Integrated Scenarios) is a recent implementation of this principle (Gibon et al. 2015). It provides insights into the "comparative environmental impacts and resource use of different electricity generation technologies" (Hertwich et al. 2015). THEMIS has four main features: (1) Its core is a nine-region integrated hybrid LC inventory model, which is a combination of foreground information on the specific technologies studied, a background LC inventory database of generic processes like materials production and transport, and MRIO tables to cover processes not contained in the LC background. (2) The historic technology mixes for electricity generation in the nine model regions were replaced with those obtained from the IEA baseline and BLUE MAP scenarios (OECD/IEA 2010) to build prospective future LC inventory models for 2030 and 2050. (3) The gradual transformation from the current to alternative future electricity mixes until 2050 was modeled with an age-cohort-based stock model of electricity generation assets, so that for every model year, the economy-wide impacts for building up new, operating, existing, and disposing of retiring electricity generation technology can be determined (Hertwich et al. 2015). (4) Exogenous scenario assumptions on the improvement of energy efficiency, capacity factors, and technology in the production of several major materials including aluminum, copper, nickel, iron and steel, and others were taken from a prospective study on efficiency improvement (ESU & IFEU 2008).

3.3 The Relation between Prospective IE Models and MFA, LCA, and I/O Analysis

The prospective IE models are built upon the established IE methods MFA, LCA, and IOA. They are integrated hybrid models of society's metabolism, in the sense that they combine a foreground system with high level of detail and strict adherence to core modeling principles such as mass balance with a generic background system. The background provides the foreground with auxiliary input, such as electricity supply for material production, and uses the products exclusively supplied by the foreground as intermediate requirements in turn.

The foreground of extended dynamic MFA comprises dynamic stock models of the materials and products studied and process models of the industries that are part of the material cycles studied. The foreground system is balanced for the products and material that are within the scope, and the background system supplies energy and other ancillary inputs to operate the stocks and processes in the foreground. Environmental impact assessment is carried out for the satellite accounts of relevant emissions from both foreground and background. Because extended dynamic MFA contains process models, considers the background economy, and uses impact assessment, one can also consider these models as macro-LCAs of the total service provided by the stocks studied, carried out as dynamic studies with scenarios for future development. The foreground model of extended dynamic MFA contains markets at all stages, preserves co-production, and contains rules for substituting secondary material for primary material. It can therefore be reformulated as combination of a physical waste-I/O model with the by-product technology assumption combined with a dynamic stock model of the products studied.

THEMIS integrates LCA and I/O modeling and combines the so-obtained hybrid model of interindustry flows with environmental impact assessment via satellite accounts. It also contains elements that are commonly found in dynamic MFA: THEMIS's foreground system is coupled to a dynamic stock model of electricity generation assets, so that material demand for building new assets and recycling of old ones is determined from the turnover of the capital stock in mass-balanced manner.

3.4   The Relation between Prospective IE Models and Consequential LCA

The desire to study the potential future consequences of a decision has been a long-standing motivation for industrial ecology research, and a few recent examples were cited above. In LCA, this desire has led to the concept of consequential life-cycle assessment (CLCA), which "is designed to generate information on the consequences of a decision" (Ekvall and Weidema 2004). While the concept of a consequential LCA is intriguing, it is also poorly defined and subject of controversy (Brandão et al. 2014; Dale and Kim 2014; Finnveden et al. 2009; Hertwich 2014; Plevin et al. 2014a; Suh and Yang 2014; Zamagni et al. 2012). CLCA was initially defined as a result of the debate on how to allocate emissions and inputs of processes with multiple products to the respective outputs. It focused on the marginal effect of producing an additional unit of output of a specific product or of recycling such a product (Ekvall and Weidema 2004). Such allocation problems are addressed through systems expansion in CLCA, so that the assessment of a product depends not only on the product system of the investigated product but also on the product systems of other products and, in particular, the production volumes of and demand for those products. This dependence on other product systems is most clearly visible when scientists, under the heading of "consequential LCA," address the question of what happens to constrained resources when a product is not produced. In the opinion of some, the consequential life-cycle emissions of a bicycle should include those of combusting the petrol that it does not use because somebody else will combust that petrol as a response of the market to the bicycle not using that petrol (Plevin et al. 2014a, b). One of us (Hertwich 2014) has questioned whether it makes sense to say that the petrol not combusted by the bicycle is part of the product system "bicycle" only because one could have used a car instead. It is also problematic to say that riding the bicycle to work causes petrol combustion somewhere else in the economy due to price elasticity. The petrol combustion is rather the consequence of somebody else's decision somewhere else in the economy.

It is of course a legitimate research questions to ask, e.g., what is the effect of the massive and intended expansion of cycling in Copenhagen on GHG emissions? The question, however, remains ill-defined until one juxtaposes the observed or planned expansion of cycling to some counterfactual possible scenario of increased car or bus transport. In addition, one needs to define the scope and functioning of the system investigated, including the causal mechanisms to be addressed. Mechanisms may or may not include the fuel market response to the petrol demand in the counterfactual scenario, the effect of the inspiration Copenhagen now provides to town planners all over the world and the effect of increased life expectancy of the cyclists on food demand and future economic development. Then one has two scenarios to compare, and one may colloquially argue the difference between the scenarios that indicates the effect of Copenhagen's cycling policy and the enthusiastic popular response it has received. Such causality is an imputed, assumed causality; the assumptions are made in the setup of the systems model and the definition of the scenarios and the imputation in the interpretation of the scenario results as showing the difference. Other system models and scenario assumptions may be equally reasonable; the true consequences are unmeasurable because we do not have a second Earth to run an experiment on.

The early developers of systems expansion as a way of addressing allocation issues fully understood that system expansion involved assumptions about other product systems and that results should be interpreted with these assumptions in mind.

We expand our above argument regarding the predictive capacity of prospective models for indeterminate systems and assert that the hypothetical CLCA approach as described by Plevin et al. (2014a) faces a dilemma: The capability of the hypothetical CLCA model to reliably predict the future outcome is the better the smaller the changes to the system and the shorter the time horizon, because fewer human actors, who are the major source of indeterminacy, are involved if changes are small and local and the inertia represented by existing stocks is larger in the near future.

Modeling on the small scale with short time horizons is the opposite of what is needed for studying strategies for a socio-metabolic transition, however, and for large-scale and long-term changes, prediction remains an illusion. Therefore, we need a practical and scientifically credible implementation of the ideal represented by the hypothetical CLCA approach.

In our opinion, prospective IE models provide such implementation. The scenario approach makes explicit the underlying exogenous assumptions that necessarily accompany any prospective model of an indeterminate system. Several authors, including Zamagni et al. (2012), Plevin et al. (2014a), and Suh and Yang (2014), acknowledge the importance of scenario modeling for the scientific assessment of decision-making in general and the questions posed by CLCA in particular. The use of a comprehensive model of society's metabolism allows us to study the system with the high level of detail and biophysical consistency that is a distinctive feature of industrial ecology methods. The combination of the scenario approach and a detailed model of society's metabolism make prospective IE models a powerful and scientifically credible approach to explore the potential consequences of decisions.

4 Prospective Modeling in Industrial Ecology: Future Development

4.1 Future Applications and Model Development of Prospective Models within Industrial Ecology

A major goal of prospective modeling is to assess bundles of mitigation and adaptation strategies and investigate whether the different strategies together can transform socioeconomic metabolism to a more sustainable state. Studying strategy bundles reveals which strategies may yield co-benefits and which ones counteract each other, which is an important information for decision-makers. Bundled assessment leads to "big picture" scenarios for a feasible future, from which environmental, economic, and social performance indicators for individual strategies can be derived. These indicators can then be monitored during real implementation to ensure that the impact of the strategies is as intended. Performance indicators may be material, product, industry sector, or region specific.

Strategy bundles affect different materials and energy carriers, which are substitutable to some extent. Flexibility in the choice of materials and energy carriers allows us to design a more resilient and potentially more sustainable SEM, but it also represents a challenge for prospective modeling, as models need to provide insights into the potential consequences of a wide spectrum of material and energy supply choices.

The cycles of different materials are tightly coupled at several places: Base minerals of different materials often occur together; they are coproduced, often with fixed ratios on certain sites. At higher stages of fabrication, materials are mixed again into compound materials and alloys, products consist of many different materials, and finally, waste streams contain material mixes. Assessments of individual metals on the small scale can neglect this coupling, as it can be assumed that the rest of the economy is able to supply or absorb ancillary flows and a credit or discredit for this service is given by allocation. In prospective modeling of society's future metabolism at full scale, however, the tight coupling between different material cycles ultimately necessitates parallel modeling of different materials across products and over time. Only then can one assess whether and how system-wide supply can meet system-wide demand for different chemical elements at different stages of the material cycles. Supply-demand imbalances may arise under business-as-usual assumptions, as studies for aluminum (Modaresi and Müller 2012) and rare earth metals (Elshkaki and Graedel 2014) show, which points out the necessity to design future material cycles from a systems' perspective.

For dynamic MFA, several trends that point toward comprehensive assessment of multi-material product portfolios are already emerging. One trend goes toward a higher level of detail of material types (alloys) and products studied to better understand quality issues in the recycling systems of different materials (Løvik et al. 2014; Ohno et al. 2014). Another trend goes toward modeling of co-occurrence, co-mining, and co-production of mineral and metal resources and production systems and energy-ore grade relationships (Graedel et al. 2013; Northey et al. 2014). Finally, there are recent advances in the modeling of the fate of the end-of-life materials from the waste management industries back into new products using Markov chains and supply-driven I/O modeling (Duchin and Levine 2013; Nakamura et al. 2014).

The trend of using I/O models for prospective assessments is also likely to continue. Service-driven modeling, or – if stocks are used as proxy for services – stock-driven modeling, can be used (a) to determine the final demand vector for I/O models (Kagawa et al. 2015) and (b) to determine the Leontief-A matrix from an age-cohort-based model of the productive capital stock (Pauliuk et al. 2015). Multilayer modeling (Schmidt et al. 2012) can be used to cover different materials in a common I/O framework, and when building I/O models of future industrial systems, the by-product technology construct can be applied to avoid allocation (Majeau-Bettez et al. 2014). A major application of the so-obtained I/O models is the prospective attributional assessment of certain quanta of final demand to measure strategy performance and derive policy targets related to specific transformation strategies.

4.2 Linking Industrial Ecology and Integrated Assessment Models (IAMs)

The most prominent contribution to prospective modeling of society's metabolism did not emerge from industrial ecology but from other disciplines, especially integrated assessment modeling. While widely successful in generating integrated scenarios of the society' future metabolism, the biosphere, and the climate system, integrated assessment models were criticized for several shortcomings (Arvesen et al. 2011; Pindyck 2013; Stern 2013). To our knowledge, a detailed criticism of IAMs from an industrial ecology perspective is still lacking; it is also beyond the scope of this chapter; however, below we list central points of critique and propose some ideas for the integration of IE principles into IAMs.

4.2.1 Integrated Assessment Models from an Industrial Ecology Perspective

In general, one can say that industrial ecologists have a more refined notion of industry as a complex system than what is currently characterized in most integrated assessment models. IAMs trace the extraction, refinement, and conversion of energy carriers but lack a description of the life cycle of the conversion technologies. IAMs model energy demand and energy efficiency opportunities in industry, buildings, and transport but lack a description of value chains and hence the interaction of different sectors. IAMs contain a description of material production as the most important industrial energy consumer but lack detail regarding the modeling of waste generation and recycling, co-production, material demand, and associated environmental impacts.

In IAMs, the level of material recycling – if recycling is considered at all – is in general not determined from the turnover of the industrial assets and products in the use phase. Either the material stocks (mostly steel and aluminum) are modeled separately from capital and product stocks and an average lifetime is used, or there is no connection at all between the extent of recycling and historic levels of material consumption. Co-production, substitution of by-products, and industrial symbiosis are only rudimentarily considered, if at all. Due to the focus on climate change, other environmental impact categories, like toxicity or acidification, are generally not taken into account in energy systems models nor are mineral resource depletion and the relation between ore grade and energy demand for extraction commonly considered.

In a nutshell, we assert that integrated assessment models lack several features that are central achievements of industrial ecology research. From the perspective of our field, the scenarios of society's future metabolism constructed by IAMs lack consistency and validity, which may compromise the credibility of the conclusions drawn.

4.2.2 The Link Between the Prospective IE Models and IAMs

Integrating IE principles into IAMs will allow the latter to construct more consistent and realistic scenarios of society's future metabolism. Moreover, this integration can increase the scope of IE research, and as a result, both research fields can move forward. We present some ideas for how this integration could happen.

The general system structure of socioeconomic metabolism acknowledges three types of processes: the industries, the markets, and the final use phase (Pauliuk et al. 2015). There are canonical models for each process type. Industries are modeled by production functions, markets by supply and demand curves, and the use phase by dynamic stock models. We discuss each process type in turn.

(1) *Industries*: Both IAMs and IE models commonly use Leontief-type production functions with fixed technical coefficients and no substitution between inputs to

describe intermediate demand. Technological detail is high in both model families, and in IAMs, industrial assets are often modeled as discrete units to simulate individual plants. Technological change is exogenous in IE models but endogenous in some IAMs. We see three ways in which IE principles can improve the modeling of production processes in IAMs: (1) mass balance consistency between industrial input and output, inclusion of waste generation and recycling; (2) consideration of multi-output processes and industrial symbiosis, that is, joint production of commodities and by-products within and across industrial sectors; (3) separate description of primary and secondary material production. Point (4) is already part of several IAMs, but the way scrap supply is modeled differs substantially across different IAMs, and not all approaches are meaningful from an IE perspective.

(2) *Marketsand products*: In IE models and IAMs, markets balance supply and demand. It is common in IE to distinguish between markets for primary products and waste and to require the models to clear both markets at the same time. Introduction of waste markets into IAMs and the development of mechanisms for how they are cleared would allow for realistic modeling of waste treatment and recycling activities. Detailed descriptions of waste handling, recycling, and substitution of secondary for primary materials are essential in understanding how a transition to a more circular economy could happen; they should therefore be an integral part of future prospective models. More details regarding the quality of materials, for example, by considering different alloys, will be necessary to build scenarios for the use of secondary materials in different products.

(3) *The final use phase* comprises in-use stocks such as products used by households, public buildings and infrastructure, and industrial assets. Both prospective IE models and IAMs use age-cohort-based dynamic models to represent in-use stocks. The age cohort technology composition of industrial assets, vehicles, or buildings determines the overall intermediate energy and material requirements to produce industrial output, drive vehicles, or heat buildings, and modeling this relation is the traditional strength of IAMs. In-use stocks, however, also represent material stocks or "urban mines." Modeling the material layer of in-use stocks along with their economic value and technical coefficients allows us to obtain a comprehensive picture of stocks and to connect the future extent of waste recovery and recycling to the physical turnover of industrial assets and other in-use stocks. The principles for this multilayer modeling of stocks are presented elsewhere (Pauliuk et al. 2015), and to implement them into IAMs, one needs to amend the description of in-use stocks by adding the material composition layer.

IAMs can also be extended regarding the *interaction of SEM with environment and society*. Due to their focus on climate change, IAMs generally focus on greenhouse gas emissions, but other types of emissions, such as particulate matter or heavy metals and other toxic substances, can be readily included (Gibon et al. 2015). IAMs contain detailed descriptions of biotic resources but should be extended to better reflect the depletion of mineral resources, especially metal ores. More details can be added on the social side, too, for example, by capturing direct relations between production activities and society, such as labor requirements at different levels of skill or labor conditions (Simas et al. 2014).

IAMs that fully respect IE principles will generate scenarios that include a wide spectrum of transformation strategies, provide a mass balance consistent and material-specific representation of SEM, and comprehensively cover interactions of SEM with the environment and society. These comprehensive and scientifically credible scenarios of society's future metabolism will then form a common basis for the analysis by different scientific disciplines.

5   Conclusion

Prospective IE models combine central features of the established IE methods into a new framework. They allow researchers to conduct comprehensive and dynamic scenario analyses of society's future metabolism and to study the potential system-wide effect of sustainable development strategies. The development of these models is motivated by the desire to study the coming socio-metabolic transition and to assess the different transformation strategies at full scale and with long-term scope, while maintaining the high level of detail and biophysical consistency that is a distinctive feature of industrial ecology methods. The intellectual framing that comes along with the prospective IE models provides a "new slang" for the field: It broadens the perspective of industrial ecology research because it gives impulses for the development of new and important research questions for further refinement and integration of assessment methods and for the development of a common, model-independent database of socioeconomic metabolism.

The recent development in industrial ecology methods necessitates a discussion about the relation between prospective IE models and IAMs since the latter are the major tool for prospective assessment of transformation strategies. We contributed to this debate by proposing how industrial ecology principles could become an integral part of integrated assessment models and how this integration could strengthen both fields and increase the relevance, robustness, and credibility of scientific assessment of transformation strategies.

Stefan Pauliuk is a postdoctoral researcher, and **Edgar G Hertwich** is a professor at the Industrial Ecology Programme and the Department of Energy and Process Engineering at the Norwegian University of Science and Technology (NTNU), Trondheim, Norway.

Acknowledgements The authors acknowledge the work of Daniel B Müller, who is the principal investigator in the development of extended dynamic MFA and who commented on an early draft of this chapter. Guillaume Majeau-Bettez pointed out the necessity for the overview presented here. The contribution of Stefan Pauliuk was funded by the Research Council of Norway under the CENSES Project (Grant number 209697). The funding source was not involved in this work.

References

Arvesen, A., Bright, R. M., & Hertwich, E. G. (2011). Considering only first-order effects? How simplifications lead to unrealistic technology optimism in climate change mitigation. *Energy Policy, 39*(11), 7448–7454.

Baccini, P., & Bader, H.-P. (1996). *Regionaler Stoffhaushalt. Erfassung, Bewertung und Steuerung* (p. 420). Heidelberg: Spektrum.

Barnosky, A. D., Hadly, E. A, Bascompte, J., Berlow, E. L., Brown, J. H., Fortelius, M., Getz, W. M., Harte, J., Hastings, A., Marquet, P. A., Martinez, N. D., Mooers, A., Roopnarine, P., Vermeij, G., Williams, J. W., … & Smith, A. B. (2012). Approaching a state shift in Earth's biosphere. *Nature, 486*(7401), 52–58.

Binder, C. R., Hinkel, J., Bots, P. W. G., & Pahl-Wostl, C. (2013). Comparison of frameworks for analyzing social-ecological systems. *Ecology and Society, 18*(4), 26.

Börjeson, L., Höjer, M., Dreborg, K.-H., Ekvall, T., & Finnveden, G. (2006). Scenario types and techniques: Towards a user's guide. *Futures, 38*(7), 723–739.

Brandão, M., Clift, R., Cowie, A., & Greenhalgh, S. (2014). The use of life cycle assessment of the support of robust (climate) policy making: Comment on "Using Attributional Life Cycle Assessment to Estimate Climate-Change Mitigation …". *Journal of Industrial Ecology, 18*(3), 461–463.

Burfisher, M. E. (2011). *Introduction to computable general equilibrium models*. New York: Cambridge University Press.

Busch, J., Steinberger, J. K., Dawson, D. A., Purnell, P., & Roelich, K. E. (2014). Managing critical materials with a technology- specific stocks and flows model. *Environmental Science & Technology, 48*(2), 1298–1305.

Cambridge Econometrics. (2014). *E3ME technical manual, version 6.0 April 2014*. Cambridge.

Cantono, S., Heijungs, R., & Kleijn, R. (2008). Environmental accounting of eco-innovations through environmental input–output analysis: The case of hydrogen and fuel cells buses. *Economic Systems Research, 20*(3), 303–318.

Daigo, I., Osako, S., Adachi, Y., & Matsuno, Y. (2014). Time-series analysis of global zinc demand associated with steel. *Resources Conservation and Recycling, 82*, 35–40.

Dale, B. E., & Kim, S. (2014). Can the predictions of consequential life cycle assessment be tested in the real world? Comment on "Using Attributional Life Cycle Assessment to Estimate Climate-Change Mitigation…". *Journal of Industrial Ecology, 18*(3), 466–467.

De Koning, A., Huppes, G., Deetman, S., & Tukker, A. (2015, February). Scenarios for a 2 °C world: A trade-linked input–output model with high sector detail. *Climate Policy*, 1–17.

De Lange, A. R. (1980). A dynamic input-output model for investigating alternative futures: Applications to the South African economy. *Technological Forecasting and Social Change, 18*, 235–245.

Duchin, F., & Levine, S. H. (2013). Embodied resource flows in a global economy. *Journal of Industrial Ecology, 17*(1), 65–78.

Earles, J. M., & Halog, A. (2011). Consequential life cycle assessment: A review. *The International Journal of Life Cycle Assessment, 16*(5), 445–453.

Ekvall, T., & Weidema, B. P. (2004). System boundaries and input data in consequential life cycle inventory analysis. *The International Journal of Life Cycle Assessment, 9*(3), 161–171.

Elshkaki, A., & Graedel, T. E. (2013). Dynamic analysis of the global metals flows and stocks in electricity generation technologies. *Journal of Cleaner Production, 59*, 260–273.

Elshkaki, A., & Graedel, T. E. (2014). Dysprosium, the balance problem, and wind power technology. *Applied Energy, 136*, 548–559.

ESU & IFEU. (2008). *LCA of background processes*. Project report of NEEDS: "New energy externalities – Development for sustainability." Zürich.

Finnveden, G., Hauschild, M. Z., Ekvall, T., Guinée, J., Heijungs, R., Hellweg, S., … & Suh, S. (2009). Recent developments in life cycle assessment. *Journal of Environmental Management, 91*(1), 1–21.

Fischer-Kowalski, M., & Haberl, H. (1998). Sustainable development: Socio-economic metabolism and colonization of nature. *International Social Science Journal, 50*(158), 573–587.

Fischer-Kowalski, M., & Haberl, H. (2007). *Socioecological transitions and global change. Trajectories of social metabolism and land use*. Cheltenham: Edward Elgar.

Fischer-Kowalski, M., & Weisz, H. (1999). Society as hybrid between material and symbolic realms: Toward a theoretical framework of society-nature interaction. *Advances in Human Ecology, 8*, 215–251.

Fischer-Kowalski, M., Krausmann, F., & Pallua, I. (2014). A sociometabolic reading of the Anthropocene: Modes of subsistence, population size and human impact on Earth. *The Anthropocene Review*.

Gallardo, C., Sandberg, N. H., & Brattebø, H. (2014). Dynamic-MFA examination of Chilean housing stock: long-term changes and earthquake damage. *Building Research and Information, 42*(3), 1–16.

Gibon, T., Hertwich, E. G., Wood, R., Bergesen, J., & Suh, S. (2015). A methodology for scenario analysis in hybrid input-output analysis: Case study on energy technologies. NTNU, Trondheim (in preparation).

Graedel, T. E., Harper, E. M., Nassar, N. T., & Reck, B. K. (2013). On the materials basis of modern society. *Proceedings of the National Academy of Sciences of the United States of America*.

Hashimoto, S., Tanikawa, H., & Moriguchi, Y. (2007). Where will large amounts of materials accumulated within the economy go?–A material flow analysis of construction minerals for Japan. *Waste Management, 27*(12), 1725–1738.

Hatayama, H., Daigo, I., Matsuno, Y., & Adachi, Y. (2009). Assessment of the recycling potential of aluminum in Japan, the United States, Europe and China. *Materials Transactions, 50*(3), 650–656.

Hatayama, H., Daigo, I., Matsuno, Y., & Adachi, Y. (2010). Outlook of the world steel cycle based on the stock and flow dynamics. *Environmental Science & Technology, 44*(16), 6457–6463.

Hawkins, T. R., Singh, B., Majeau-Bettez, G., & Strømman, A. H. (2013). Comparative environmental life cycle assessment of conventional and electric vehicles. *Journal of Industrial Ecology, 17*(1), 53–64.

Hertwich, E. G. (2005). Consumption and the rebound effect: An industrial ecology perspective. *Journal of Industrial Ecology, 9*(1–2), 85–98. Retrieved February 23, 2015, from http://mit-press.mit.edu/jie

Hertwich, E. G. (2014). Understanding the climate mitigation benefits of product systems: Comment on "Using Attributional Life Cycle Assessment to Estimate Climate-Change Mitigation….". *Journal of Industrial Ecology, 18*(3), 464–465.

Hertwich, E. G., Gibon, T., Bouman, E. A., Arvesen, A., Suh, S., Heath, G. A., … & Shi, L. (2015). Integrated life-cycle assessment of electricity-supply scenarios confirms global environmental benefit of low-carbon technologies. *Proceedings of the National Academy of Sciences, 112*(20), 6277–6282.

Idenburg, A. M., & Wilting, H. C. (2000). *DIMITRI: A dynamic input-output model to study the impacts of technology related innovations* (pp. 1–18). Macerata: University of Macerata.

Igarashi, Y., Daigo, I., Matsuno, Y., & Adachi, Y. (2007). Estimation of the change in quality of domestic steel production affected by steel scrap exports. *ISIJ International, 47*(5), 753–757.

Kagawa, S., Nakamura, S., Kondo, Y., Matsubae, K., & Nagasaka, T. (2015). Forecasting replacement demand of durable goods and the induced secondary material flows. *Journal of Industrial Ecology, 19*(1), 10–19.

Kleijn, R., Huele, R., & van der Voet, E. (2000). Dynamic substance flow analysis: The delaying mechanism of stocks, with the case of PVC in Sweden. *Ecological Economics, 32*(2), 241–254.

Krausmann, F. (2011). *The socio-metabolic transition. Long term historical trends and patterns in global material and energy use.* Vienna: Institute of Social Ecology.

Krausmann, F., & Fischer-Kowalski, M. (2013). Global socio-metabolic transitions. In S. J. Singh, H. Haberl, M. Chertow, M. Mirtl, & M. Schmid (Eds.), *Long term socio-ecological research* (Human-envi, pp. 339–365). Dordrecht: Springer.

Leontief, W. W., & Duchin, F. (1986). *The future impact of automation on workers.* New York: Oxford University Press.

Levine, S. H., Gloria, T. P., & Romanoff, E. (2007). A dynamic model for determining the temporal distribution of environmental burden. *Journal of Industrial Ecology, 11*(4), 39–49.

Loulou, R., Remne, U., Kanudia, A., Lehtila, A., & Goldstein, G. (2005). *Documentation for the TIMES model* (pp. 1–78). Paris: Energy Technology Systems Analysis Programme (ETSAP).

Løvik, A. N., Modaresi, R., & Müller, D. B. (2014). Long-term strategies for increased recycling of automotive aluminum and its alloying elements. *Environmental Science & Technology, 48*(8), 4257–4265.

Lundie, S., Peters, G. M., & Beavis, P. C. (2004). Life cycle assessment for sustainable metropolitan water systems planning. *Environmental Science & Technology, 38*(13), 3465–3473.

Majeau-Bettez, G., Wood, R., & Strømman, A. H. (2014). Unified theory of allocations and constructs in life cycle assessment and input-output analysis. *Journal of Industrial Ecology, 18*(5), 747–770.

Milford, R. L., Pauliuk, S., Allwood, J. M., & Müller, D. B. (2013). The roles of energy and material efficiency in meeting steel industry CO2 targets. *Environmental Science & Technology, 47*(7), 3455–3462.

Modaresi, R., & Müller, D. B. (2012). The role of automobiles for the future of aluminum recycling. *Environmental Science & Technology, 46*(16), 8587–8594.

Modaresi, R., Pauliuk, S., Løvik, A. N., & Müller, D. B. (2014). Global carbon benefits of material substitution in passenger cars until 2050 and the impact on the steel and aluminum industries. *Environmental Science & Technology, 48*(18), 10776–10784.

Müller, D. B. (2006). Stock dynamics for forecasting material flows – Case study for housing in The Netherlands. *Ecological Economics, 59*(1), 142–156.

Müller, D. B., Bader, H.-P., & Baccini, P. (2004). Long-term coordination of timber production and consumption using a dynamic material and energy flow analysis. *Journal of Industrial Ecology, 8*(3), 65–87.

Müller, E., Hilty, L. M., Widmer, R., Schluep, M., & Faulstich, M. (2014). Modeling metal stocks and flows – A review of dynamic material flow analysis methods. *Environmental Science & Technology, 48*(4), 2102–2113.

Murakami, S., Oguchi, M., Tasaki, T., Daigo, I., & Hashimoto, S. (2010). Lifespan of commodities, Part I – The creation of a database and its review. *Journal of Industrial Ecology, 14*(4), 598–612.

Nakamura, S., Nakajima, K., Kondo, Y., & Nagasaka, T. (2007). The waste input-output approach to materials flow analysis concepts and application to base metals. *Journal of Industrial Ecology, 11*(4), 50–63.

Nakamura, S., Kondo, Y., Kagawa, S., Matsubae, K., Nakajima, K., & Nagasaka, T. (2014). MaTrace: Tracing the fate of materials over time and across products in open-loop recycling. *Environmental Science & Technology, 48*(13), 7207–7214.

Northey, S., Mohr, S., Mudd, G. M., Weng, Z., & Giurco, D. (2014). Modelling future copper ore grade decline based on a detailed assessment of copper resources and mining. *Resources, Conservation and Recycling, 83*, 190–201.

OECD/IEA. (2010). *Energy technology perspectives : Scenarios and strategies to 2050.* Paris: International Energy Agency.

Ohno, H., Matsubae, K., Nakajima, K., Nakamura, S., & Nagasaka, T. (2014). Unintentional flow of alloying elements in steel during recycling of end-of-life vehicles. *Journal of Industrial Ecology, 18*(2), 242–253.

Pauliuk, S. (2013). *The role of stock dynamics in climate change mitigation*. PhD thesis, NTNU, Trondheim, Norway.

Pauliuk, S., & Hertwich, E. G. (2015). *Socioeconomic metabolism as paradigm for studying the biophysical basis of human society*. Trondheim: NTNU. Under review with Ecological Economics.

Pauliuk, S., Milford, R. L., Müller, D. B., & Allwood, J. M. (2013a). The steel scrap age. *Environmental Science & Technology, 47*(7), 3448–3454.

Pauliuk, S., & Müller, D. B. (2014). The role of in-use stocks in the social metabolism and in climate change mitigation. *Global Environmental Change, 24*, 132–142.

Pauliuk, S., Sjöstrand, K., & Müller, D. B. (2013b). Transforming the Norwegian dwelling stock to reach the 2 degrees celsius climate target. *Journal of Industrial Ecology, 17*(4), 542–554.

Pauliuk, S., Wang, T., & Müller, D. B. (2012). Moving toward the circular economy: The role of stocks in the Chinese steel cycle. *Environmental Science & Technology, 46*(1), 148–154.

Pauliuk, S., Wood, R., & Hertwich, E. G. (2015). Dynamic models of fixed capital stocks and their application in industrial ecology. *Journal of Industrial Ecology, 19*(1), 104–116.

Pauliuk, S., Majeau-Bettez, G., & Müller, D. B. (2015). A general system structure and accounting framework for socioeconomic metabolism. *Journal of Industrial Ecology* (forthcoming).

Pindyck, R. S. (2013). Climate change policy: What do the models tell us? *Journal of Economic Literature, 51*(3), 860–872.

Plevin, R. J., Delucchi, M. A., & Creutzig, F. (2014a). Using attributional life cycle assessment to estimate climate-change mitigation benefits misleads policy makers. *Journal of Industrial Ecology, 18*(1), 73–83.

Plevin, R. J., Delucchi, M., & Creutzig, F. (2014b). Response to comments on "Using Attributional Life Cycle Assessment to Estimate Climate-Change Mitigation …". *Journal of Industrial Ecology, 18*(3), 468–470.

Sandberg, N. H., & Brattebø, H. (2012). Analysis of energy and carbon flows in the future Norwegian dwelling stock. *Building Research and Information, 40*(2), 123–139.

Sartori, I., Bergsdal, H., Müller, D. B., & Brattebø, H. (2008). Towards modelling of construction, renovation and demolition activities: Norway's dwelling stock, 1900–2100. *Building Research & Information, 36*(5), 412–425.

Schaffartzik, A., Mayer, A., Gingrich, S., Eisenmenger, N., Loy, C., & Krausmann, F. (2014). The global metabolic transition: Regional patterns and trends of global material flows, 1950–2010. *Global Environmental Change, 26*, 87–97.

Schmidt, J., Merciai, S., Delahaye, R., Vuik, J., Heijungs, R., de Koning, A., & Sahoo, A. (2012). *EU-CREEA project. Deliverable no. 4.1. Recommendation of terminology, classification, framework of waste accounts and MFA, and data collection guideline*. Aalborg: 2.0 LCA consultants.

Scholz, R. W., & Binder, C. R. (2011). *Environmental literacy in science and society: From knowledge to decisions*. Cambridge: Cambridge University Press.

Sieferle, R. P., Krausmann, F., Schandl, H., & Winiwarter, V. (2006). *Das Ende der Fläche*. Cologne: Böhlau & Cie.

Simas, M., Golsteijn, L., Huijbregts, M., Wood, R., & Hertwich, E. G. (2014). The "Bad Labor" footprint: Quantifying the social impacts of globalization. *Sustainability, 6*(11), 7514–7540.

Spielmann, M., Scholz, R. W., Tietje, O., & Haan, P. D. (2005). Scenario modelling in prospective LCA of transport systems – Application of formative scenario analysis. *The International Journal of Life Cycle Assessment, 10*(5), 325–335.

Stern, N. (2013). The structure of economic modeling of the potential impacts of climate change: Grafting gross underestimation of risk onto already narrow science models. *Journal of Economic Literature, 51*(3), 838–859.

Suh, S., & Yang, Y. (2014). On the uncanny capabilities of consequential LCA. *The International Journal of Life Cycle Assessment, 19*, 1179–1184.

Suh, S., Lenzen, M., Treloar, G. J., Hondo, H., Horvath, A., Huppes, G., ... Norris, G. (2004). System boundary selection in life-cycle inventories using hybrid approaches. *Environmental Science & Technology, 38*(3), 657–664.

Tanikawa, H., Hashimoto, S., & Moriguchi, Y. (2002). Estimation of material stock in urban civil infrastructures and buildings for the prediction of waste generation. In *5th International conference on ecobalance* (pp. 806–809). Tokyo: Society for Non-Traditional Technology.

Van der Voet, E., Kleijn, R., Huele, R., Ishikawa, M., & Verkuijlen, E. (2002). Predicting future emissions based on characteristics of stocks. *Ecological Economics, 41*(2), 223–234.

Whitefoot, K. S., Grimes-Casey, H. G., Girata, C. E., Morrow, W. R., Winebrake, J. J., Keoleian, G. A., & Skerlos, S. J. (2011). Consequential life cycle assessment with market-driven design. *Journal of Industrial Ecology, 15*(5), 726–742.

Zamagni, A., Guinée, J., Heijungs, R., Masoni, P., & Raggi, A. (2012). Lights and shadows in consequential LCA. *The International Journal of Life Cycle Assessment, 17*(7), 904–918.

Chapter 3
Life Cycle Sustainability Assessment: What Is It and What Are Its Challenges?

Jeroen Guinée

Abstract Environmental life cycle assessment (LCA) has developed fast over the last three decades. Today, LCA is widely applied and used as a tool for supporting policies and performance-based regulation, notably concerning bioenergy. Over the past decade, LCA has broadened to also include life cycle costing (LCC) and social LCA (SLCA), drawing on the three-pillar or 'triple bottom line' model of sustainability. With these developments, LCA has broadened from merely environmental assessment to a more comprehensive life cycle sustainability assessment (LCSA). LCSA has received increasing attention over the past years, while at the same time, its meaning and contents are not always sufficiently clear. In this chapter, we therefore addressed the question: what are LCSA practitioners actually doing in practice? We distinguished two sub-questions: which definition(s) do they adopt and what challenges do they face? To answer these questions, LCSA research published over the past half decade has been analysed, supplemented by a brief questionnaire to researchers and practitioners. This analysis revealed two main definitions of LCSA. Based on these two definitions, we distinguished three dimensions along which LCSA is expanding when compared to environmental LCA: (1) broadening of impacts, LCSA = LCA + LCC + SLCA; (2) broadening level of analysis, product-, sector- and economy-wide questions and analyses; and (3) deepening, including other than just technological relations, such as physical, economic and behavioural relations. From this analysis, it is clear that the vast majority of LCSA research so far has focused on the 'broadening of impacts' dimension. The challenges most frequently cited concern the need for more practical examples of LCSA, efficient ways of communicating LCSA results and the need for more data and methods particularly for SLCA indicators and comprehensive uncertainty assessment. We conclude that the three most crucial challenges to be addressed first are developing quantitative and practical indicators for SLCA, life cycle-based approaches to evaluate scenarios for sustainable futures and practical ways to deal with uncertainties and rebound effects.

J. Guinée (✉)
Faculty of Sciences, Institute of Environmental Sciences (CML), Department of Industrial Ecology, Leiden University, P.O.box 9518, 2300 RA Leiden, The Netherlands
e-mail: Guinee@cml.leidenuniv.nl

© The Author(s) 2016
R. Clift, A. Druckman (eds.), *Taking Stock of Industrial Ecology*,
DOI 10.1007/978-3-319-20571-7_3

45

Keywords Life cycle assessment, LCA • Life cycle sustainability assessment, LCSA • Life cycle scenario assessment • Rebound • Uncertainty • Social life cycle assessment, SLCA • Life cycle costing, LCC

1 Introduction

Environmental life cycle assessment (LCA) has developed fast over the last three decades. The first studies that are now recognised as (partial) LCAs date from the late 1960s and early 1970s, a period in which environmental issues like resource and energy efficiency, pollution control and solid waste became issues of broad public concern. One of the first studies quantifying the resource requirements, emission loadings and waste flows of different beverage containers was conducted by Midwest Research Institute (MRI) for the Coca Cola Company in 1969. Similar but independent studies were conducted in Europe by Sundström (1971) and by Basler and Hofman (1974). Together with several follow-ups, this marked the beginning of the development of LCA as we know it today (Guinée 1995; Hunt and Franklin 1996; Baumann and Tillman 2004; Guinée et al. 2011).

The period 1970–1990 comprised the decades of conception of LCA with widely diverging approaches, terminologies and results. There was a clear lack of international scientific discussion and exchange platforms for LCA. LCAs were performed using different methods and without a common theoretical framework. The obtained results differed greatly, even when the objects of the study were the same (Guinée et al. 1993).

The 1990s saw a remarkable growth of scientific and coordination activities worldwide, which among other things is reflected in the number of LCA guides and handbooks produced (ILV et al. 1991; Lindfors 1992; Grieshammer et al. 1991; Heijungs et al. 1992; Vigon et al. 1993; Lindfors et al. 1995; Curran 1996; Hauschild and Wenzel 1998). Also the first scientific journals appeared with LCA as their key topic or one of their main key topics. The period 1990–2000 showed convergence and harmonisation of methods through SETAC's coordination and ISO's standardisation activities, providing a standardised framework and terminology, and platforms for debate and harmonisation of LCA methods. During this period, LCA also became increasingly part of policy documents and legislation, particularly focusing on packaging. It is also the period that the scientific field of industrial ecology (IE) emerged, with life cycle thinking and LCA as one of its key tools (Graedel 1996; see Chap. 1).

The first decade of the twenty-first century has shown an ever-increasing attention to LCA resulting in new textbooks (e.g. Guinée et al. 2002; Baumann and Tillman 2004; EC 2010; Curran 2012; Klöpffer and Grahl 2014). LCA was increasingly used as a tool for supporting policies and (bioenergy) performance-based regulation. Life cycle-based carbon footprint standards were established worldwide in this period. During this period, LCA methods were elaborated in further detail, which unfortunately resulted in divergence in methods again. New approaches were developed with respect to system boundaries and allocation methods (e.g. consequential LCA; see also Chap. 2 of this book), dynamic LCA, spatially differentiated LCA, environmental input–output-based LCA (EIO-LCA) and hybrid LCA. On top of this, various life cycle costing (LCC) and social life cycle assessment (SLCA) approaches were proposed and/or developed. This broadening of environmental LCA to LCC and SLCA draws on the three-pillar (or triple bottom line, TBL) model of sustainability, distinguishing environmental, economic and social impacts of product systems along their life cycle. The original conception of LCA only dealt with the environmental or ecological component, whereas with these latter developments, LCA broadened itself from a merely environmental LCA to a more comprehensive life cycle sustainability assessment (LCSA).[1] This broadening is consistent with developments in IE, for which sustainability and the three-pillar model are principal motivations (Allenby 1999; Graedel and Allenby 1999).

As a matter of course, a subject section was formed within the International Society for Industrial Ecology (ISIE) in 2011, to focus on life cycle assessment (LCA) as currently existing and on life cycle sustainability analysis (LCSA) as a direction in which LCA was developing. Meanwhile several journals have opened up special sections on LCSA, clearly confirming that we are in the middle of the 'LCSA age'.

While several researchers have proposed definitions and methods for LCSA over the past recent years, many practitioners are still left in confusion on what LCSA exactly is, what its methods are and when to apply what. An interesting question therefore is what are LCSA practitioners doing in practice? We distinguished two sub-questions:

- Which definition(s) do they adopt?
- What challenges do they face?

In this chapter, these questions will be addressed by first discussing two different definitions of LCSA and the interpretations of sustainability that these definitions are grounded in and then analysing the LCSA research published over the past half decade replenished by inputs from members of the ISIE-LCSA section[2] on adopted definitions of LCSA and main challenges faced. We will conclude with our top three of the main challenges.[3]

---

[1] Sometimes LCSA is taken as life cycle sustainability analysis. For a discussion on the different reasons for adopting assessment or analysis, we here refer to Zamagni et al. (2009) and Sala et al. (2013b). Here we adopt assessment to stay close to the ISO definition of LCA.

[2] In order to learn what exactly the understanding of LCSA is by members of the ISIE-LCSA section, a questionnaire was issued. All members were invited through the section's electronic platform and through e-mail invitations to provide their views on:

1. Their (preferred) definition of life cycle sustainability assessment
2. Their top three of (scientific and/or practical) challenges for LCSA

Seventeen people reacted on this invitation and their inputs are gratefully used below.

[3] Note that this does not imply that there are no other challenges; it just reflects the author's top three.

2   Definitions of LCSA

Definitions of LCSA are not yet carved in a stone. The first use of the term LCSA was by Zhou et al. (2007), but they only addressed climate change and resource depletion impacts in their LCA and combined it with an LCC, which doesn't fully comply with the three-pillar model. Shortly after Zhou et al. (2007), Klöpffer and Renner (2007; see also Klöpffer 2008) provided a definition of LCSA, and later on, Guinée et al. (2011) built on that definition. Thus today, at least two definitions of LCSA exist:

*   Klöpffer and Renner (2007; see also Klöpffer 2008): 'Given the widespread acceptance of the [triple bottom line] model, it is rather straightforward to propose the following scheme for Life Cycle Sustainability Assessment (LCSA): LCSA = LCA + LCC + SLCA, where LCA is the SETAC/ISO environmental Life Cycle Assessment, LCC is an LCA-type ('environmental') Life Cycle Costing assessment and SLCA stands for societal or social Life Cycle Assessment'. According to this definition, LCSA thus broadens ISO-LCA to also include economic and social aspects adopting a life cycle approach. Klöpffer (2003) already argued for combining LCA with LCC and SLCA, but he did not use the term LCSA at that time. As mentioned by Klöpffer and Renner, this TBL-based life cycle approach was earlier introduced by the German Oeko-Institut in a method called 'Produktlinienanalyse' in 1987 (Projektgruppe ökologische Wirtschaft 1987).
*   Guinée et al. (2011): LCSA links 'life cycle sustainability questions to knowledge needed for addressing them, identifying available knowledge and related models, knowledge gaps and defining research programs to fill these gaps. [...] It *broadens* the scope of current LCA from mainly environmental impacts only to covering all three dimensions of sustainability (people, planet and prosperity). It also *broadens* the object (or level) of analysis from predominantly product-related questions (product level) to questions related to sector (sector level) or even economy-wide levels (economy level). In addition, it *deepens* current LCA to also include other than just technological relations, e.g. physical relations (including limitations in available resources and land), economic and behavioural relations, etc. [...] LCSA is a trans-disciplinary *framework* for integration of models rather than a model in itself. LCSA works with a plethora of disciplinary models and guides selecting the proper ones, given a specific sustainability question'.

Guinée et al. (2011) basically adopted the definition by Klöpffer and Renner (2007) but added two dimensions and called it a framework rather than a method in itself. Based on these two definitions, we can thus distinguish between three dimensions along which LCSA expands when compared to (environmental) LCA:

1. Broadening of impacts: LCSA = LCA + LCC + SLCA
2. Broadening level of analysis: product-, sector- and economy-wide questions and analyses

3. Deepening: including other than just technological relations, such as physical, economic and behavioural relations

To better understand the different LCSA definitions above, we need to discuss the interpretations of 'sustainability' that the different definitions of LCSA are grounded in.

3 Sustainability

As mentioned above, the Projektgruppe ökologische Wirtschaft (1987) firstly introduced a life cycle approach including all three dimensions of sustainability. The year of publication of their 'Produktlinienanalyse' coincided with the year of publication of the Brundtland report 'Our Common Future' (WCED 1987). The Projektgruppe ökologische Wirtschaft obviously did not yet use the term 'sustainability'.

Klöpffer (2008; English version of Klöpffer and Renner (2007); see also Klöpffer (2003)) extensively discusses what exactly they mean by LCSA. They adopted the 'triple bottom line' (Elkington 1998) or the 'three-pillar' interpretation of sustainability, referred to as 'people, planet and prosperity' at the World Summit on Sustainable Development in Johannesburg in 2002. The triple bottom line approach basically says that for achieving more sustainable futures, environmental, economic as well as social impacts of activities have to be taken into account. In the World Summit on Sustainable Development in Johannesburg in 2002, also life cycle analysis (http://www.un-documents.net/jburgpln.htm) was introduced, and thus, Klöpffer (2003) argues that 'any environmental, economic, or social assessment method for products has to take into account the full life cycle from raw material extraction, production to use and recycling or waste disposal. In other words, a systems approach has to be taken'. The background for the LCSA definition by Klöpffer (2008) and Klöpffer and Renner (2007) is thus the 'triple bottom line' or 'three-pillar' interpretation of sustainability, which is a very common interpretation (e.g. Mitchell et al. 2004; Blewitt 2008) adopting a system approach.

The Guinée et al. (2011) LCSA framework is based on the work done as part of the EU FP6 CALCAS (Co-ordination Action for innovation in Life Cycle Analysis for Sustainability) project (http://www.calcasproject.net/). The interpretation of sustainability is similar to Klöpffer (2008) and Klöpffer and Renner (2007), but two additions were made: broadening of the level of analysis and deepening the analysis itself. The rationales behind these two additions originate from:

(a) An analysis of the bioenergy debate and the role of LCA in this debate (Zamagni et al. 2009)
(b) The simple observation that although huge efforts have been made to improve the environmental performance of products applying LCA, little or no progress has been made improving the environmental sustainability of the global economy as a whole (Rockström et al. 2009; EPA2013; PBL 2013)

The bioenergy debate showed that LCAs may show some fundamental flaws when applied as a tool for supporting bioenergy performance-based regulation (PBR). We distinguish between flaws related to differences in methods applied between studies (e.g. related to attributional vs. consequential analysis, data sources, gaps and uncertainties, choices of functional unit, allocation method, impact categories and characterisation method) and flaws in impacts and mechanisms considered for the systems analysed. For PBRs, LCA results should be robust and 'lawsuit proof', implying that the freedom of methodological choices for the handling of such issues as biogenic carbon balances and allocation should be reduced to an absolute minimum, uncertainties should be properly dealt with and it should be realised that there may be a gap between the translation of results based on a functional unit of a litre of biofuel to real-world improvements for millions of litres. There are huge differences between LCA studies on bioenergy systems as identified by Voet et al. (2010). Besides these methodological differences, most of these LCA studies have been limited to considering only environmental impacts and not taking into account system effects and consequences such as indirect land use, rebound effects and market mechanisms. These all play a role in how a large-scale production of bioenergy could affect the food market, scarcity, social structure, land use, nature and other conditions that are important for society. Large-scale policies to stimulate bioethanol in the USA and Europe have led to consequences which were not really foreseen and were barely considered in the preparatory LCA-type studies (Zamagni et al. 2009). A framework for deepened analysis – including more of these mechanisms – was lacking so far.

The fact that we may improve the environmental performance of products while still increasing the global pressure on the environment implies that we cannot simply focus on single product systems only, but also have to broaden our life cycle-based analyses to baskets of products, sectors and whole economies. Referring to the well-known IPAT equation (Ehrlich and Holdren 1971), which decomposes environmental impact (I) into the separate effects of population size (P), affluence (A) and technology (T), LCAs so far have focused on the pollution per functional unit of product or service. This basically is no more than a 'supermicro' analysis of T. If the total consumption of products and services (increasing affluence) and the size of the population keep increasing meanwhile, we may not achieve any improvement in (macro) global sustainability despite significant progresses in (micro) sustainability of (a number of) individual products and services.

Both these arguments resulted in the LCSA definition by Guinée et al. (2011) which added two dimensions to the definition by Klöpffer (2008) and Klöpffer and Renner (2007).

4   LCSA Definitions Adopted in Practice

In order to find out which definition of LCSA practitioners adopt in practice, a bibliometric analysis was carried out of the ISI Web of Science (WoS) published by Thomson Reuters. The keywords used under 'topic' for searching 'all databases' were 'life cycle sustainability assessment*' OR 'life cycle sustainability analys*' for the time span = 2000–2014 (accessed on 24/11/2014). The result of this bibliometric analysis is shown in Table 3.1. References basically covering the same topic and originating from the same research institute were grouped together. For example, Heijungs et al. (2010), Guinée and Heijungs (2011) and Guinée et al. (2011) basically cover the same topic (presenting an LCSA framework covering all three dimensions of the LCSA definition) and originate from the same research institute (CML). In addition, references that despite the use of LCSA had little or no connection to LCSA and the two questions posed here were eliminated from the results. Put more precisely, a reference was excluded from further analysis if it could not comply with one or more of the following criteria:

- The term LCSA was used to refer to one of the two (revised or otherwise) definitions of LCSA discussed above.
- If the reference focused on broadening of impacts, it should include analyses of all three pillars (e.g. LCA + LCC + SLCA).
- If the reference focused on broadening of the level of analysis and/or deepening the analysis, it should do so as part of LCSA.

The resulting (groups of) references were then analysed on their coverage of the three dimensions mentioned above (see also Table 3.1).

The bibliometric analysis resulted in about 30 articles covering the topic of LCSA (Table 3.1). Table 3.1 shows that almost all of the LCSA studies published so far focus on the 'broadening of impacts' dimension: LCSA = LCA + LCC + SLCA. Among these studies are many case studies. In addition, explorations have been made to widen the scope of the three pillars to include, for example, cultural aspects (Pizzirani et al. 2014). Along a similar line, Jørgensen et al. (2013) argue that when fully adopting the WCED (1987) definition of sustainability, LCA and SLCA in particular 'should be expanded to better cover how product life cycles affect poverty and produced capital'. Only a few studies report on the 'broadening of the level of analysis' and/or 'deepening' dimensions; most of these studies are reviews or methodological by nature.

The main keywords popping up among the ISIE-LCSA membership from the response concerning the question on their preferred definition of LCSA are 'environmental-social-economic' besides 'product', 'sustainability' and 'assessment'.

From both the bibliometric analysis and the brief questionnaire, it becomes obvious that the vast majority of LCSA articles have focused on the 'broadening of impacts' dimension: LCSA = LCA + LCC + SLCA. However, this may rather be a limitation of our bibliometric analysis since we only searched for articles including the terms life cycle sustainability assessment(s) or life cycle sustainability analysis(es), while many articles in the 'broadening of the level of analysis' (like IOA) and 'deepening' (like rebound modelling and uncertainty analysis) domains may not use these terms in their topical descriptions. This immediately touches upon a problem of too encompassing or too strict definitions: the Guinée et al. (2011) definition of LCSA includes broadening of the level of analysis and deepen-

Table 3.1 LCSA references as a result from the bibliometric analysis of the Thomson Reuters ISI Web of Science (WoS) databases on 'life cycle sustainability assessment*' OR 'life cycle sustainability analys*' for the time span = 2000–2014 (accessed on 24/11/2014), classified on their coverage of the three dimensions of LCSA

References	Case (C) or methodology/ review (M) study	Broadening impacts	Broadening analysis	Deepening
Klöpffer (2008) and Klöpffer and Renner (2007)	M	Y	N	N
Finkbeiner et al.(2010)	C	Y	N	N
Moriizumi et al. (2010)	C	Y	N	N
Heijungs et al. (2010), Guinée and Heijungs (2011), and Guinée et al. (2011)	M	Y	Y	Y
Halog and Manik (2011)	M/C	Y	Y	Y
Manzardo et al. (2012)	M	Y	N	N
Menikpura et al. (2012)	C	Y	N	N
Stamford and Azapagic (2012)	C	Y	N	N
Traverso et al. (2012a, b)	M/C	Y	N	N
Zamagni (2012)	M	Y	Y	Y
Bachmann (2013)	M	Y	N	N
Cinelli et al. (2013)	M[a]	Y	Y	Y
Giesen et al. (2013)	M	Y	Y	Y
Hu et al. (2013)	M/C	Y	Y	Y
Jørgensen et al. (2013)	M	Y	N	N
Kucukvar and Tatari (2013)	C	Y	N	N
Pesonen and Horn (2013)	M	Y	N	N
Sala et al. (2013a, b)	M	Y	Y	Y
Vinyes et al. (2013)	C	Y	N	N
Zamagni et al. (2013)	M	Y	Y	Y
Onat et al. (2014) and Kucukvar et al. (2014a, b)	C	Y	Y	N
Ostermeyer et al. (2013)	C	Y	N	N
Stefanova et al. (2014)	M/C	Y	Y	Y
Heijungs et al. (2014)	C/M	N	Y	Y

Y Yes, *N* No
[a]This reference is a workshop report

ing, while research in these dimensions is often developed as specific approaches rather than topics under the umbrella of LCSA.

5   Main Challenges Identified in LCSA Studies So Far

The references in Table 3.1 were then analysed on the challenges faced. In Annex 1, these challenges are summarised for the different (groups of) references. Scanning through these references and generalising the challenges identified results in the following interpretation of the main challenges:

- The need for data and methods, particularly the lack of (proper and quantitative) SLCA indicators (Klöpffer 2008; Finkbeiner et al. 2010; Traverso et al. 2012a; Hu et al. 2013; Ostermeyer et al. 2013; Kucukvar and Tatari 2013; Vinyes et al. 2013; Zamagni et al. 2013).
- The need for practical (case study) examples (how to put LCSA in practice?) (Cinelli et al. 2013; Giesen et al.. 2013; Hu et al. 2013; Zamagni et al. 2013).
- How to communicate LCSA results (Finkbeiner et al. 2010; Traverso et al. 2012a, b; Bachmann 2013; Pesonen and Horn 2013)?
- The need for comprehensive methods dealing with all relevant uncertainties related to life cycle-based approaches (Zamagni 2012; Pesonen and Horn 2013; Kucukvar and Tatari 2013 and Kucukvar et al. 2014a, b).
- How to deal with technological, economic and political mechanisms at different levels of analysis (Cinelli et al. 2013; Sala et al. 2013a, b; Zamagni 2012; Zamagni et al. 2013)?
- The need for more dynamic models (Ostermeyer et al. 2013; Onat et al. 2014).
- How to deal with value choices and subjectivity in, particularly, the weighting step (Stamford and Azapagic 2012; Traverso et al. 2012b; Bachmann 2013; Manzardo et al. 2012; Sala et al. 2013a, b; Vinyes et al. 2013)?
- The need for further development of life cycle-based scenario evaluations (Zamagni 2012; Heijungs et al. 2014).
- How to deal with benefits (beneficial impacts), particularly in SLCA (Bachmann 2013)?
- How to avoid double counting (inconsistent application) between LCA, LCC and SLCA (Zamagni 2012; Bachmann 2013)?
- How to deal with different perspectives (producer, customer, societal) on costs in LCC (Finkbeiner et al. 2010)?
- How to (practically) relate (disciplinary) models to different types of life cycle sustainability questions (Guinée et al. 2011; Zamagni et al. 2013; Stefanova et al. 2014)?

From this list of challenges, those most frequently cited concern the 'broadening of impacts' dimension in general, the need for more practical examples of LCSA, efficient ways of communicating LCSA results and the need for more data and methods particularly for SLCA indicators and comprehensive uncertainty assessment. Note that with respect to SLCA, there are many more authors that identified these challenges (e.g. Jørgensen et al. 2008), but their references were excluded due to the limitations of our bibliometric analysis (see above).

The number of indicators that the various studies adopt for addressing the three pillars of sustainability in a life cycle perspective varies from a few (e.g. Moriizumi et al. limit their LCSA of two mangrove management systems in Thailand to just three indicators, one for each dimension of the 'triple bottom line') to several dozen indicators (e.g. Stamford and Azapagic adopted 43 indicators to address the same three pillars in their LCSA on electricity options for the UK). The challenges faced by studies adopting only a few indicators obviously include how to broaden the number of indicators. The challenges for studies adopting dozens of indicators include how to communicate their results to decision-makers and/or how to further weight (evaluate) and aggregate the indicator results, for example, applying (multi-criteria) decision analysis.

The topic of 'deepening' is addressed less by the studies listed in Annex 1. Nevertheless, several references mention (e.g. Cinelli et al. 2013; Sala et al. 2013b; Zamagni 2012; Zamagni et al. 2013; Pesonen and Horn 2013; Kucukvar and Tatari 2013; Kucukvar et al. 2014a, b) and some even address (Hertwich et al. 2014) typical 'deepening' topics such as the need for comprehensive uncertainty assessment and methods for dealing with rebound effects. But, again, these references were excluded from Table 3.1 and Annex 1 due to the limitations of our bibliometric analysis (see above). However, we feel that deepening discussions are very important as part of maturing LCSA approaches. We illustrate this by the example of modelling rebound effects in a life cycle perspective, which has been addressed by several authors (Hertwich 2005; Hofstetter et al. 2006; Thiesen et al. 2008; Girod et al. 2011; Druckman et al. 2011; Font Vivanco and Voet 2014).

Hertwich (2005) defines the rebound effect as 'a behavioural or other systemic response to a measure taken to reduce environmental impacts that offsets the effect of the measure. As a result of this secondary effect, the environmental benefits of eco-efficiency measures are lower than anticipated (rebound) or even negative (backfire)'. For example, the positive effect of more efficient cars has largely been offset by an overall shift to larger and heavier cars (see Chap. 18). Similarly, the introduction of high-efficient light bulbs has been combined with an expansion of the number of light points. Recently, Font Vivanco and Voet (2014) performed a review describing the state of the art in incorporating the rebound effect into LCA-based studies and analysed their main strengths and weaknesses. Their literature review identified a total of 42 relevant scientific documents, from which 17 provided quantitative estimates of the rebound effect using LCA-based approaches. It appeared that 'the inclusion of the rebound effect into LCA-based studies is still one of the most relevant unresolved issues in the field; […] only few studies provide quantitative estimates (mostly for carbon dioxide and global warming […])'. Font Vivanco and Voet concluded that 'while a number of LCA-based studies have considered such effects […], no generally applicable guidelines have been developed so far; […] consequently, a panoply of non-consensual definitions and analytical approaches have arisen within the LCA community, and rebound effects have been both unevenly and inconsistently incorporated into LCA-based studies'. The results of this reviews show that, while incorporating the rebound effect into LCA studies is recognised as a very important topic and has received some attention, there is still no generally applicable and/or comprehensive method for dealing with rebound in a life cycle perspective. A similar conclusion is valid with respect to the challenge of incorporating comprehensive though practical uncertainty assessments into LCA (see, e.g. Gregory et al. 2013; Harst and Potting, 2013; Henriksson et al. 2014, 2015; Mendoza et al. 2014).

Finally, the bibliographic analysis showed that there are an increasing number of LCSA studies (e.g. Giesen et al. 2013; Hu et al. 2013; Manzardo et al. 2012; Stefanova et al. 2014; Heijungs et al. 2014) dealing with scenarios.[4] The studies explore possible configurations of emerging new technologies, product systems or consumption baskets, comparing their potential impacts to alternative technologies, product systems and consumption baskets. Such studies are very relevant, particularly if performed ex ante or parallel to the technology development trajectory, as in that way LCSA is able to advise the technology developer whether developments are on the 'right' track while identifying hot spots for improvement. Considering this increase and the relevance of scenarios for evaluating possible more sustainable futures, we might even consider changing the meaning of the abbreviation LCSA from life cycle sustainability assessment to life cycle-based scenario assessment.

The results from our brief questionnaire among the ISIE-LCSA membership largely support the challenges discussed above while particularly adding challenges as communication with and involvement of stakeholders in the LCSA process, education and standardisation of LCSA methods.

6   Conclusions

Adopting the Guinée et al. (2011) definition of LCSA while not underestimating other challenges, we see the following challenges as crucial to address first (one challenge for each dimension of LCSA):

1. *Broadening of impacts*: proper, preferably quantitative and practical indicators for SLCA.
2. *Broadening the level of analysis*: develop, implement and apply life cycle-based approaches to evaluate scenarios for sustainable futures.
3. *Deepening*: develop and implement ways to deal with uncertainties and rebound effects as comprehensively and practically as possible.

The challenge to develop proper, preferably quantitative and practical, indicators for SLCA has been present ever since SLCA was proposed as a possible approach.

---

[4] See also Spielmann et al. (2005), Hertwich et al. (2014) and Koning et al. (2015) that report on scenario-based life cycle modelling but not as part of an LCSA framework. See also Chap. 2 of this book.

Many proposals have been developed in this area (see Jørgensen et al. 2008), but the range of methods proposed and developed differs widely. They also often face implementation problems. The bottom line is that there is not at present 'anything resembling an agreed approach or methodology' (Clift 2014). Most efforts so far have focused on finding and developing ways to include social impacts using impact categories and indicators, similar to environmental LCA. Considering the challenges identified in Annex 1 and the period over which discussions on SLCA's challenges have continued, one may wonder 'whether it is really appropriate to model social LCA on environmental LCA' and whether or not 'Social LCA is more likely to develop as a useful tool if it is not forced into the mould of environmental LCA' (Clift 2014). This is not a new discussion since Udo de Haes (see Klöpffer 2008) already argued in 2008 that 'social indicators do not fit in the structure of LCA' because developing 'a quantitative relationship of the indicator to the functional unit' or properly handling the high spatial dependency of the indicator is problematic when trying to squeeze such impacts into environmental LCA. To prevent progress on SLCA coming to a dead end, fundamental re-examination of SLCA's paradigm seems necessary eventually leading to increased applicability and a more comprehensive coverage of social benefits and impacts of life cycles. Since a platform for this discussion seems to be lacking, the ISIE-LCSA section could offer this.

Life cycle-based approaches have an important role to play in assessing scenarios on how to feed, fuel and fibre about nine billion people – all longing for the 'good' life – in a sustainable way in 2050 (cf. Frosch and Gallopoulos 1989). We need to develop approaches and tools within the LCSA framework for evaluating the sustainability of scenarios for such a future. One of the sub-challenges is to make sensible and proper use of the different modes of LCA and LCSA available. The key challenge is to effectively combine backcasting LCSA[5] (BLCSA; Heijungs et al. 2014) with forecasting LCSA (FLCSA) approaches (e.g. Hertwich et al. 2014; Koning et al. 2015) and eventually also product LCA (CLCA as well as ALCA) in such a way that policies and transitions towards a more sustainable future can be properly supported and monitored.[6]

All our life cycle tools should be accompanied with proper ways of dealing with uncertainties of data, methodological choices, assumptions and scenarios and pref-

---

[5] Heijungs et al. (2014) defined backcasting LCSA as exploring ways, in a life cycle perspective, to stay within normatively defined sustainability levels (e.g. planetary boundaries) through adapted affluence, population growth and/or technologies.

[6] Note that we make a distinction between supporting policy *development* and monitoring *developed* policies. It's our belief that we need different tools for supporting policy development (e.g. CLCA; see also Chap. 2 of this book) and for monitoring accepted policy (e.g. ALCA for monitoring bioenergy performance-based regulation through carbon footprint studies). For policy development, we need to analyse all possible direct and indirect consequences of potential policy options using life cycle-based scenario analysis for which CLCA, BLCSA, FLCSA and other scenario-based life cycle approaches (e.g. Spielmann et al. 2005; Hertwich et al. 2014; Koning et al. 2015) are best suited. For monitoring existing, accepted policies, we need clear black and white answers and no scenario-based ranges of answers; for this, ALCA seems better suited.

erably also with proper ways of handling rebound effects. This is particularly important since the results of our tools are increasingly supporting public policies and performance-based regulations. However, most of our studies still present their results as point values, suggesting that life cycle tools produce black and white results with no uncertainties while all experienced practitioners of these tools know better than this. Thus, in order to maintain and increase the credibility of our life cycle decision-support tools, we need to develop, as a matter of priority, approaches to properly and transparently deal with uncertainties associated with data, models, choices and assumptions of all life cycle-based methods (LCA, LCC, SLCA, IOA, hybrid LCA, etc.). Several methods have been proposed for this (see above), but the main remaining challenge is to harmonise them to be comprehensive (e.g. covering all types of uncertainty for all phases of LCA in a common approach, covering all types of rebound effects for complete life cycles in a common approach) and implement them (through, e.g. data and software tools) in the daily practice of practitioners. Similar reasoning is valid for rebound effects.

Finally, as mentioned above, one of the sub-challenges is to make sensible and proper use of the different modes of LCA and LCSA available. For LCA and LCSA, we currently have at least the following modes of analysis at our disposal: attributional (ALCA/ALCSA), backcasting (BLCA/BLCSA), consequential (CLCA/CLCSA), decision or dynamic (DLCA/DLCSA), exergy (ELCA/ELCSA) and potentially resulting in A–Z LCA/LCSA. We should thus pay due attention to relating sustainability questions to the most appropriate tools of our industrial ecology toolbox. The alternative is to throw the dic

e.

Acknowledgements We thank Göran Finnveden, Sheetal Gavankar, Wenjie Liao, Aleksandar Lozanovski, Sergio Pacca, Stefania Pizzirani, Richard Plevin, Anne Ventura and Bo Weidema for providing their views on their preferred definition of life cycle sustainability assessment and their top three of scientific and/or practical challenges for LCSA. Furthermore, we are grateful for the inputs that Roland Clift provided on previous versions of this manuscript.

Annex 1: Challenges Faced in the LCSA References from the Bibliometric Analysis

References	Short topic description	Challenges faced
Klöpffer (2008) and Klöpffer and Renner (2007)	Discussion of possible definitions for LCSA of products	Quantitative methods needed for decision-making if different solutions are offered; quantification difficult challenge for SLCA
Finkbeiner et al.(2010)	The 'life cycle sustainability dashboard' and the 'life cycle sustainability triangle' are presented as examples for communication tools for both experts and non-expert stakeholders	Real and substantial implementation of the sustainability concept
		Comprehensive, yet understandable presentation of LCSA results
		Different possible perspectives (producer, customer, societal) when considering the life cycle costs
		Data availability
		To unambiguously determine and measure sustainability performance for processes and products
		Trade-off between validity and applicability
Moriizumi et al. (2010)	To assess the sustainability of two mangrove management systems in Thailand using a life cycle approach and using three indicators: net global warming emissions (CO_2-eq.; environmental indicator), amount of employment created in local communities (social indicator) and the value of cash flow generated (economic indicator)	Other environmental indicators; estimation of spillover impacts of production activity; combining the results with decision-support tools

Heijungs et al. (2010), Guinée and Heijungs (2011), and Guinée et al. (2011)	Proposal of a framework for LCSA	Structuring, selecting and making the plethora of disciplinary models practically available in relation to different types of life cycle sustainability questions
		To derive consistent criteria for implementing methods (e.g. attributional, consequential and scenario-based modelling of systems and related time frames, including aspects of unpredictability of emerging systems, complex adaptive systems, etc.)
		LCSA research methods and practical examples
		Linking approaches to questions
		Backcasting LCA
Halog and Manik (2011)	Integrated methodology for LCSA by capitalising the complementary strengths of different methods used by industrial ecologists and biophysical economists	n.a.
Manzardo et al. (2012)	To develop a 'grey-based' group decision-making methodology for the selection of the best renewable energy technology using an LCSA (LCA+LCC+SLCA) perspective and addressing the issue of uncertainty	Subjectivity involved in qualitative evaluations
Menikpura et al. (2012)	Development of a method for comprehensive sustainability assessment on most of the critical environmental, economic and social impacts starting from LCA	n.a.
Stamford and Azapagic (2012)	To identify the most sustainable options for the future UK electricity mix applying a sustainability assessment framework developed previously by the same authors	The influence of priorities and preferences of different stakeholders on the outcomes of the sustainability assessment of electricity options
Traverso et al. (2012a)	The integration of LCSA and the dashboard of sustainability into a so-called life cycle sustainability dashboard (LCSD) and its first application to a group of hard floor coverings	Having (quantitative) data for the (particularly SLCA) indicators considered
		How to handle qualitative data that can particularly be meaningful in the social assessment, in a basically quantitative method

(continued)

References	Short topic description	Challenges faced
Traverso et al. (2012b)	Application of life cycle sustainability assessment (LCSA) and the life cycle sustainability dashboard (LCSD) for comparing different PV modules	Selection of social LCA indicators
		Weighting sets needed for the LCSD
		More case studies needed, for example, to calibrate indicators and weights
Zamagni (2012)	Editorial on a new section on LCSA	How can the LCSA framework be consistently applied, considering different degrees of maturity of LCA, LCC and SLCA?
		Is adopting the same system boundary for LCA, LCC and SLCA always feasible and conceptually correct?
		What role does scenario modelling play in the LCSA framework?
		What approaches exist for including mechanisms in the analysis?
		How can different domains, normative positions (values) and empirical knowledge be dealt with?
		How can future changing structures of the economy be accounted for?
		How can uncertainty be accommodated and managed?
Bachmann (2013)	Ranking of power generation technologies by means of (1) the total cost approach, adding private and external costs, and (2) a multi-criteria decision analysis (MCDA) integrating social, economic and environmental criteria	How to separate the different life cycle-based assessments into environmental, economic or social in particular to avoid double counting?
		The inclusion of risks
		Dealing with benefits
		Dealing with value choices
Cinelli et al. (2013)	Workshop on LCSA: the state-of-the-art and research needs – November 26, 2012, Copenhagen, Denmark	How to communicate LCSA results?
		How to put LCSA into practice?
		Dealing with technological, economic, political relations at different scales of analysis
		Theoretical roots of LCSA and frameworks

		How can LCSA be approached practically?
Giesen et al. (2013) and Hu et al. (2013)	Putting the LCSA framework of Guinée et al. (2011) into practice by five operational steps: (1) broad system definition, (2) making scenarios, (3) defining sub-questions for individual tools, (4) application of the tools and (5) interpreting the results in an LCSA framework	Only one social impact indicator could be modelled in the process-based LCA structure
		More case study examples needed
Pesonen and Horn (2013)	Streamlined rapid assessment tool – the sustainability SWOT – and empirical testing of its impact on the corporate world analysing whether or not it leads changes in either strategic or operative-level activities	Development of approaches to quicken the resource-consuming inventory and assessment phases of LCSA; easy-to-understand communication of results; streamlined approach for managing uncertainties of all types with transparency and competence
Sala et al. (2013a, b)	Review of main challenges posed to sustainability assessment methodologies and related methods in terms of ontology, epistemology and methodology of sustainability science	A framework for SA should be able to better deal with externalities, interrelations, different applications, multiple stakeholder needs and multiplicity of legitimate perspectives of stakeholders, to deal with nonlinearities, normative choices, uncertainties and risks
		LCSA should be developed ([further] in order to:
		Guarantee a holistic perspective in the assessment
		Be hierarchically different from LCA, eLCC and sLCA. It should represent the holistic approach integrating (and not substituting) the reductionist approach of the single part of the analysis
		Enhance transparency and scientific robustness
		Tailor the assessment for local/specific impact
		Encourage and systematise the interaction among stakeholders involved in the
		development, application and use of the LCSA results
		Widen the goal of the integrated assessment (e.g. including not only negative but also positive impacts)

(continued)

References	Short topic description	Challenges faced
Vinyes et al. (2013)	LCSA comparison of three domestic collection systems for used cooking oil (UCO) to determine which systems should be promoted for the collection of UCO in cities in Mediterranean countries	Quantitative indicators for SLCA
		Relating social indicators/impacts to the functional unit
		How to restrict social indicators proposed to a manageable and comparable number
		Weighting methods
Zamagni et al. (2013)	From LCA to LCSA: concept, practice and future directions. Introductory article to a special issue on LCSA	Limited number of LCSA applications, the majority of which focus on the interface of environmental and economic aspects; social aspects are less addressed
		SLCA data and indicators
		Weak understanding at the conceptual level of SLCA and LCSA
		What is the appropriate scale of LCSA: products, enterprises, communities or nations?
Kucukvar and Tatari (2013)	To quantify the overall environmental, economic and social impacts of the US construction sectors using an economic input–output-based sustainability assessment framework	Lack of comprehensive data sets for all three pillars
		Uncertainty assessment
Onat et al. (2014)	Integrating several social and economic indicators demonstrating the usefulness of IO modelling for quantifying sustainability impacts, providing an economy-wide analysis and a macro-level LCSA	Dynamic system approach
Kucukvar et al. (2014b)	To develop a triple bottom line sustainability assessment model evaluating the environmental and socio-economic impacts of pavements	Uncertainty assessment; weighting of different impacts
Kucukvar et al. (2014a)	Adding fuzzy multi-criteria decision-making method to the approach above	Uncertainty of LCA results including weighting of sustainability indicators and limitation of the EIO method should be included in decision-making

Ostermayer et al. (2013)	The application and potential of LCSA in the built environment, focusing on refurbishment of residential buildings, LCA, LCC and limited social assessment, and applying a multidimensional Pareto optimisation method	Dynamic modelling of energy mixes and related LCI data sets; dynamic modelling of discount rates and energy price scenarios for LCC; suitable indicators for SLCAs
		Extreme dependence and differentiation of SLCA indicators on regional and cultural conditions; lack of data for SLCA; discussion needed on what aspects need to be implemented into SLCA at least
Stefanova et al. (2014)	An approach to structure the goal and scope phase of LCSA to identify the relevant mechanisms [deepening] to be further modelled for a case study on a new technology for the production of high-purity hydrogen from biomass to be used in automotive fuel cells	Structured identification of [deepening] mechanisms to be modelled in a specific case
Heijungs et al. (2014)	Using IO tables, planetary boundaries and minimum consumption levels to backcast directions to 'safe operating spaces'	Improving the backcasting models to including more impact categories, dynamics, definition of a welfare function, allocation of surplus consumption to consumption categories, etc.

References

Allenby, B. R. (1999). *Industrial ecology: Policy framework and implementation*. Upper Saddle River: Prentice-Hall, Inc.

Bachmann, T. M. (2013). Towards life cycle sustainability assessment: drawing on the NEEDS project's total cost and multi-criteria decision analysis ranking methods. *International Journal of Life Cycle Assessment, 18*(9), 1698–1709.

Basler & Hofman Ingenieure und Planer. (1974). *Studie Umwelt und Volkswirtschaft: Vergleich der Umweltbelastung von Behältern aus PVC, Glas, Blech und Karton*. Bern: Eidgenössisches Amt für Umweltschutz.

Baumann, H., & Tillman, A.-M. (2004). *The hitchhiker's guide to LCA: An orientation in life cycle assessment methodology and application*. Lund: Studentlitteratur.

Blewitt, J. (2008). *Understanding sustainable development*. London: Earthscan.

Cinelli, M., Coles, S. R., Jørgensen, A., Zamagni, A., Fernando, C., & Kirwan, K. (2013). Workshop on life cycle sustainability assessment: The state of the art and research needs-November 26, 2012, Copenhagen, Denmark. *International Journal of Life Cycle Assessment, 18*(7), 1421–1424.

Clift, R. (2014, November). *Social life cycle assessment: What are we trying to do?* Paper presented at the International Seminar on Social LCA, Montpellier.

Curran, M. A. (1996). *Environmental life-cycle assessment*. New York: McGraw-Hill.

Curran, M. A. (Ed.). (2012). *Life cycle assessment handbook: A guide for environmentally sustainable products*. Beverly: Scrivener Publishing.

Druckman, A., Chitnis, M., Sorrell, S., & Jackson, T. (2011). Missing carbon reductions? Exploring rebound and backfire effects in UK households. *Energy Policy, 39*, 3572–3581.

Ehrlich, P. R., & Holdren, J. P. (1971). Impact of population growth. *Science, 171*, 1212–1217.

Elkington, J. (1998). *Cannibals with forks: The triple bottom line of 21st century business*. Gabriola Island: New Society Publishers.

EPA. (2013). *Global greenhouse gas emissions data*. Retrieved March 15, 2015, from http://www.epa.gov/climatechange/ghgemissions/global.html

European Commission – Joint Research Centre – Institute for Environment and Sustainability. (2010). *International reference life cycle data system (ILCD) handbook: General guide for life cycle assessment*. Luxembourg: Publications Office of the European Union.

Finkbeiner, M., Schau, E. M., Lehmann, A., & Traverso, M. (2010). Towards life cycle sustainability assessment. *Sustainability, 2*, 3309–3322.

Font Vivanco, D., & van der Voet, E. (2014). The rebound effect through industrial ecology's eyes: A review of LCA-based studies. *International Journal of Life Cycle Assessment, 19*(12), 1933–1947.

Frosch, R. A., & Gallopoulos, N. E. (1989). Strategies for manufacturing. *Scientific American, 261*(3), 144–152.

Girod, B., de Haan, P., & Scholz, R. W. (2011). Consumption-as-usual instead of ceteris paribus assumption for demand: Integration of potential rebound effects into LCA. *International Journal of Life Cycle Assessment, 16*(1), 3–11.

Graedel, T. E. (1996). On the concept of industrial ecology. *Annual Review of Energy and the Environment, 21*, 69–98.

Graedel, T. E., & Allenby, B. R. (1999). *Industrial ecology* (2nd ed.). Upper Saddle River: Pearson Education, Inc.

Gregory, J. R., Montalbo, T. M., & Kirchain, R. E. (2013). Analyzing uncertainty in a comparative life cycle assessment of hand drying systems. *International Journal of Life Cycle Assessment, 18*(8), 1605–1617.

Grieshammer, R., Schmincke, E., Fendler, R., Geiler, N., & Lütge, E. (1991). *Entwicklung eines Verfahrens zur ökologischen Beurteilung und zum Vergleich verschiedener Wasch- und Reinigungsmittel* (Band 1 und 2). Berlin: Umweltbundesamt.

Guinée, J. B. (1995). *Development of a methodology for the environmental life-cycle assessment of products, with a case study on margarines.* Dissertation, Leiden University. Retrieved March 15, 2015, from http://openaccess.leidenuniv.nl/handle/1887/8052

Guinée, J. B., & Heijungs, R. (2011). Life cycle sustainability analysis: Framing questions to approaches. *Journal of Industrial Ecology, 15*(5), 656–658.

Guinée, J. B. (Ed.), Gorrée, M., Heijungs, R., Huppes, G., Kleijn, R., Koning, A. de, Oers, L. van, Wegener Sleeswijk, A., Suh, S., Udo de Haes, H. A., Bruijn, J. A. de, Duin, R. van, & Huijbregts, M. A. J. (2002). *Handbook on life cycle assessment: Operational guide to the ISO standards.* Series: Eco-efficiency in industry and science (Vol. 7). Dordrecht: Springer.

Guinée, J. B., Udo de Haes, H. A., & Huppes, G. (1993). Quantitative life cycle assessment of products: Goal definition and inventory. *Journal of Cleaner Production, 1*(1), 3–13.

Guinée, J. B., Heijungs, R., Huppes, G., Zamagni, A., Masoni, P., Buonamici, R., Ekvall, T., & Rydberg, T. (2011). Life cycle assessment: Past, present and future. *Environmental Science and Technology, 45*(1), 90–96.

Halog, A., & Manik, Y. (2011). Advancing integrated systems modelling framework for life cycle sustainability assessment. *Sustainability, 3*(2), 469–499.

Hauschild, M., & Wenzel, H. (1998). *Environmental assessment of products. Volume 1: Methodology, tools and case studies in product development – Volume 2: Scientific background.* London: Chapman & Hall.

Heijungs, R., Guinée, J. B., Huppes, G., Lankreijer, R. M., Udo de Haes, H. A., Wegener Sleeswijk, A., Ansems, A. M. M., Eggels, P. G., van Duin, R., & de Goede, H. P. (1992). *Environmental life cycle assessment of products. Guide & backgrounds – October 1992.* Leiden: Leiden University, Centre of Environmental Science.

Heijungs, R., Huppes, G., & Guinée, J. B. (2010). Life cycle assessment and sustainability analysis of products, materials and technologies. Toward a scientific framework for sustainability life cycle analysis. *Polymer Degradation and Stability, 95*(3), 422–428.

Heijungs, R., de Koning, A., & Guinée, J. B. (2014). Maximising affluence within the planetary boundaries. *International Journal of Life Cycle Assessment, 19*(6), 1331–1335.

Henriksson, P. J. G., Guinée, J. B., Heijungs, R., de Koning, A., & Green, D. M. (2014). A protocol for horizontal averaging of unit process data – Including estimates for uncertainty. *International Journal of Life Cycle Assessment, 19*(2), 429–436.

Henriksson, P. J. G., Heijungs, R., Dao, H. M., Phan, L. T., de Snoo, G. R., & Guinée, J. B. (2015). Product carbon footprints and their uncertainties in comparative decision contexts. *PloS One.* doi:10.1371/journal.pone.0121221.

Hertwich, E. G. (2005). Consumption and the rebound effect: An industrial ecology perspective. *Journal of Industrial Ecology, 9*(1–2), 85–98.

Hertwich, E. G., Gibon, T., Bouman, E. A., Arvesen, A., Suh, S., Heath, G. A., Bergesen, J. D., Ramirez, A., Vega, M. I., & Shi, L. (2014). Integrated life-cycle assessment of electricity-supply scenarios confirms global environmental benefit of low-carbon technologies. *PNAS, 112*(20), 6277–6282.

Hofstetter, P., Madjar, M., & Ozawa, T. (2006). Happiness and sustainable consumption: Psychological and physical rebound effects at work in a tool for sustainable design. *International Journal of Life Cycle Assessment, 11*(SI1), 105–115.

Hu, M., Kleijn, R., Bozhilova-Kisheva, K. P., & Di Maio, F. (2013). An approach to LCSA: The case of concrete recycling. *International Journal of Life Cycle Assessment, 18*(9), 1793–1803.

Hunt, R. G., & Franklin, W. E. (1996). LCA- how it came about – Personal reflections on the origin and LCA in the USA. *International Journal of Life Cycle Assessment, 1*(1), 4–7.

ILV., GVM., & IFEU. (1991). *Umweltprofile von Packstoffen und Packmitteln, Methode* (Entwurf). München, Wiesbaden & Heidelberg: Fraunhofer-Institut für Lebensmitteltechnologie und Verpackung (ILV), Gesellschaft für Verpackungsmarktforschung Wiesbaden (GVM) & Institut für Energie- und Umweltforschung Heidelberg (IFEU).

Jørgensen, A., Le Bocq, A., Nazurkina, L., & Hauschild, M. (2008). Methodologies for social life cycle assessment. *International Journal of Life Cycle Assessment, 13*(2), 96–103.

Jørgensen, A., Herrmann, I. T., & Bjørn, A. (2013). Analysis of the link between a definition of sustainability and the life cycle methodologies. *International Journal of Life Cycle Assessment, 18*(8), 1440–1449.

Klöpffer, W. (2003). Life-cycle based methods for sustainable product development. *International Journal of Life Cycle Assessment, 8*(3), 157–159.

Klöpffer, W. (2008). Life cycle sustainability assessment of products. *International Journal of Life Cycle Assessment, 13*(2), 89–95.

Klöpffer, W., & Grahl, B. (2014). *Life cycle assessment (LCA) – A guide to best practice.* Weinheim: Wiley-VCH.

Klöpffer, W., & Renner, I. (2007). Lebenszyklusbasierte Nachhaltigkeitsbewertung von Produkten. *TaTuP Zeitschrift des ITAS zur Technikfolgenabschätzung, 16*(3), 32–38.

Koning, A. de, Huppes, G., Deetman, S., & Tukker, A. (2015). Scenarios for a 2 °C world: a trade-linked input–output model with high sector detail. *Climate Policy*, (in press). doi:10.1080/146 93062.2014.999224

Kucukvar, M., & Tatari, O. (2013). Towards a triple bottom-line sustainability assessment of the U.S. construction industry. *International Journal of Life Cycle Assessment, 18*(5), 958–972.

Kucukvar, M., Gumus, S., Egilmez, G., & Tatari, O. (2014a). Ranking the sustainability performance of pavements: An intuitionistic fuzzy decision making method. *Automation in Construction, 40*, 33–43.

Kucukvar, M., Noori, M., Egilmez, G., & Tatari, O. (2014b). Stochastic decision modeling for sustainable pavement designs. *International Journal of Life Cycle Assessment, 19*(6), 1185–1199.

Lindfors, L.-G. (Ed.) (1992). Product life cycle assessment – Principles and methodology. *Nord*, 1992, 9. Copenhagen: Nordic Council of Ministers.

Lindfors, L.-G., Christiansen, K., Hoffman, L., Virtanen, Y., Juntilla, V., Hanssen, O. J., Rønning, A., Ekvall, T., & Finnveden, G. (1995). Nordic guidelines on life-cycle assessment. *Nord*, 1995, 20. Copenhagen: Nordic Council of Ministers.

Manzardo, A., Ren, J., Mazzi, A., & Scipioni, A. (2012). A grey-based group decision-making methodology for the selection of hydrogen technologies in life cycle sustainability perspective. *International Journal of Hydrogen Energy, 37*(23), 17663–17670.

Mendoza Beltran, A., Guinée, J., & Heijungs, R. (2014, October). *A statistical approach to deal with uncertainty due to the choice of allocation method in LCA.* Paper presented at the 9th international conference on LCA of food, San Francisco. Retrieved March 15, 2015, from http://lcafood2014.org/papers/163.pdf

Menikpura, S. N. M., Gheewala, S. H., & Bonnet, S. (2012). Framework for life cycle sustainability assessment of municipal solid waste management systems with an application to a case study in Thailand. *Waste Management and Research, 30*(7), 708–719.

Mitchell, C. A., Carew, A. L., & Clift, R. (2004). The role of the professional engineer and scientist in sustainable development. In A. Azapagic, R. Clift, & S. Perdan (Eds.), *Sustainable development in practice: Case studies for engineers and scientists* (pp. 29–55). Chichester: Wiley.

Moriizumi, Y., Matsui, N., & Hondo, H. (2010). Simplified life cycle sustainability assessment of mangrove management: A case of plantation on wastelands in Thailand. *Journal of Cleaner Production, 18*(16–17), 1629–1638.

Onat, N. C., Kucukvar, M., & Tatari, O. (2014). Integrating triple bottom line input–output analysis into life cycle sustainability assessment framework: The case for US buildings. *International Journal of Life Cycle Assessment, 19*(8), 1488–1505.

Ostermeyer, Y., Wallbaum, H., & Reuter, F. (2013). Multidimensional Pareto optimization as an approach for site-specific building refurbishment solutions applicable for life cycle sustainability assessment. *International Journal of Life Cycle Assessment, 18*(9), 1762–1779.

PBL (Netherlands Environmental Assessment Agency). (2013). *Trends in global CO₂emissions: 2013 report*. ISBN: 978-94-91506-51-2, PBL publication number: 1148, JRC Technical Note number: JRC83593, EUR number: EUR 26098 EN, The Hague.

Pesonen, H.-L., & Horn, S. (2013). Evaluating the Sustainability SWOT as a streamlined tool for life cycle sustainability assessment. *International Journal of Life Cycle Assessment, 18*(9), 1780–1792.

Pizzirani, S., McLaren, S. J., & Seadon, J. K. (2014). Is there a place for culture in life cycle sustainability assessment? *International Journal of Life Cycle Assessment, 19*(6), 1316–1330.

Projektgruppe ökologische Wirtschaft (Ed.). (1987). *Produktlinienanalyse: Bedürfnisse, Produkte und ihre Folgen*. Köln: Kölner Volksblattverlag.

Rockström, J., Steffen, W., Noone, K., Persson, Å., Chapin, F. S., Lambin, E. F., Lenton, T. M., Scheffer, M., Folke, C., Schellnhuber, H. J., Nykvist, B., de Wit, C. A., Hughes, T., van der Leeuw, S., Rodhe, H., Sörlin, S., Snyder, P. K., Costanza, R., Svedin, U., Falkenmark, M., Karlberg, L., Corell, R. W., Fabry, V. J., Hansen, J., Walker, B., Liverman, D., Richardson, K., Crutzen, P., & Foley, J. A. (2009). A safe operating space for humanity. *Nature, 461*, 472–475.

Sala, S., Farioli, F., & Zamagni, A. (2013a). Progress in sustainability science: Lessons learnt from current methodologies for sustainability assessment (part 1). *International Journal of Life Cycle Assessment, 18*(9), 1653–1672.

Sala, S., Farioli, F., & Zamagni, A. (2013b). Life cycle sustainability assessment in the context of sustainability science progress (part 2). *International Journal of Life Cycle Assessment, 18*(9), 1686–1697.

Spielmann, M., Scholz, R. W., Tietje, O., & de Haan, P. (2005). Scenario modelling in prospective LCA of transport systems: Application of formative scenario analysis. *International Journal of Life Cycle Assessment, 10*(5), 325–335.

Stamford, L., & Azapagic, A. (2012). Life cycle sustainability assessment of electricity options for the UK. *International Journal of Energy Research, 36*(14), 1263–1290.

Stefanova, M., Tripepi, C., Zamagni, A., & Masoni, P. (2014). Goal and scope in life cycle sustainability analysis: The case of hydrogen production from biomass. *Sustainability, 6*(8), 5463–5475.

Sundström, G. (1971). *Investigation of energy requirements from raw material to garbage treatment for four Swedish beer and packaging alternatives*. Malmö: Rigello Pak AB.

Thiesen, J., Christensen, T. S., Kristensen, T. G., Andersen, R. D., Brunoe, B., Gregersen, T. K., Thrane, M., & Weidema, B. P. (2008). Rebound effects of price differences. *International Journal of Life Cycle Assessment, 13*(2), 104–114.

Traverso, M., Asdrubali, F., Francia, A., & Finkbeiner, M. (2012a). Towards life cycle sustainability assessment: An implementation to photovoltaic modules. *International Journal of Life Cycle Assessment, 17*(8), 1068–1079.

Traverso, M., Finkbeiner, M., Jørgensen, A., & Schneider, L. (2012b). Life cycle sustainability dashboard. *Journal of Industrial Ecology, 16*(5), 680–688.

van der Giesen, C., Kleijn, R., Kramer, G. J., & Guinée, J. (2013). Towards application of life cycle sustainability analysis. *Revue de Métallurgie, 110*, 31–38.

van der Harst, E., & Potting, J. (2013). Variation in LCA results for disposable polystyrene beverage cups due to multiple data sets and modelling choices. *Environmental Modelling and Software, 51*, 123–135.

van der Voet, E., Lifset, R. J., & Luo, L. (2010). Life cycle assessment of biofuels, convergence and divergence. *Biofuels, 1*(3), 435–449.

Vigon, B. W., Tolle, D. A., Cornaby, B. W., Latham, H. C., Harrison, C. L., Boguski, T. L., Hunt, R. G., & Sellers, J. D. (1993). *Life-cycle assessment: Inventory guidelines and principles* (EPA/600/R-92/245). Washington, DC: Environmental Protection Agency.

Vinyes, E., Oliver-Sola, J., Ugaya, C., Rieradevall, J., & Gasol, C. M. (2013). Application of LCSA to used cooking oil waste management. *International Journal of Life Cycle Assessment, 18*(2), 445–455.

WCED – World Commission on Environment and Development. (1987). *Our common future.* Oxford: Oxford University Press.

Zamagni, A. (2012). Life cycle sustainability assessment. *International Journal of Life Cycle Assessment, 17*(4), 373–376.

Zamagni, A., Buttol, P., Buonamici, R., Masoni, P., Guinée, J. B., Huppes, G., Heijungs, R., van der Voet, E., Ekvall, T., & Rydberg, T. (2009). *Blue paper on life cycle sustainability analysis; deliverable 20 of the CALCAS project, 2009.* Available at www.calcasproject.net/

Zamagni, A., Pesonen, H.-L., & Swarr, T. (2013). From LCA to life cycle sustainability assessment: Concept, practice and future directions. *International Journal of Life Cycle Assessment, 18*(9), 1637–1641.

Zhou, Z., Jiang, H., & Qin, L. (2007). Life cycle sustainability assessment of fuels. *Fuel, 86,* 256–263.

Chapter 4
Industrial Ecology and Cities

Christopher A. Kennedy

Abstract The study of cities, or urban systems, in Industrial Ecology has a peculiar history. In the 1960s, there was a false dawn for green cities in the United States under the Experimental City project, the unfulfilled plans for which included numerous aspects of Industrial Ecology (IE). When IE eventually began to form as a discipline in the 1990s, cities or urban systems were at best a fringe topic, although their importance was recognized by thought leaders in the field. The development of research on cities as a theme within IE perhaps followed with the broadening of IE to include Social Ecology. Then the study of urban metabolism, which had its own separate literature, arguably became one of the three metabolisms within IE – along with industrial and socio-economic. In this review of work on IE and cities, a Scopus search of ISI-rated publications finds over 200 papers on the topic, many of which are in the *Journal of Industrial Ecology*. Amongst the common themes are papers on urban industrial symbiosis, urban infrastructure frameworks, transportation, waste, energy, greenhouse gas emissions, other urban contaminants, metals, phosphorus and food in cities. The great ongoing challenge for work on IE and cities remains to understand the environmental impacts related to urban metabolism and attempt to reduce them. More specific examples of possible future work include determining potentials for city-scale industrial symbiosis and uncovering how much is occurring and exploring theoretical limits to the sustainability of cities using non-equilibrium thermodynamics.

Keywords Urban systems • Urban metabolism • Industrial symbiosis • Material flow analysis • Energy • Water • Waste • Urban infrastructure

C.A. Kennedy (✉)
Department of Civil Engineering, University of Toronto,
35 St. George Street, Toronto, ON M5S 1A4, Canada
e-mail: christopher.kennedy@utoronto.ca

© The Author(s) 2016
R. Clift, A. Druckman (eds.), *Taking Stock of Industrial Ecology*,
DOI 10.1007/978-3-319-20571-7_4

1   A False Dawn

Between 1966 and 1973, a sustained attempt was made to initiate the construction of an experimental green city in rural Minnesota, USA (Wildermuth 2008). Promoted as a solution to the problems of inner city decay and growing concerns about environmental pollution, a site for the space age city was selected, in 1972, near the small community of Swatara, 130 miles north of Minneapolis. The vision for the city, stemming largely from its influential brainchild Athelstan Spilhaus, contained concepts that are today central to Industrial Ecology. The Experimental City would not generate any wastes, as all residuals would become useful substances. The city regulations would exclude polluting industries. Transport would be by a guideway network system, eliminating the need for cars. Furnace sewers would pump away any polluted air produced by delivery trucks in the underground service roads. Water would be completely recycled. Even buildings and infrastructure would be constructed from precast modular parts that could be adapted or reused to meet the changing needs of future generations. Energy supply for the city would be provided by a fossil-fuel-free power plant, likely nuclear, located at the city centre. No single part of the city was ever constructed – it was a utopian dream with many failings – but it was backed by powerful politicians and received hundreds of thousands of dollars in supporting grants over the 7 years of planning.

The initial instigator of the Experimental City was a scientist of the highest standing (Wildermuth 2008). Athelstan Spilhaus (1911–1998) was a South-African born geophysicist and oceanographer who is credited with developing the bathythermograph while working with the Woods Hole Oceanographic Institution in Massachusetts. The device, which provides measurements of ocean depth and temperatures from moving vessels, was important for the United States in submarine warfare during WWII and helped establish Spilhaus as a scientist of notoriety. In January 1949, at age 37, he was appointed Dean at the University of Minnesota's Institute of Technology. It was at Minnesota that his interest in future cities first began, but Spilhaus was also engaged in a broad range of scientific issues. In 1957, he began a scientific comic strip known as "Our New Age" which was published in 102 US and 19 foreign newspapers. From 1954 to 1958, he was US representative on the executive board of UNESCO and was the commissioner of a hugely successful popular science exhibit at the 1962 World Fair in Seattle. Following publication of an article on The Experimental City in *Science* magazine (Spilhaus 1968), he became President of the American Association for the Advancement of Science in 1969.

The aspects of Industrial Ecology in the Experimental City link back to Spilhaus' role as the Chair of the Committee on Pollution of the National Academy of Sciences and National Research Council. As early as 1961, Spilhaus had observed that "Waste … is some useful substance that we don't know how to use or we don't yet know how to use economically" (quoted in Wildermuth 2008: 59). The national committee on pollution, which Spilhaus chaired from 1963, recognized that growing environment challenges in the United States were linked to increasing population and urbanization. A report by the committee published in 1967 had some bold solutions; it proposed "closed system" manufacturing, under which goods were "designed in the first place with the return to the factory for remaking and reuse in mind" (Wildermuth 2008: 64; Committee on Pollution 1966: 06). Spilhaus' later writings on the Experimental City recognized the dematerialization strategy of selling services rather than products: "Instead of owning things in the future, you will own the services that the things will give you" (Wildermuth 2008: 113). In an article on the Experimental City, he also notes "There are examples of industrial symbiosis where one industry feeds off, or at least neutralizes, the wastes of another…" and he provides several examples (Spilhaus 1967: 1131). Recognizing American cities to be the hot spots for pollution, Spilhaus' committee had at first decided to conduct a case study in the highly urbanized Delaware River Basin. Getting data on pollution for the urban area proved, however, to be too challenging due to the myriad of state, county and municipal governments involved (a common challenge for urban metabolism studies). It was this experience with the frustrating institutional structures within existing cities that to a large extend encouraged Silhaus to pursue his vision of a full-scale experimental new city.

There were other powerful actors supporting the Experimental City idea (Wildermuth 2008). Back in Minneapolis, working in parallel to Spilhaus at first, newspaper executive Otto Silha had taken an interest in the idea of enclosed cities that had been popularized by Buckminster Fuller. Silha had strong connections to government and industry – including the US Vice President Hubert Humphrey, who was a former mayor of Minneapolis. Keen to see Minnesota lead the country on building new cities, Silha assembled a powerful steering committee and co-opted Spilhaus to be the front man for the project. Other members of the steering committee included a retired four-star general, a former economic advisor to President Kennedy, the personal physician of President Johnson, and further representatives from government, industry, labour unions, foundations and academia, as well as Buckminster Fuller. In May 1966, Silha's committee received $248,000 in federal funding to kick-start the planning of the city, with more to follow. Federal support for the Experimental City later waned, but then the state government took a stronger role. On June 4, 1971, Minnesota Governor Wendell Anderson even signed legislation to create a new state agency known as the Minnesota Experimental City Authority.

The great plans for the Experimental City came to nothing, however, due to a variety of reasons. First Spilhaus' vision of the city was utopian, and he was inflexible with many details. The utopian aspects were not so much the ambitious technological features, although some no doubt still needed to be worked out but lay with the social and cultural dimensions. Spilhaus argued that the city should be designed for a fixed population of 250,000 people, no more or less, and that potential residents would have to be screened to achieve the right balance of required professions. Spilhaus saw the city as being a perfectible machine and ignored the aspirations of regular people. Even after Spilhaus withdrew from the implementation efforts, Silha's committee continued to make mistakes. In choosing a possible site for the Experimental City, they failed to suitably engage environmentalists or local people, the majority of whom saw urbanization of their rural surroundings to be deplorable. Ultimately, the effort failed on both cultural and environmental grounds. As one of the local citizen's group concluded: "the idea of relocating population masses rather than treating and solving problems of cities where they exist is just as repugnant as … timbering without reforestation…" (Wildermuth 2008: 231). With the failure of the Experimental City project, opportunities to make progress on both green cities and Industrial Ecology were lost.

2 Formative Years of IE

When Industrial Ecology did emerge in North America around the 1990s, the subject of green cities or urban systems was at first largely absent. This can be seen from examining the contents of early books and conference proceedings in the formative years of Industrial Ecology (Table 4.1). Georg Winter's 1988 book on *Business and the Environment* considers IE from a corporate perspective; it does have advice for companies with respect to buildings, but gives no mention to cities. Contributors to a broad purview of intersections between technology and the environment (Ausubel and Sladovich 1989) mention issues of air pollution in cities and the management of municipal solid waste, but none give specific focus to cities. On May 20–21, 1991, Kumar Patel organized a colloquium on Industrial Ecology at the US National Academy of Sciences; the 23 papers from the meeting, none of which were on cities, were published in Proceedings of the National Academy of Sciences. A further early publication in which cities are omitted is a collection of papers edited by Brad Allenby and Deanna Richards (1994). There are also examples of IE principles being applied earlier than the 1990s outside of North America – such as in Japan in the early 1970s (Erkman 1997). The report from a US-Japan Workshop on Industrial Ecology, held March 1–3, 1993, in Irvine, California, however, has no mention of urban issues either from Japanese or American perspectives (Richards and Fullerton 1994). Of course, it may be the case that cities were considered in other early literature pertaining to IE in Japan, or elsewhere for that matter, but clearly cities were low priority.

The subject of cities did get on the agenda for the 1992 Snowmass conference on Industrial Ecology and Global Change, which was important for the development of the discipline (Socolow et al. 1994). Approximately 50 people attended the meeting in Colorado, supported by the Office for Interdisciplinary Earth Studies. One of the 32 papers in the proceedings was a study quantifying methane emissions for an American city by Robert Harris of the University of New Hampshire (Harris 1994).

In Tom Graedel and Brad Allenby's textbook on Industrial Ecology, cities are recognized in a chapter on the future of industrial activities (Graedel and Allenby 1995). The section entitled *The Ecologically Planned City* suggests that engineers might consider cities "as systems to be optimized for sustainability" (p. 336), but discussion is limited to just two paragraphs. The reason for such little detail on cities perhaps becomes apparent in Allenby's later text on policy frameworks and implementation of IE – which indicates that the science has yet to be done. Under the subtitle *Physical Models of Communities*, Allenby (1999: 20) writes:

Table 4.1 Coverage of cities or urban systems in early books or conference proceedings on industrial ecology

Book or conference proceeding	Content on cities or urban systems
Winter (1988). Business and the environment: a handbook of industrial ecology with 22 checklists for practical use	No chapters, or checklists, specific to cities
Ausubel and Sladovich (1989). Technology and environment	Cursory mention of urban air pollution and waste management
National academy of science colloquium on industrial ecology published in PNAS (1992)	No papers specific to cities
Allenby and Richards (1994). The greening of industrial ecosystems	No chapters or cases on cities
Socolow et al. (1994). Industrial ecology and global change	One paper on reducing urban sources of methane
Richards and Fullerton (1994). Industrial ecology: U.S. –Japan perspectives	No urban content
Graedel and Allenby (1995). Industrial ecology	Two paragraphs on ecologically planned cities
Ayres and Ayres (1996). Industrial ecology: towards closing the materials cycle	Brief mention of a proposed industrial ecosystem including a municipal waste treatment plant
Lowe et al. (1997). Discovering industrial ecology: an executive briefing and sourcebook	Section on applications of IE for local government
Allenby (1999). Industrial ecology: policy framework and implementation	One paragraph describing the need for research on physical models of communities
Ayres and Ayres (2002). A handbook of industrial ecology	Three chapters on: urban material flows, urban planning; and municipal waste
Bourg and Erkman (2003). Perspectives on industrial ecology	One chapter on urban transportation and IE

Note the list of books examined here may not be comprehensive and is influenced by availability at the University of Toronto library and online access

It is increasingly chic for some communities to call themselves "sustainable communities" yet the science to understand what such assertions really mean has yet to be done. Developing integrated models of urban communities, including small, relatively self-contained cities, larger cities with surrounding suburbs, and large megalopolises with decayed centres and most business activity decentralized throughout the suburbs will be a necessary step in achieving such an understanding. Urban centres in developing and developed nations should also be modelled and compared. Such models would include: transportation, physical infrastructure, food energy, material stocks and flows, and other systems. Both direct and embedded impacts (…) should be included. This information would facilitate identification of major sources of environmental impacts, patterns of activities that give rise to them, and potential environmentally preferable technological or mitigation options. It would also provide a necessary basis for comparing the environmental impacts of different kinds of communities, as well as creating higher level, integrated regional models.

Further recognition of the topic of cities is given in *Discovering Industrial Ecology* by Ernest Lowe, John Warren and Stephen Moran (1997). The front cover shows a night-time city panorama superimposed on to a view of Earth from Space.

Inside the authors include a city public works director in a conversation amongst six professionals about the challenges of implementing IE. A chapter on opportunities for governments in applying IE discusses the need to develop integrated models of urban communities. Several pages on local government then describe how practical applications of IE in cities can lead to waste reduction, extended capacity to local infrastructure, development of resource exchange markets and related economic development – all of which should be pursued within strategic community green planning. As part of this planning, Lowe et al. also encourage the study of urban metabolism (although they call it the industrial metabolism of the community).

In *A Handbook of Industrial Ecology* (Ayres and Ayres 2002), cities are explicitly recognized as being relevant to IE – at least to a small extent – by the inclusion of 3, out of 46, chapters relating to urban issues. The first of these, by Ian Douglas and Nigel Lawson (2002), shows sensible understating of the dominant material flow processes being entitled *Material flows due to Mining and Urbanization*. With respect to the urban end, Douglas and Lawson distinguish between the urban fabric of buildings and infrastructure with slow turnover of materials and the other materials such as food, clothing, packaging, water and energy that flow relatively rapidly through cities. They also note how, over centuries, the in situ deposition of construction and demolition waste in cities gradually raises the elevation of cities. The authors also make reference to a limited number of ecological footprint studies and urban material flow balances, but concede like others previously, that these are challenging to conduct. They resort to presenting a few measures of Earth moved or materials placed in major tunnels, airports runways or similar projects in the United Kingdom or otherwise present national level data.

In *Industrial Ecology and Spatial Planning*, Clinton Andrews (2002a) explains some of the important linkages between industrial ecology and the activities of urban planners. He begins by recognizing that geography matters because it brings details of scale and the level of analytical resolution to IE. Challenges in applying IE to cities sometime occur, however, because of misalignment between natural and political boundaries and the existence of multiple levels of government. Andrews notes the wide variation in physical characteristics of cities worldwide and also comments that planning practices are just as varied – e.g. ranging from pure regulatory roles to those who are active in shaping urban form. He then discusses several intersections between spatial planning and IE: (1) connection between urban patterns and the environmental performance of cities; (2) the significant role for planners in eco-industrial developments, which are subject to a several regulatory issues; (3) urban waste management; (4) several ways in which the design of the built environment impacts eco-efficiency; (5) the use of non-toxic, biocompatible materials in cities; and (6) the potential for industrial ecologists to study behavioural attributes of citizens, linked to policy interventions. Andrews concludes with a warning that IE should be careful to learn from some of the ill-conceived, utopian interventions that urban planners implemented in previous decades.

> Like planning, prescriptive IE is in danger of cycling from hubris to despair before discovering humility and effectiveness.....Until the grand visions of IE are tempered by implementation experience they will be a poor basis for public decisions. (Andrews 2002a: 487)

A second contribution from Andrews in the *Handbook* examines municipal solid waste management through the combined lens of IE and political economy. Andrews (2002b) notes that methods of IE have been used by US waste management policy analysts since the 1970s for mass flow analysis and since the 1980s for life cycle assessment. The perspective of political economy is required to address issues beyond IE – such as trans-boundary transport, deregulation and environmental justice. Andrews provides integrative analysis of the predominant actors at each stage of the product life cycle. This leads to an examination of the basis for government intervention and a summary of lessons from implementation.

A further collection of papers on IE edited by Dominique Bourg and Suren Erkman (2003) includes one urban flavoured chapter, out of 28, this being on Urban Transportation and IE by Tom Graedel and Michael Jensen (2003). The short paper reviews the costs and benefits of automotive transportation and considers three different approaches to optimizing the net benefits. An urban system is proposed comprising trams, high-speed networks and feeder lines, with automobile restricted to low-density suburbs and new ways of handling personal cargo developed.

This review of formative literature on IE shows relatively sparing attention to urban systems, but it would be wrong to conclude that the early protagonists of IE failed to understand the significance of cities in environmental challenges. For example, both Andrews and Graedel were attendees of the Snowmass 1992 summit and were (and remain) influential within IE. Andrews was coeditor of the Snowmass proceedings and went on to become a Chair of Planning at Rutgers University, where he published several papers pertaining to IE and cities (some discussed below here). Graedel, the first Professor of Industrial Ecology, and the first President of the International Society for Industrial Ecology (ISIE), examines metal stocks and flows in cities within his wider work at other scales. Another who worked on cities was Jesse Ausubel, coeditor of a National Academy of Engineering book on *Cities and Their Vital Systems* (Ausubel and Herman 1988). Like Andrews and Graedel, he attended the formative meeting of the ISIE in 2000 and also coedited the proceeding of a National Academy of Engineering meeting on Energy and Environment (Ausubel and Sladovich 1989). A further attendee at the Snowmass meeting – and early influence in IE – was Arnulf Grübler – who had a profound understanding of the role of cities on global environmental change. In his text *Technology and Global Change* (Grübler 1998) – which arguably should be the first textbook read by students of IE – Grübler clearly describes how the phenomena of urbanization result from improvements in agricultural productivity.

3 Into the Twenty-First Century

Moving into the twenty-first century, there is the emergence of an academic research literature specifically concerned with applications of IE to cities. This can be examined through use of the search function on Scopus, although many caveats are required. A first set of search results, shown in Fig. 4.1, is for papers that explicitly

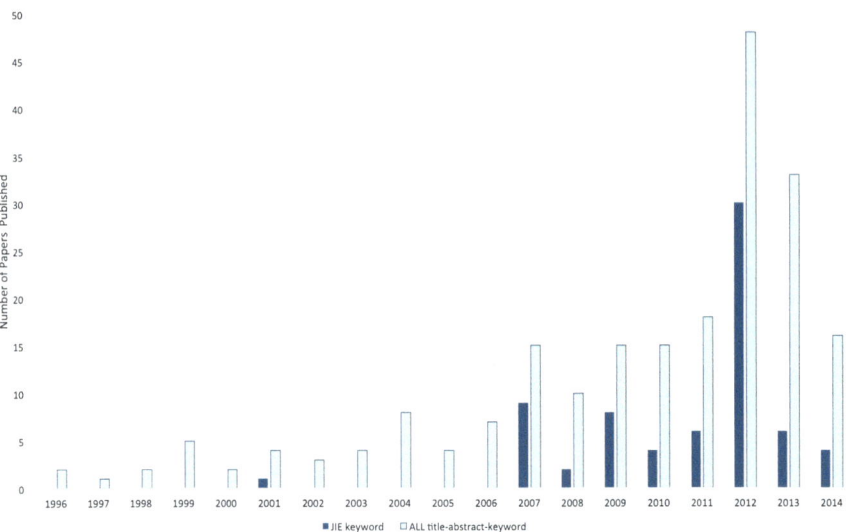

Fig. 4.1 Papers published (1996–2014) in: (1) Scopus-tracked journals including industrial ecology and cities, city, urban or urbanization in the title, abstract or keywords; (2) the Journal of Industrial Ecology, including the keywords cities, city, urban or urbanization

use the term "industrial ecology" and any one of four city-related words: "cities", "city", "urban" or "urbanization" in the title, abstract or keywords. The search results are modified to remove obvious anomalies as well as articles in press for 2014. Still, the search results include some papers that perhaps do not really reflect the essence of applying IE at the city scale while missing others that do. To give some examples, several of the early papers discuss eco-industrial parks as a strategy for sustainable cities, or eco-parks which just happen to be in cities, without really reflecting the idea that industrial symbiosis might be more broadly applied on a city scale. Similarly, the search picks up a few papers that are focused on household consumption within cities, which, while contributing to urban metabolism, might miss industrial and larger-scale urban processes in cities. Another scale-related problem is that a few papers are included that primarily discuss energy or material flows at national and regional levels, with only brief mention of processes at the urban scale. Of the 212 papers recorded in the first search perhaps as many as half might be rejected if the search were restricted to research at the city scale. That said, the search terms exclude a potentially large literature on energy and material flows in cities that does not use the term "industrial ecology". Omitted papers include, for example, a material flow study of paper in a city (Leach et al. 1997) and an overview of strategies for the ecological restructuring of cities (Hahn and Simonis 1991). In more recent years, industrial ecologists have made numerous studies of greenhouse gas emissions associated with cities (e.g. Kennedy et al. 2009), or urban sustainability more broadly (e.g. Baynes and Wiedmann 2012), many of which are missed.

Most significantly, the search misses many papers on urban metabolism – now seen as central to urban IE – but also used in other fields; this will be addressed in a separate section below.

A few reoccurring themes can be observed within the Scopus-tracked papers in the first search. Starting with a paper by Robbins and Kumar (1999), several authors discuss the application of industrial symbiosis with an exchange of residuals occurring broadly at a city scale (Cerceau et al. 2014; Liu and Chen 2013; Chen et al. 2012; Giurco et al. 2011; Van Berkel et al. 2009a, b). Amongst these are two papers by Pierre Desrochers which suggest that industrial symbiosis has a long history in cities (Desrochers 2002; Desrochers and Leppälä 2010). A couple of papers analysing the use of biogas or sewage gas for transportation in Swedish cities might also be considered examples of city-scale industrial symbiosis (Fallde and Eklund 2015; Vernay et al. 2013). The topic of waste has been addressed in several papers, including work focussed on e-waste (Leigh et al. 2012), healthcare waste (Soares et al. 2013) and construction and demolition waste (Bohne et al. 2008), as well as city-specific waste studies (Murphy and Pincetl 2013; Chertow and Eckelman 2009). Urban infrastructure systems have been researched in both broad and narrow contexts. In a broad sense, several sustainable infrastructure or resource frameworks have been published (Ramaswami et al. 2012a, b; Hodson et al. 2012; Agudelo-Vera et al. 2012). Narrower sector-specific infrastructure studies include those on water systems, including drinking water, wastewater and storm water (Venkatesh 2013; De Sousa et al. 2012; Venkatesh and Brattebø 2012; Kenway et al. 2011; Pasqualino et al. 2011; Venkatesh et al. 2009). At least two papers have examined the historical use of energy in cities (Reiter and Marique 2012; Baynes and Bai 2012). Also related to energy, amongst other activities, is a large number of papers on greenhouse gas emissions for cities (Ramaswami et al. 2012a, b; Mohareb and Kennedy 2012a, b; Chavez et al. 2012; Sugar et al. 2012; Feng et al. 2012; Shi et al. 2012; Bullock et al. 2011; Kraines et al. 2010; Lebel et al. 2007; VandeWeghe and Kennedy 2007), although this is just a subset of a larger literature. Other contaminants featured include heavy metals (Batzias et al. 2011) and PBDEs (Vyzinkarova and Brunner 2013). There is also a distinct grouping of papers on the theme of metals in cities, including urban mining (Kral et al. 2014; Zhang et al. 2012; Klinglmair and Fellner 2010; Månsson et al. 2009; Drakonakis et al. 2007; Harper et al. 2006). Similarly, several papers address stocks and flows of phosphorus in cities (Ma et al. 2013; Kalmykova et al. 2012; Li et al. 2012; Fu et al. 2012), while others have addressed food more broadly (Broeze et al. 2011; Neset et al. 2006; Waggoner 2006). Over 200 papers on cities and IE are picked up in the first Scopus search, only a few of which have been mentioned here in order to demonstrate the dominant themes of the literature.

Much of the recent upward trend in publications on IE and cities is due to papers in the Journal of Industrial Ecology (JIE); it accounts in particular for the steps up in publications in 2007 and 2012 (Fig. 4.1). The second search in Fig. 4.1 is for all titles in the JIE for the keywords cities, city, urban or urbanization. In 2007, JIE had a quasi special issue on *Industrial Ecology and the Global Impacts of Cities*, edited by Xuemei Bai, with nine articles (picked up by the keyword search). Since then there have been between 2 and 8 urban systems papers per year in JIE, with the exception of 2012 when 30 papers are recorded. This was due to the publication of a full special issue on *Sustainable Urban Systems* with Larry Baker, Shobhakar Dhakal, Anu Ramaswami and myself as coeditors.

In addition to the journal articles on IE and cities, the past decade has also seen publication of textbooks on the topic. Amongst these is Sustainable Urban Metabolism by Paulo Ferrão and John Fernández (2013). Another that explicitly recognizes IE is Vortex cities to sustainable cities: Australia's urban challenge by Phil McManus (2005).

4   Urban Metabolism

The interdisciplinary field of Industrial Ecology can today perhaps be broadly understood as the study of the three types of metabolism – industrial, socio-economic and urban. When IE first began to establish as a discipline in the 1990s, it primarily did so with an initial focus on industrial metabolism. Inclusion of urban metabolism within IE arguably only occurred once IE had broadened to include Social Ecology (Fischer-Kowalski and Hüttler 1998); urban metabolism can be seen as a scale-delineated component of socio-economic metabolism. The term urban metabolism, however, seems to predate the term industrial metabolism, which is usually attributed to Ayres (1994). Abel Wolman's influential paper on the Metabolism of Cities was published in Scientific American in 1965, while the notion that cities have a metabolism can perhaps be traced back further to Marx (Newell and Cousins 2014). Interestingly, Wolman was a member of the Spilhaus' Committee on Pollution in the early 1960s, as was Alan Kneese who co-authored the highly influential paper on material flows in an economy with Bob Ayres in the late 1960s (Ayres and Knesse 1969). So the study of urban metabolism and industrial metabolism do have some shared history.

The topic of urban metabolism nonetheless has its own literature, originally separate from Industrial Ecology. Starting from the 1970s, quantitative studies of the metabolism of actual cities have been conducted by researchers from several disciplines including civil engineering, chemical engineering, ecology and urban planning (Kennedy et al. 2007). Significant contributions to urban metabolism in the early years included works by Newcombe et al. (1978), Odum (1983), Baccini and Brunner (1991) and Girardet (1992). The first urban metabolism study to recognize the term industrial ecology appears to be Peter Newman's study of Sydney in 1999. A review of the urban literature with discussion of applications to urban planning and design is given by Kennedy et al. (2011), with an extension to include a growing number of Chinese studies by Zhang (2013). Barles (2010) explores the origins of urban metabolism in relation to sustainable urban development. More recent review papers by Broto et al. (2012) and Newell and Cousins (2014) have examined relations between studies of urban metabolism in the fields of industrial ecology, urban ecology and political ecology.

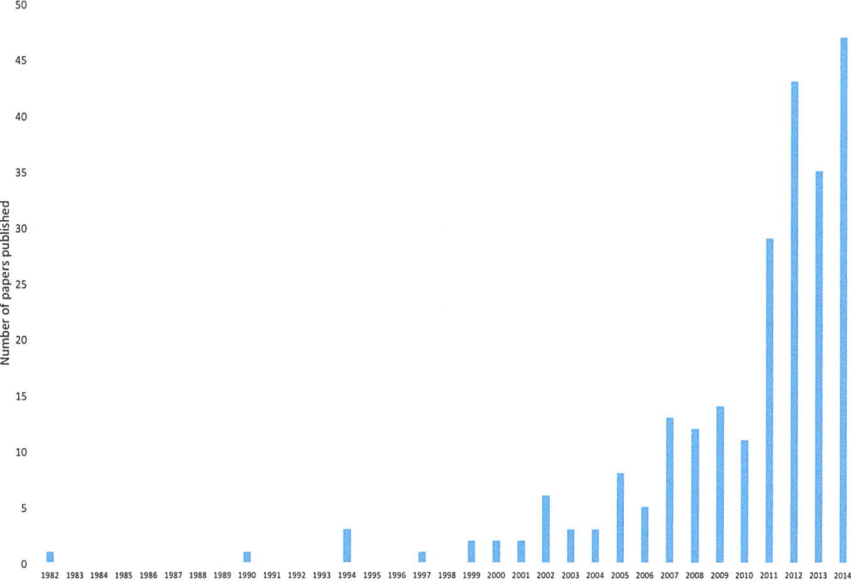

Fig. 4.2 Papers published (1982–2014) in Scopus-tracked journals including urban metabolism or metabolism of cities in the title, abstract or keywords. Wolman's 1965 publication off the chart

A further search using Scopus shows an increasing number of publications on urban metabolism over the past 15 years or so (Fig. 4.2). Around the turn of the century, there were typically two papers on urban metabolism published each year, but the number increased to 47 in 2014. The results for early years in this search should be treated cautiously, as some works recognized today as urban metabolism studies do not register. Notably, none of the works in the 1970s, between Wolman's paper and one by Olson in 1982, were picked up in the search. Nonetheless, it is clear that research on urban metabolism is accelerating.

5   Future Directions

The grand challenge for IE and cities is to understand the environmental impacts of the urban metabolism and pursue plans and strategies to reduce them. These aspirations are to some extent the same as those of Spilhaus and the Experimental City project 50 years ago, but necessarily tempered with large doses of practicality, as raised by authors such as Andrews. Unlike Spilhaus' efforts, the applications of IE have to first be in today's existing cities, many of which are rapidly growing. It might be possible in some countries or cultures that new cities are built on green-field sites, possibly on the edge of existing cities – and principles of IE might be incorporated to some degree. A small example of an experimental green city does exist in the case of Mazda, UAE; important lessons have been learnt at Mazda, such as the cost of building a city with the pedestrian realm one storey up above a ground floor dedicated to service vehicles and personal rapid transit pods. Nonetheless, sustainable transformation of today's existing and expanding cities is where the challenge lies.

Although not complete, the study of cities in Industrial Ecology has come a long way since Allenby's (1999) observation that the science had yet to be done. Progress has been made with a combination of broad holistic analyses of urban metabolism – the scale upon which differences are measured – and more detailed studies of specific resource flows or infrastructure systems – which reveal the inner workings of the city. As more cities, hopefully, move towards greater practical application of IE, then attention to both scales will be important. There is a need for further study of material flows in cities; few studies have quantified material stocks and flows with as much detail as those in Lisbon (Niza et al. 2009) or Paris (Barles 2009). More refined understanding of material flows will be necessary to address questions in a couple of related future directions discussed below: increased application of industrial symbiosis at the city scale and examination of the thermodynamics of urban metabolism.

Just how much potential is there for industrial symbiosis to be conducted at the city scale and how much of such sharing or recycling of residuals is already taking place? Examples such as the case of Kawasaki, Japan, where 565,000 tonnes of potential waste per year are diverted through seven key material exchanges hint at significant potential for industrial symbiosis in cities (Van Berkel et al. 2009a, b). Drawing upon four examples of urban regions where waste exchange is practiced, and citing several nineteenth and early twentieth century authors, Desroches (2002) argues that urban industrial symbiosis used to be relatively common. "The fact that cities or regional economies… have probably always exhibited localized inter-industry recycling linkages seems highly plausible" (Desroches 2002: 35). He suggests that industrial symbiosis is a form of agglomeration effect that occurs due to the high volumes and close proximity of waste-producing activities in cities. As well as further empirical studies, perhaps there is potential to develop theoretical economic models that describe such agglomeration effects. Further research might also seek to determine the limits to which industrial symbiosis or other notions of the circular economy can practically be applied in cities. Some categories of materials cannot be recycled or require so much energy as to be undesirable (Ayres 1997; Allenby 1999; Allwood 2014).

A final challenging topic, which will also inform the questions on industrial symbiosis, is the development of improved theoretical understanding of the urban metabolism using thermodynamics. This is important for addressing concerns over possible limits to the notion of sustainable cities. If cities were to pursue high levels of efficiency, and greater closing of material loops through increased industrial symbiosis, what would be the repercussions, feedbacks or rebound effects? For example, if today's cities were able to cut their consumption of fossil-fuel energy use in half, might that just result in the saved fuels being used to build more cities? Nonequilibrium thermodynamics, as used by Bristow and Kennedy (2015) to understand the growth of cities, might tentatively offer insights into such questions. A

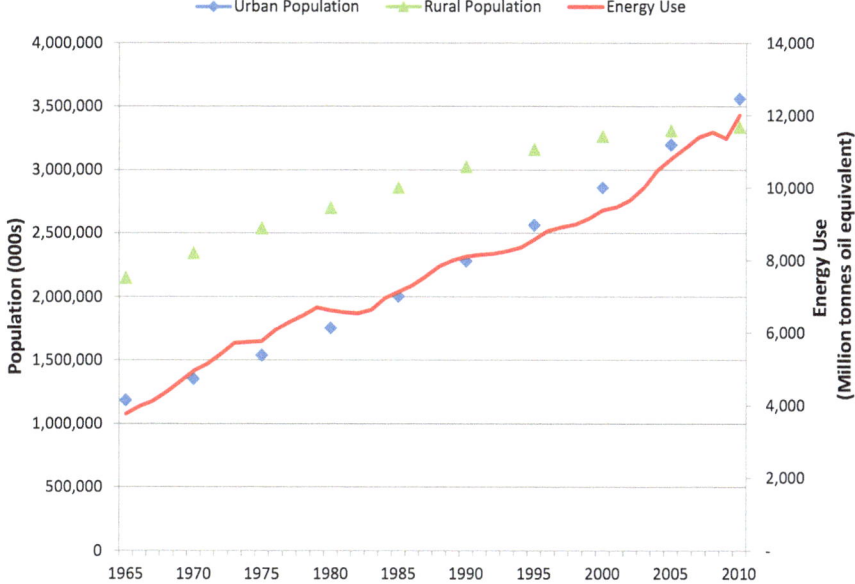

Fig. 4.3 Global energy use, urban and rural population, 1965–2010 (Figure 4.3 from Bristow and Kennedy 2015)

possibly important finding in this regard is the strong relationship showing that global energy use is directly proportional to global urban population (Fig. 4.3). Three areas for future research on the thermodynamics of cities are proposed: (1) development of nonsteady, nonequilibrium thermodynamic models specific to cities or systems of cities; (2) more studies of exergy gradients for cities; and (3) exploration of the possible intersection of thermodynamic and economic theories of urban growth (Bristow and Kennedy 2015).

References

Agudelo-Vera, C. M., Mels, A., Keesman, K., & Rijnaarts, H. (2012). The urban harvest approach as an aid for sustainable urban resource planning. *Journal of Industrial Ecology, 16*(6), 839–850.

Allenby, B. R. (1999). *Industrial ecology: Policy framework and implementation*. Upper Saddle River: Prentice Hall.

Allenby, B. R., & Richards, D. J. (Eds.). (1994). *The greening of industrial ecosystems*. Washington, DC: National Academy Press.

Allwood, J. M. (2014). Squaring the circular economy: The role of recycling within a hierarchy of material management strategies. In E. Worrell & M. A. Reuter (Eds.), *Handbook of recycling* (pp. 445–477). Waltham: Elsevier.

Andrews, C. (2002a). Industrial ecology and spatial planning. In R. U. Ayres & L. W. Ayres (Eds.), *A handbook of industrial ecology* (pp. 476–487). Cheltenham: Edward Elgar.

Andrews, C. (2002b). Municipal solid waste management. In R. U. Ayres & L. W. Ayres (Eds.), *A handbook of industrial ecology* (pp. 542–553). Cheltenham: Edward Elgar.

Ausubel, J. H., & Herman, R. (1988). *Cities and their vital systems.*

Ausubel, J. H., & Sladovich, A. E. (Eds.). (1989). *Technology and environment, National Academy of Engineering.* Washington, DC: National Academy Press.

Ayres, R. U. (1994). Industrial metabolism: Theory and policy. In R. U. Ayres & U. K. Simonis (Eds.), *Industrial metabolism: Restructuring for sustainable development* (pp. 3–20). Tokyo: United Nations University Press.

Ayres, R. U. (1997). *Industrial metabolism: Work in progress.* Retrieved February 23, 2015, from https://flora.insead.edu/fichiersti_wp/inseadwp1997/97-09.pdf

Ayres, R. U., & Ayres, L. W. (1996). *Industrial ecology: Towards closing the materials cycle.* Cheltenham: Edward Elgar.

Ayres, R. U., & Ayres, L. W. (2002). *A handbook of industrial ecology.* Cheltenham: Edward Elgar.

Ayres, R., & Kneese, A. (1969). Production, consumption, and externalities. *American Economic Review, 59*(3), 282–297.

Baccini, P., & Brunner, P. H. (1991). *Metabolism of the anthroposphere.* Berlin: Springer.

Barles, S. (2009). Urban metabolism of Paris and its region. *Journal of Industrial Ecology, 13*(6), 898–913.

Barles, S. (2010). Society, energy and materials: The contribution of urban metabolism studies to sustainable urban development issues. *Journal of Environmental Planning and Management, 53*(4), 439–455.

Batzias, F., Politi, D., & Sidiras, D. (2011). Heavy metals pollution abatement within a framework of industrial ecology. In *Recent advances in fluid mechanics and heat and mass transfer – Proceedings of the 9th IASME/WSEAS International Conference on Fluid Mechanics and Aerodynamics,* FMA'11, pp. 251–256.

Baynes, T. M., & Bai, X. (2012). Reconstructing the energy history of a city: Melbourne's population, urban development, energy supply and use from 1973 to 2005. *Journal of Industrial Ecology, 16*(6), 862–874.

Baynes, T., & Wiedmann, T. (2012). General approaches to assessing urban environmental sustainability. *Current Opinion in Environmental Sustainability, 4,* 458–464.

Bohne, R. A., Brattebø, H., & Bergsdal, H. (2008). Dynamic eco-efficiency projections for construction and demolition waste recycling strategies at the city level. *Journal of Industrial Ecology, 12*(1), 52–68.

Bourg, D., & Erkman, S. (Eds.). (2003). *Perspectives on industrial ecology.* Sheffield: Greenleaf.

Bristow, D., & Kennedy, C. A. (2015). Why do cities grow? Insights from non-equilibrium thermodynamics at the urban and global scales. *Journal of Industrial Ecology, 19*(2), 211–221.

Broeze, J., Simons, A., & Smeets, P. (2011). Sustainable agro-food production concept: Food clusters. In *6th international CIGR technical symposium – Towards a sustainable food chain: Food process, bioprocessing and food quality management,* 4 p.

Broto, V. C., Allen, A., & Rapoport, E. (2012). Interdisciplinary perspectives on urban metabolism. *Journal of Industrial Ecology, 16*(6), 851–861.

Bullock, S. H., Escoto-Rodríguez, M., Smith, S. V., & Hinojosa, A. (2011). Carbon flux of an urban system in México. *Journal of Industrial Ecology, 15*(4), 512–526.

Cerceau, J., Mat, N., Junqua, G., Lin, L., Laforest, V., & Gonzalez, C. (2014). Implementing industrial ecology in port cities: International overview of case studies and cross-case analysis. *Journal of Cleaner Production, 74,* 1–16.

Chavez, A., Ramaswami, A., Nath, D., Guru, R., & Kumar, E. (2012). Implementing trans-boundary infrastructure-based greenhouse gas accounting for Delhi, India: Data availability and methods. *Journal of Industrial Ecology, 16*(6), 814–828.

Chen, X., Fujita, T., Ohnishi, S., Fujii, M., & Geng, Y. (2012). The impact of scale, recycling boundary, and type of waste on symbiosis and recycling: An empirical study of Japanese eco-towns. *Journal of Industrial Ecology, 16*(1), 129–141.

Chertow, M. R., & Eckelman, M. J. (2009). Using material flow analysis to illuminate long-term waste management solutions in Oahu, Hawaii. *Journal of Industrial Ecology, 13*(5), 758–774.

Committee on Pollution. (1966). *Waste management and control: A report to the Federal Council for Science and Technology, National Academy of Sciences – National Research Council.* Washington, DC: National Academy Press.

De Sousa, M. R. C., Montalto, F. A., & Spatari, S. (2012). Using life cycle assessment to evaluate green and grey combined sewer overflow control strategies. *Journal of Industrial Ecology, 16*(6), 901–913.

Desrochers, P. (2002). Cities and industrial symbiosis: Some historical perspectives and policy implications. *Journal of Industrial Ecology, 5*(4), 29–44.

Desrochers, P., & Leppälä, S. (2010). Industrial symbiosis: Old wine in recycled bottles? Some perspective from the history of economic and geographical thought. *International Regional Science Review, 33*(3), 338–361.

Douglas, I., & Lawson, N. (2002). Material flows due to mining and urbanization. In R. U. Ayres & L. W. Ayres (Eds.), *A handbook of industrial ecology* (pp. 351–364). Cheltenham: Edward Elgar.

Drakonakis, K., Rostkowski, K., Rauch, J., Graedel, T. E., & Gordon, R. B. (2007). Metal capital sustaining a North American city: Iron and copper in New Haven, CT. *Resources, Conservation and Recycling, 49*(4), 406–420.

Erkman, S. (1997). Industrial ecology: An historical view. *Journal of Cleaner Production, 5*(1–2), 1–10.

Fallde, M., & Eklund, M. (2015). Towards a sustainable socio-technical system of biogas for transport: The case of the city of Linköping in Sweden. *Journal of Cleaner Production, 98*, 17–28.

Feng, K., Siu, Y. L., Guan, D., & Hubacek, K. (2012). Analyzing drivers of regional carbon dioxide emissions for China: A structural decomposition analysis. *Journal of Industrial Ecology, 16*(4), 600–611.

Ferrão, P., & Fernández, J. E. (2013). *Sustainable urban metabolism.* Cambridge, MA: The MIT Press.

Fischer-Kowalski, M., & Hüttler, W. (1998). Society's metabolism. *Journal of Industrial Ecology, 2*(4), 107–136.

Fu, Y., Yuan, Z., Wu, H., & Zhang, L. (2012). Anthropogenic phosphorus flow analysis of Hanshan County in Anhui Province. *Shengtai Xuebao/Acta Ecologica Sinica, 32*(5), 1578–1586.

Girardet, H. (1992). *The Gaia atlas of cities.* London: Gaia Books Limited.

Giurco, D., Bossilkov, A., Patterson, J., & Kazaglis, A. (2011). Developing industrial water reuse synergies in Port Melbourne: Cost effectiveness, barriers and opportunities. *Journal of Cleaner Production, 19*(8), 867–876.

Graedel, T. E., & Allenby, B. R. (1995). *Industrial ecology.* Englewood Cliffs: Prentice Hall.

Graedel, T. E., & Jensen, M. (2003). Urban transportation and industrial ecology. In D. Bourg & S. Erkman (Eds.), *Perspectives on industrial ecology* (pp. 283–290). Sheffield: Greenleaf.

Grübler, A. (1998). *Technology and global change.* Cambridge/New York: Cambridge University Press.

Hahn, E., & Simonis, U. E. (1991). Ecological urban restructuring: Method and action. *Environmental Management and Health, 2*(2), 12–19.

Harper, E. M., Johnson, J., & Graedel, T. E. (2006). Making metals count: Applications of material flow analysis. *Environmental Engineering Science, 23*(3), 493–506.

Harris, R. (1994). Reducing urban sources of methane: An experiment in industrial ecology. In R. Socolow et al. (Eds.), *Industrial ecology and global change* (pp. 173–182). Cambridge/New York: Cambridge University Press.

Hodson, M., Marvin, S., Robinson, B., & Swilling, M. (2012). Reshaping urban infrastructure: Material flow analysis and transitions analysis in an urban context. *Journal of Industrial Ecology, 16*(6), 789–800.

Kalmykova, Y., Harder, R., Borgestedt, H., & Svanäng, I. (2012). Pathways and management of phosphorus in urban areas. *Journal of Industrial Ecology, 16*(6), 928–939.

Kennedy, C. A., Cuddihy, J., & Engel Yan, J. (2007). The changing metabolism of cities. *Journal of Industrial Ecology, 11*(2), 43–59.

Kennedy, C., Steinberger, J., Gasson, B., Hillman, T., Havránek, M., Hansen, Y., Pataki, D., Phdungsilp, A., Ramaswami, A., & Villalba Mendez, G. (2009). Greenhouse gas emissions from global cities. *Environmental Science and Technology, 43*, 7297–7302.

Kennedy, C. A., Pincetl, S., & Bunje, P. (2011). The study of urban metabolism and its applications to urban planning and design. *Journal of Environmental Pollution, 159*(8–9), 1965–1973.

Kenway, S., Gregory, A., & McMahon, J. (2011). Urban water mass balance analysis. *Journal of Industrial Ecology, 15*(5), 693–706.

Klinglmair, M., & Fellner, J. (2010). Urban mining in times of raw material shortage: Copper management in Austria during World War I. *Journal of Industrial Ecology, 14*(4), 666–679.

Kraines, S. B., Ishida, T., & Wallace, D. R. (2010). Integrated environmental assessment of supply-side and demand-side measures for carbon dioxide mitigation in Tokyo, Japan. *Journal of Industrial Ecology, 14*(5), 808–825.

Kral, U., Lin, C.-Y., Kellner, K., Ma, H.-W., & Brunner, P. H. (2014). The copper balance of cities: Exploratory insights into a European and an Asian city. *Journal of Industrial Ecology, 18*(3), 432–444.

Leach, M. A., Bauen, A., & Lucas, N. (1997). A systems approach to materials flow in sustainable cities a case study of paper. *Journal of Environmental Planning and Management, 40*(6), 705–724.

Lebel, L., Garden, P., Ma Banaticla, R. N., Lasco, R. D., Contreras, A., Mitra, A. P., Sharma, C., Nguyen, H. T., Ooi, G. L., & Sari, A. (2007). Integrating carbon management into the development strategies of urbanizing regions in Asia: Implications of urban function, form, and role. *Journal of Industrial Ecology, 11*(2), 61–81.

Leigh, N. G., Choi, T., & Hoelzel, N. Z. (2012). New insights into electronic waste recycling in metropolitan areas. *Journal of Industrial Ecology, 16*(6), 940–950.

Li, G.-L., Bai, X., Yu, S., Zhang, H., & Zhu, Y.-G. (2012). Urban phosphorus metabolism through food consumption: The case of China. *Journal of Industrial Ecology, 16*(4), 588–599.

Liu, G.-F., & Chen, F.-D. (2013). NISP-based research on the system structure of urban symbiosis network in China. *Applied Mechanics and Materials, 427–429*, 2923–2927.

Lowe, E. A., Warren, J. L., & Moran, S. R. (1997). *Discovering industrial ecology: An executive briefing and sourcebook*. Columbus: Battelle.

Ma, D., Hu, S., Chen, D., & Li, Y. (2013). The temporal evolution of anthropogenic phosphorus consumption in China and its environmental implications. *Journal of Industrial Ecology, 17*(4), 566–577.

Månsson, N., Bergbäck, B., & Sörme, L. (2009). Phasing out cadmium, lead, and mercury effects on urban stocks and flows. *Journal of Industrial Ecology, 13*(1), 94–111.

McManus, P. (2005). *Vortex cities to sustainable cities: Australia's urban challenge*. Sydney: University of New South Wales Press.

Mohareb, E., & Kennedy, C. (2012a). Greenhouse gas emission scenario modeling for cities using the PURGE model: A case study of the greater Toronto area. *Journal of Industrial Ecology, 16*(6), 875–888.

Mohareb, E., & Kennedy, C. (2012b). Gross direct and embodied carbon sinks for urban inventories. *Journal of Industrial Ecology, 16*(3), 302–316.

Murphy, S., & Pincetl, S. (2013). Zero waste in Los Angeles: Is the emperor wearing any clothes? *Resources, Conservation and Recycling, 81*, 40–51.

Neset, T.-S. S., Bader, H.-P., & Scheidegger, R. (2006). Food consumption and nutrient flows: Nitrogen in Sweden since the 1870s. *Journal of Industrial Ecology, 10*(4), 61–75.

Newcombe, K., Kalma, J., & Aston, A. (1978). The metabolism of a city: The case of Hong Kong. *Ambio, 7,* 3–15.

Newell, J. P., & Cousins, J. J. (2014). The boundaries of urban metabolism: Towards a political–Industrial ecology. *Progress in Human Geography,* 1–27.

Niza, S., Rosado, L., & Ferrao, P. (2009). Urban metabolism methodological advances in urban material flow accounting based on the Lisbon case study. *Journal of Industrial Ecology, 13*(3), 384–405.

Odum, H. T. (1983). *Systems ecology, an introduction.* New York: Wiley-Interscience.

Olson, S. (1982). Urban metabolism and morphogenesis. *Urban Geography, 3*(2), 87–109.

Pasqualino, J. C., Meneses, M., & Castells, F. (2011). Life cycle assessment of urban wastewater reclamation and reuse alternatives. *Journal of Industrial Ecology, 15*(1), 49–63.

Ramaswami, A., Weible, C., Main, D., Heikkila, T., Siddiki, S., Duvall, A., Pattison, A., & Bernard, M. (2012a). A social-ecological-infrastructural systems framework for interdisciplinary study of sustainable city systems: An integrative curriculum across seven major disciplines. *Journal of Industrial Ecology, 16*(6), 801–813.

Ramaswami, A., Chavez, A., & Chertow, M. (2012b). Carbon footprinting of cities and implications for analysis of urban material and energy flows. *Journal of Industrial Ecology, 16*(6), 783–785.

Reiter, S., & Marique, A.-F. (2012). Toward low energy cities: A case study of the urban area of Liége, Belgium toward low energy cities. *Journal of Industrial Ecology, 16*(6), 829–838.

Richards, D. J., & Fullerton, A. B. (1994). *Industrial ecology: U.S. – Japan perspectives.* Washington, DC: National Academy Press.

Shi, F., Huang, T., Tanikawa, H., Han, J., Hashimoto, S., & Moriguchi, Y. (2012). Toward a low carbon-dematerialization society: Measuring the materials demand and CO2 emissions of building and transport infrastructure construction in China. *Journal of Industrial Ecology, 16*(4), 493–505.

Soares, S. R., Finotti, A. R., Prudêncio da Silva, V., & Alvarenga, R. A. F. (2013). Applications of life cycle assessment and cost analysis in health care waste management. *Waste Management, 33*(1), 175–183.

Socolow, R., Andrews, C., Berkhout, F., & Thomas, V. (Eds.). (1994). *Industrial ecology and global change.* Cambridge/New York: Cambridge University Press.

Spilhaus, A. (1967). The experimental city. *Daedalus, 96*(4), 1129–1141.

Spilhaus, A. (1968). The experimental city. *Science, 159*(3816), 710–715.

Sugar, L., Kennedy, C., & Leman, E. (2012). Greenhouse gas emissions from Chinese cities. *Journal of Industrial Ecology, 16*(4), 552–563.

Van Berkel, R., Fujita, T., Hashimoto, S., & Geng, Y. (2009a). Industrial and urban symbiosis in Japan: Analysis of the Eco-Town program 1997–2006. *Journal of Environmental Management, 90*(3), 1544–1556.

Van Berkel, R., Fujita, T., Hashimoto, S., & Fujii, M. (2009b). Quantitative assessment of urban and industrial symbiosis in Kawasaki, Japan. *Environmental Science and Technology, 43*(5), 1271–1281.

VandeWeghe, J. R., & Kennedy, C. (2007). A spatial analysis of residential greenhouse gas emissions in the Toronto census metropolitan area. *Journal of Industrial Ecology, 11*(2), 133–144.

Venkatesh, G. (2013). An analysis of stocks and flows associated with water consumption in Indian households. *Journal of Industrial Ecology, 17*(3), 472–481.

Venkatesh, G., & Brattebø, H. (2012). Assessment of environmental impacts of an aging and stagnating water supply pipeline network: City of Oslo, 1991–2006. *Journal of Industrial Ecology, 16*(5), 722–734.

Venkatesh, G., Hammervold, J., & Brattebø, H. (2009). Combined MFA-LCA for analysis of wastewater pipeline networks: Case study of Oslo, Norway. *Journal of Industrial Ecology, 13*(4), 532–550.

Vernay, A.-L., Mulder, K. F., Kamp, L. M., & De Bruijn, H. (2013). Exploring the socio-technical dynamics of systems integration-the case of sewage gas for transport in Stockholm, Sweden. *Journal of Cleaner Production, 44*, 190–199.

Vyzinkarova, D., & Brunner, P. H. (2013). Substance flow analysis of wastes containing polybrominated diphenyl ethers: The need for more information and for final sinks. *Journal of Industrial Ecology, 17*(6), 900–911.

Waggoner, P. E. (2006). How can EcoCity get its food? *Technology in Society, 28*(1–2), 183–193.

Wildermuth, T. A. (2008). *Yesterday's city of tomorrow: The Minnesota experimental city and green urbanism*. PhD dissertation, University of Illinois at Urbana Champaign.

Winter, G. (1988). *Business and the environment: A handbook of industrial ecology with 22 checklists for practical use and a concrete example of the integrated system of Environmentalist Business Management (the Winter model)*. Das umweltbewusste Unternehmen. English Hamburg/New York: McGraw-Hill Book Co.

Zhang, Y. (2013). Urban metabolism: A review of research methodologies. *Environmental Pollution, 178*, 463–473.

Zhang, L., Yuan, Z., & Bi, J. (2012). Estimation of copper in-use stocks in Nanjing, China. *Journal of Industrial Ecology, 16*(2), 191–202.

Chapter 5
Scholarship and Practice in Industrial Symbiosis: 1989–2014

Marian Chertow and Jooyoung Park

Abstract Industrial symbiosis, a subfield of industrial ecology, engages traditionally separate industries and entities in a collaborative approach to resource sharing that benefits both the environment and the economy. This chapter examines the period 1989–2014 to "take stock" of industrial symbiosis. First, we look at the earliest days to discuss what inspired industrial symbiosis both in the scholarly literature and in practice. Next, we draw attention to certain dilemmas and sharpen the distinctions between industrial symbiosis and some related concepts such as eco-industrial parks and environmentally balanced industrial complexes. With regard to dissemination of industrial symbiosis ideas, we found that at the country level, China has now received the most attention in industrial symbiosis academic research and this continues to grow rapidly.

The final section looks at both theory (conceptual knowledge largely from academia) and practice (on-the-ground experience of public, not-for-profit, and private organizations working to implement industrial symbiosis) as both are essential to industrial symbiosis. A bibliometric analysis of the scholarly work, capturing 391 articles indexed in Scopus and Web of Science for 20 years between 1995 and 2014, is used to define and track the types of articles, how the mix of articles has changed over time, and what the most popular journals are. Taking a closer look at the research literature, distinct themes are identified and discussed such as the scale of industrial symbiosis, whether industrial symbiosis is based on planning or self-organization, the role of social factors, and what is known about the actual performance of industrial symbiosis. To assess important issues with regard to practice, we compile a list of industrial symbiosis-related events from database searches of reports, media, and key consulting and business organizations and examine trends, mechanisms, and motivations of industrial symbiosis practice by surveying key practitioners and academics.

M. Chertow (✉)
Yale School of Forestry and Environmental Studies, Yale University,
195 Prospect Street, New Haven, CT 06511, USA
e-mail: marian.chertow@yale.edu

J. Park
School of Management, Universidad de los Andes, Calle 21 No. 1-20, Bogotá, Colombia
e-mail: jy.park@uniandes.edu.co

© The Author(s) 2016   87
R. Clift, A. Druckman (eds.), *Taking Stock of Industrial Ecology*,
DOI 10.1007/978-3-319-20571-7_5

Since 1989, there has been significant uptake of industrial symbiosis around the world as shown by the increasing number of journal articles and also events on the ground. Industrial symbiosis has become more geographically and institutionally diverse, as more organizations in more countries learn about the ideas and diffuse regionally specific versions. This presents additional opportunities to understand the phenomenon, but also makes the search to embrace a coherent framework more immediate.

Keywords Eco-industrial park • Industrial ecosystem • By-product reuse • Circular economy • Industrial ecology

1 Introduction

This chapter examines the period 1989–2014 during which many people associated with the industrial ecology community have participated in building the knowledge base of industrial symbiosis. At the heart of industrial symbiosis is cooperative resource sharing of water, energy, and material by-products and wastes across organizations for both environmental and economic benefit. Taking stock of industrial ecology includes looking back at the earliest days and threads of industrial symbiosis, honoring the catalytic role that the industrial symbiosis in Kalundborg, Denmark, has played, and compiling some of the lessons learned about how to describe this phenomenon and its many manifestations as experience and understanding evolve. Industrial symbiosis is designated as a subfield of industrial ecology where both theory (conceptual knowledge largely from academia) and practice (on-the-ground experience of public, not-for-profit, and private organizations working to implement industrial symbiosis) are highly valued. The chapter includes a bibliometric analysis of the scholarly work, a survey of practitioners and projects, and commentary from the authors and many other colleagues who contributed ideas for the chapter.

2 Part I: Why People Sometimes Equate Industrial Symbiosis with Industrial Ecology—Frosch and Gallopoulos, Kalundborg, and Beyond

While there are numerous antecedents to industrial ecology as Erkman (1997) and others have well demonstrated, many agree that modern industrial ecology was greatly inspired by, or even began with, the seminal 1989 article in *Scientific American*, "Strategies for Manufacturing." Written by two members of the research and development staff at General Motors, this article laid out the conceptual groundwork for industrial ecology with the idea of following material flows through "industrial ecosystems" wherein "the consumption of energy and material is optimized, waste generation is minimized and the effluents from one process serve as the raw material for another" (Frosch and Gallopoulos 1989). The notion of an industrial ecosystem has played an especially large ideation role in industrial symbiosis. This construct not only denotes a space where industrial symbiosis can occur but also does so through the introduction of a compelling tie back to natural systems. Indeed, we now know that the authors had proposed an alternative title that the editors rejected, but may have offered even more prominence to what would become industrial ecology ideas. This title was "Manufacturing: The Industrial Ecosystem View."

As industrial ecology developed, industrial symbiosis became known as a sub-field of industrial ecology, distinct from other branches of the new field in that it straddles both theory and practice. Not only were Frosch and Gallopoulos employed by one of the largest automobile companies in the world, but the date of their article, 1989, is the convergent year that the industrial cluster in the City of Kalundborg, Denmark, which had been developing since the 1960s, began to come into much view given the extensive network of cooperating industrial operations there. Industrial ecology lore cites 1989 as the year the label "industrial symbiosis" was applied to Kalundborg by Inge Christensen, a pharmacist, and her husband, Valdemar Christensen, the Kalundborg power plant manager, to describe what was happening in the Kalundborg industrial ecosystem (Hewes and Lyons 2008). The inviting imagery of "the effluents from one process" serving "as the raw material for another" and the recognition of the extensive and interconnected resource sharing network of Kalundborg have proven so powerful for reimagining sustainable industrial development that many people, even today, mistake industrial symbiosis as what defines all of industrial ecology.

In Kalundborg, 1972 is regarded as the year of the first real interfirm symbiosis which, it should be noted, also brought a new enterprise to town. The gypsum board company was established in Kalundborg, in part because of the availability of excess butane from the nearby oil refinery. The butane transaction lasted nearly 30 years, and while it is now inactive, the gypsum board company remains in place under its third set of owners. Having officially celebrated 40 years with a newly published booklet (Kalundborg Symbiosis 2014), there is much to look back upon, and it is hard not to be impressed with the way this icon has continued to evolve and change, even with the generational passage to new management (Ehrenfeld and Chertow 2002; Ehrenfeld and Gertler 1997; Gertler 1995; Jacobsen 2006; Kalundborg Symbiosis 2014). In addition to substantial participation from the long-term partners, several new organizations have joined the symbiosis with 33 identified instances of interfirm resource sharing (Fig. 5.1).

There were foundational changes over the years in Kalundborg that could have shaken the entire system if it were fragile. Instead, this industrial ecosystem has continuously weathered many varieties of disturbance, from changes in the fuel type at the power plant to significant ownership and organizational changes within the companies such as the doubling in size of the oil refinery and the splitting up of the pharmaceutical operation into two separate companies. Even the official City of Kalundborg expanded from 20,000 to 50,000 residents during this period based on reorganization of municipal boundaries. While some have been concerned about

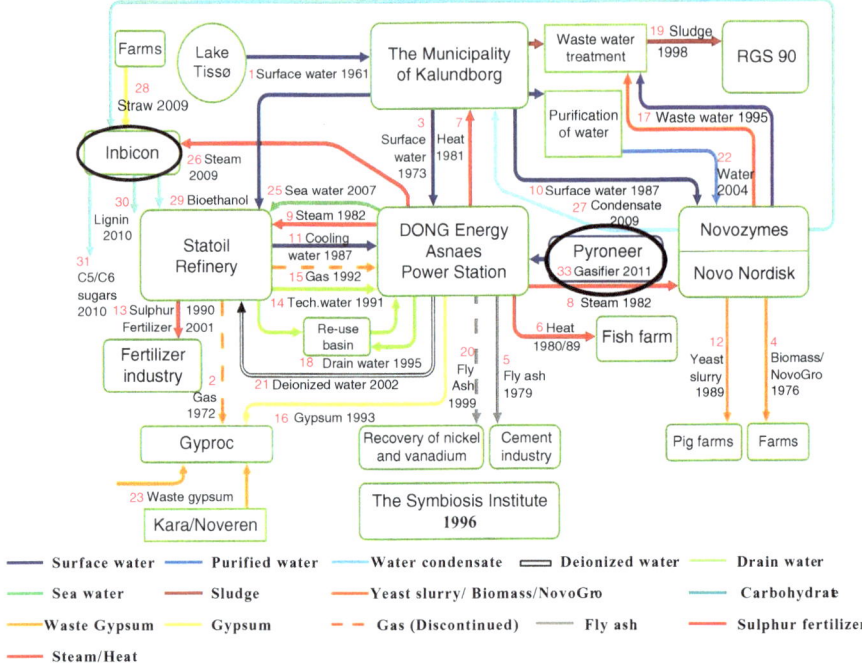

Fig. 5.1 Kalundborg Symbiosis as of 2012 with two new power plants circled. Both use biomass rather than fossil fuel (Source: http://symbiosecenter.dk)

technology lock-in and path dependence (Reuter et al. 2005; Sagar and Frosch 1997), two of the newest additions to the industrial ecosystem, both biomass power plants, suggest further breakthroughs in adaptation and regional sustainability (Fig. 5.1). Previously, with only one, albeit large, coal-fired power plant, the Kalundborg system was less resilient to perturbations. Matching the numbers here to Fig. 5.1, the Inbicon Biomass Refinery, begun in 2009, uses local straw (28) for conversion to bioethanol (29) and also generates lignin pellets (30) and molasses (31). DONG Energy's Pyroneer plant, begun in 2011, is a 6 MW demonstration facility that gasifies local biomass (33). Diversifying energy sources has reduced reliance on fossil fuel toward increased use of bioenergy.

While Kalundborg is a familiar story for most industrial ecologists, the true legacy of Kalundborg comprises all that it spawned geographically and intellectually. While it is not at all the sole narrative for physically connected enterprises, in a surprising number of cases evaluated, Kalundborg is directly cited as an influencing factor, whether the projects envisioned succeeded or failed. Perhaps most interesting, even with some common reference points such as Frosch and Gallopoulos and Kalundborg, is how extensive and varied the evolutionary experiments that embed industrial symbiosis have come to be. The range extends from North American eco-industrial parks (Cohen-Rosenthal and Musnikow 2003; Côté and Hall 1995; Lowe and Evans 1995) to Southeast Asian industrial estates (Panyathanakun et al. 2013;

Van Ha et al. 2009) and even to broader concepts such as Japan's "resource-circulating society" (Morioka et al. 2011) and China's "ecological civilization" (Hu 2012).

3 Part II: Bounding Industrial Symbiosis in Time and Space—Distinctions and Differences

As a truly multi-, inter-, and transdisciplinary focus of study, industrial symbiosis easily becomes far-flung and sometimes misunderstood. The diverse communities of industrial symbiosis, from those who pursue scholarship to those creating policy to those facing the reality of business development, do not often overlap in ideas and audiences, as most people see just a part (Fig. 5.2). Some "big data" colleagues complain that examining industrial symbiosis is too much work because the information sources related to material and energy flows reveal only part of a multifaceted story that cannot be told without some human intervention to sort it out (Nikolic 2013, July 1, Personal communication). This section emphasizes a few of the ongoing questions raised and distinctions made as industrial symbiosis has progressed through time and space.

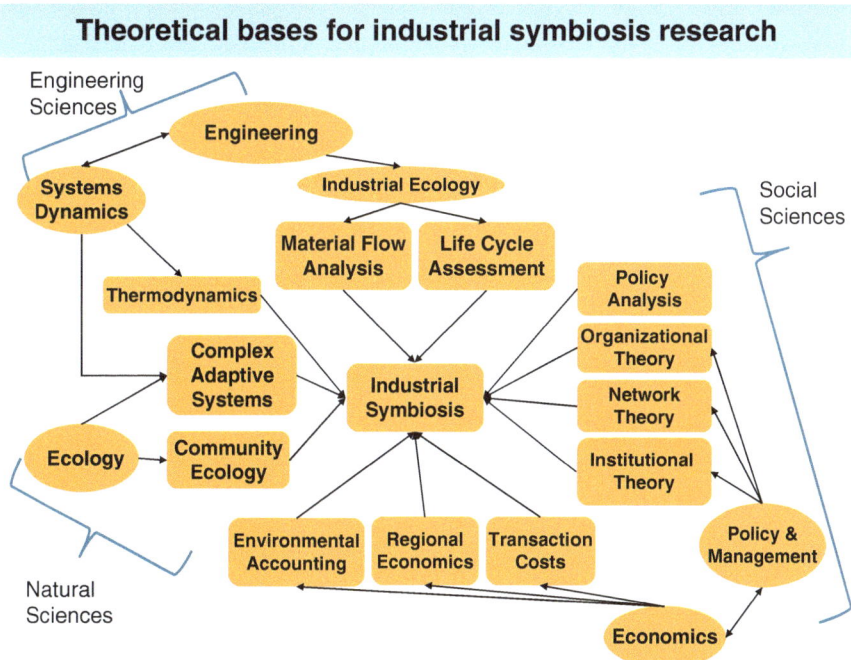

Fig. 5.2 Industrial symbiosis examined through many research fields (Source: Ashton and Chertow, Yale Center for Industrial Ecology, updated 2015)

3.1 Industrial Symbiosis: Old, New, or Hidden

One of the distinctive crosscutting themes of industrial ecology is the way it attends to "hidden" flows such as (1) the often uncounted overburden left behind from mining extraction that becomes part of mass flow analysis; (2) the "missing" lead or mercury or other elements that dissipate into air, water, and soil; and, (3) in industrial symbiosis, resource exchanges that may have been implemented long ago, but are not known or recognized other than among the participating economic actors as part of a formal or informal localized supply chain. In each of these classic industrial ecology instances, something formerly hidden or unknown becomes attended to and counted. By identifying these hidden flows and giving them more careful attention, both economic and environmental implications are clarified. Environmental consequences can be negative, as with escaping lead, but can often be positive, as with repurposed sulfur, wastewater, or cogenerated steam characteristic of industrial symbiosis. The nature of economic activities that fly under the radar of visibility and consciousness is a phenomenon within industrial symbiosis first discussed by Schwarz and Steininger (1997) and helps to explain the surprise that often accompanies forays into industrial ecosystems.

In the early 2000s, geographer Pierre Desrochers identified industrial symbiosis transactions as rearrangements of practices that had come before (Desrochers 2000). Indeed, as long as there have been people, products, and wastes—from animal hides to flower pollen to metal scraps—such items have been exchanged if the recipient finds value in the material being received, whether or not there is a special name for it. Desrochers (2001, 2004) cited numerous industrial examples of resource sharing over hundreds of years including the role that cities have played in agglomerating industries in common locations and, in the process, enabling the creation of reuse and recycling among players in those industries. Desrochers has investigated reuse of by-products, for example, in Victorian England, and identified some very interesting historical accounts such as an 1862 volume titled *Waste Products and Undeveloped Substances*, whose authors attributed the ingenuity they saw around them to opportunities to earn additional revenue from what appeared to have no value at all.

Such reflections help to stimulate thinking about industrial symbiosis and claims about it by its observers. When a phenomenon has a strong economic basis—such as the revenue earning aspect of industrial symbiosis—then there is motivation and impetus to pursue it. To a large extent, much of industrial symbiosis has been concealed in the broader realm of economic exchanges. What we recognize and indeed can quantify from today's version of industrial symbiosis is that alongside the economic benefits, environmental ones are generated as well. These benefits may be hidden, but can be demonstrated to the broader community in the form of reduced emissions and waste and jobs created through reuse and recycling. Once these industrial symbiosis benefits become recognized more broadly and further steps are taken to continue them, then the activities of the business cluster where they occur can be classified as a distinct environmental phenomenon beyond what happens in other economic agglomeration networks (Chertow and Ehrenfeld 2012).

3.2 Single Industry Dominated vs. Multiple Industry Involvement

The 1990s was a time when great effort went into creating bounded systems to include resource sharing across organizations in eco-industrial parks and through related constructs such as zero-waste projects, by-product synergy, or integrated biosystems (IBS). Just as there was a convergence in 1989 with the article by Frosch and Gallopoulos and the revelation of Kalundborg, convergent in 1995 were environmental engineer Nelson Nemerow's book, *Zero Pollution for Industry: Waste Minimization Through Industrial Complexes*, and the first peer-reviewed journal article relating to industrial symbiosis appearing in the scientific literature, "Industrial Ecology and Industrial Ecosystems" by Ernest Lowe and Laurence Evans, in the *Journal of Cleaner Production*. Nemerow imagined the creation of a system of environmentally balanced "industrial complexes" where companies in the complex consume each other's by-products as a means of increasing production efficiency and reducing waste. Since 1992, Lowe and colleagues at Indigo Development along with Professor Ray Côté and colleagues at Dalhousie University in Nova Scotia and Cornell University's former Work and Environment Initiative led by Ed Cohen-Rosenthal formalized the concept of the eco-industrial park. By 1994, the US EPA had hired Lowe and Indigo, along with the Research Triangle Institute staff, to develop the concept further.

Notably, the examples of Nemerow's environmentally balanced industrial complexes were rooted in individual industries such as pulp and paper, sugar refining, and textile complexes. In these bounded systems, Nemerow described the wealth of potential resource sharing opportunities that could be usefully implemented by related industrial operations. In contrast, Lowe and colleagues were envisioning business clusters populated by many unrelated firms thereby creating opportunities for sharing the inputs and outputs from a diverse array of facilities in eco-industrial parks that would foster economic cooperation, environmental improvement, and community benefit. It was understood that if firms had a high diversity of inputs and outputs rather than uniformity, there would be numerous new business opportunities based on reuse of by-products. This distinction of "single-industry dominated clusters" and "multiple industry clusters" is an important one. What Kalundborg, Lowe, and other projects and colleagues demonstrated was the potential involvement of many, diverse industries and outcomes in one geography.

3.3 Industrial Symbiosis and Eco-industrial Parks (EIPs)

While the section above discusses different lineages of more traditional single-industry dominated industrial complexes and the broader diversity within eco-industrial parks, resource sharing occurs in both. Another important element in sorting out concepts of industrial symbiosis is reexamining its relationship to eco-industrial parks, clusters, and/or estates. When ideas come together from many different places, traditions, and cultures, there are larger and smaller variations in content and emphasis even as many institutions are embracing the notion of inter-firm resource sharing. The international business community has leaned toward other expressions such as by-product synergy, a keystone of the US Business Council for Sustainable Development (US BCSD) with focus not only on interfirm resource reuse but also on intra-firm reuse (Mangan 2015, January 19, Personal communication). The Ellen MacArthur Foundation has greatly popularized the notion of the "circular economy" especially in business circles.

One of the early EIP pioneers, Ray Côté of Dalhousie University, recently expressed his concern with "the number of researchers and practitioners who continue to equate eco-industrial parks with industrial symbiosis or seem to believe that once industrial symbiosis has occurred within an industrial park, the latter can be called an eco-industrial park" (Côté 2015, January 16, Personal Communication). Côté perceives the EIP beginning with the land itself and the necessary understanding of the ecological services of the area where the park is to be situated. In this way, changes to those services can be appropriately addressed—for example, if wetlands will be compromised, planners could incorporate constructed or engineered wetlands. Industrial symbiosis, a less bounded spatial concept than EIP, focuses on resource reuse and how to achieve it technically, economically, and behaviorally, more than the physicality of any particular site, which of course is also very important (Lowitt 2015, January 20, Personal communication with Director of Devens Enterprise Center).

According to Côté, even as the park management looks for symbiotic opportunities involving water, energy, and materials, other key elements of EIPs include, for example, standards or guidelines for green building features, water cycling, reuse, and landscaping. Especially when there are many small- and medium-sized companies in an EIP, then "scavenger and decomposer" operations can be encouraged to fill many reuse niches across firms through "repair, rent, restore, reclaim, remanufacture, and recycle," which become important components "of local circular economies" (Côté 2015, January 16, Personal Communication; Geng and Côté 2002).

3.4 Diffusion of Industrial Symbiosis

Now in its twelfth year, the annual Industrial Symbiosis Research Symposium, building on smaller regional fora, has linked academics and practitioners in 1–2-day meetings to discuss and exchange ideas in numerous locations around the globe (Table 5.1). In addition, based on the bibliometric analysis presented in Part III, it is possible to track, at least from the perspective of peer-reviewed scientific journals, those articles in which individual countries received some of the focus of industrial symbiosis papers (Fig. 5.3). By this measure, China has now received the most attention in industrial symbiosis academic research and continues to grow rapidly. The strong interest in China appears to be policy driven following establishment of the program creating the Chinese National Demonstration EIPs and circular

Table 5.1 The annual Industrial Symbiosis Research Symposium

Year	Hosting location	Host or contact person
2004	New Haven, Connecticut, USA	Marian Chertow
2005	Stockholm, Sweden	Noel Jacobsen
2006	Birmingham, UK	Peter Laybourn
2007	Toronto, Canada	Ray Côté
2008	Devens, Massachusetts, USA	Peter Lowitt
2009	Kalundborg, Denmark	Jørgen Christensen
2010	Kawasaki, Japan	Tsuyoshi Fujita (with Chinese, Korean, and Japanese sponsorship)
2011	San Francisco, California, USA	Marian Chertow
2012	Tianjin, China	Shi Han, Yuyan Song
2013	Ulsan, South Korea	Hung-Suck Park
2014	Melbourne, Australia	Robin Branson, Biji Kurup
2015	Lausanne, Switzerland	Guillaume Massard, Suren Erkman

economy policy. The second highest number of articles is about the USA, followed by Australia, Denmark, the UK, Finland, Japan, and South Korea, many of which have or had national industrial symbiosis-related programs and initiatives or have representative industrial clusters such as Kalundborg and Kwinana. The increasing focus on developing countries can be seen at the bottom of Fig. 5.3 with each having one peer-reviewed paper, thus entering the broader conversation. Many more papers and conference proceedings appear in the "gray literature" found in search engines and the offices of professors, consultants, and project officials, but this was not fully surveyed here.

3.5   Understanding Industrial Symbiosis in a Chinese Context

Given the prominence of industrial symbiosis in China (Fig. 5.3), it is important to describe its development in more depth. Professor Shi Lei of Tsinghua University put together some background information, noting that the emergence of industrial symbiosis and eco-industrial parks in China stems from earlier types of industrial development following the passage of the 1978 policy on Reform and Opening Up. In 1984, the first Economic-Technological Development Areas (ETDAs) began which, according to Pi and Wang (2004), were "designed to break the ice of the planned economy" and subsequently began to focus on attracting foreign direct investment. These were followed by the development of Hi-Tech Parks (HTP) in the early 1990s. By the turn of the new century, China had shaped a large industrial system based on more than 6,000 parks located across China. With the high concentration of industrial activities, resource and environmental problems became increasingly serious, which ignited the rise of EIPs (Shi 2015, January 18, Personal communication).

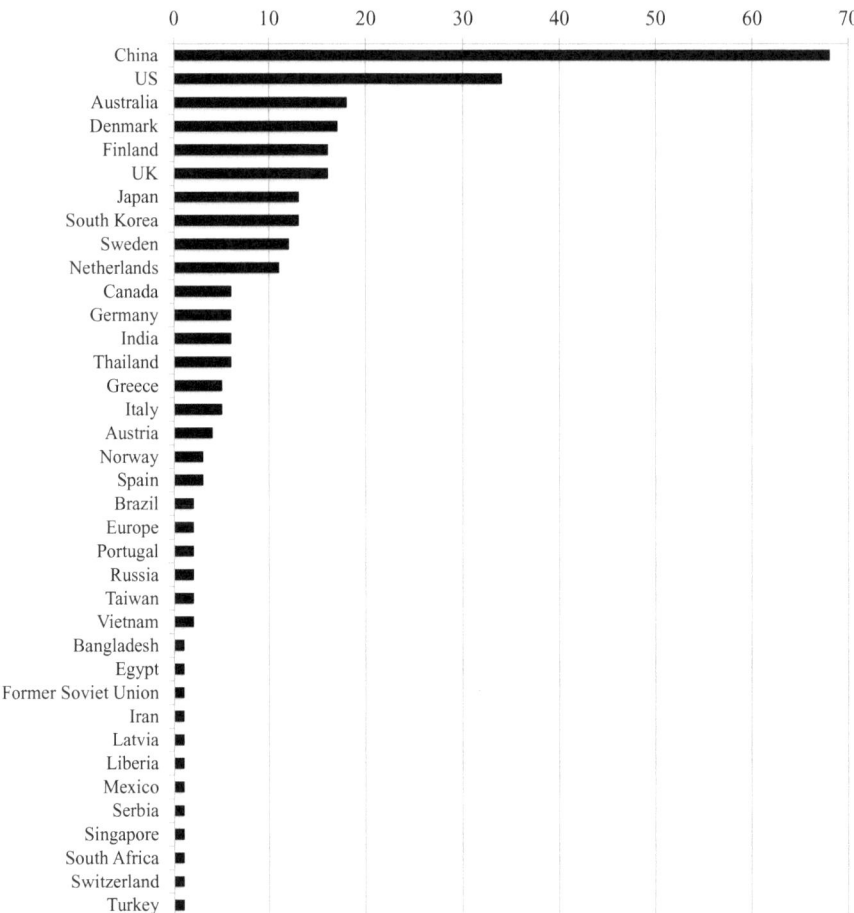

Fig. 5.3 Countries featured in 286 academic papers on industrial symbiosis (out of total reviewed of 391) from 1995 to 2014

The first experimentation happened in single-industry dominated, sector-specific parks, such as chemicals, metals, and sugarcane, then in ETDAs and HTPs, and finally in waste-recycling parks. Along with the spread of EIPs, industrial symbiosis research also has become increasingly diverse in China. The early research mainly focused on EIP conceptual frameworks motivated by industrial ecology metaphors, planning methods/tools based on planning theories and frameworks, and symbiosis system integration from system engineering perspectives. With more and more cases arising, two additional academic communities joined in industrial symbiosis research: environment and ecology and economics and management science. The former mainly focused on win-win solutions, environmental performance evaluation, and formulation of EIP guidelines. The latter mainly focused on cost-benefit analysis, experience identification, and case studies. More recently, research has begun to uncover the mechanisms and processes of industrial symbiosis from complex theory and network science (Shi and Shi 2014).

Overall, the role of the Chinese government has been crucial in financing and promoting EIPs and opening up space for experimentation and exchange. With the currently perceived need for updating and "eco-transformation," Yu et al. (2014b) found that the planned EIP model is most useful in the early stage of development, but later it should be combined with a facilitation model to achieve long-term goals with more flexibility. Further, it has become clear that industrial parks cannot succeed without being effectively embedded into the capital, labor, and resources of their local regions. China continues to work toward turning eco-industrial niches into mainstream development, both through practice and through industrial symbiosis research (Gibbs 2009; Shi and Yu 2014).

3.6 Organizational Drivers and Barriers

An ongoing quest in industrial symbiosis is the search for a more definitive understanding of what drives and what hinders it. Since 2011, there have been three review articles with emphasis on the social science side (Boons et al. 2011; Jiao and Boons 2014; Walls and Paquin 2015). Walls and Paquin analyzed 121 industrial symbiosis articles focused on organizational and institutional issues. Using content analysis, they analyzed the factors that authors mention most frequently to explain (1) what facilitates industrial symbiosis in the first place (antecedents), (2) what factors generally help industrial symbiosis to grow over time (lubricants), (3) what inhibits industrial symbiosis over time (limiters), and (4) what are the outcomes of industrial symbiosis (consequences). Table 5.2 lists the top six factors determined by Walls and Paquin in each of the four categories, illuminating where organizational and institutional research in industrial symbiosis has focused. Walls and Paquin's review highlights the role of the social sciences to contribute to a deeper theory of industrial symbiosis, by taking into account the complex interactions, motivations, and dynamics that occur within and among organizations. A greater understanding of these aspects and dynamics is increasingly important for facilitating and developing robust industrial symbiosis, especially in the policy realm, as industrial symbiosis grows more diverse in typology, geography, and organizational contexts (Jiao and Boons 2014).

4 Part III: Industrial Symbiosis in Both Scholarship and Practice

This section covers scholarship and practice, two integrally related strains of industrial symbiosis. At one level, the rise of bibliometrics has given us a convenient way to analyze the rapidly developing academic area of focus where peer-reviewed

Table 5.2 Drivers and barriers of industrial symbiosis

Antecedents	Lubricants
Co-location, proximity	Intermediaries, coordinators, and champions
Government regulation	Trust, openness
Anchors, scavengers, other roles	Knowledge creation or sharing
Diversity of actor's involvement	Embeddedness
Common strategic vision, beliefs, and alignment	Culture or mind-set
Economic reasons	Social and network ties
Limiters	Consequences
Power, status, asymmetries	Innovation
Too much diversity	Co-benefits: environmental and economic
Exit of player, personnel, or change in flows	Learning
Cost, risk	Resilience
Environmental regulation too restrictive	Lock-in, domino effect
Lack of trust	Social capital

Source: Walls and Paquin (2015)

articles increased sixfold over 10 years and doubled since 2010. Some of the findings of the bibliometric analysis are captured here. At another level, however, it is through practice that we see actual projects and can assess the extent of the contribution of industrial symbiosis to on-the-ground industrial ecosystems, global resource management, and collaborative business behavior. Broadly, practitioners include government actors, private sector actors, and NGO representatives. Of course, there is a great deal of crossover, and ideally, scholarship can eventually be tested in practice, and on-the-ground performance can be measured through academic analysis. Many academics serve as idea brokers and also as evaluators and have been linked to specific projects such as Baas and Boons in Rotterdam, the Netherlands; Park in Ulsan, Korea; Ashton in Puerto Rico; and Salmi on the Kola Peninsula in Russia. Academics also bring projects to light as we have seen in Styria, Austria, TEDA in China, and Östergötland in Sweden. Process engineers, too, have been important bridges between theory and practice.

4.1   Section A: Industrial Symbiosis in Scholarship

Two questions drove the bibliometric analysis of articles from 1995 to 2014:

- What is the intellectual structure of this field based on a review of the academic literature?
- How have frameworks, concepts, and theories advanced in the last 20 years?

This analysis relied on Scopus and ISI Web of Knowledge searches for bibliographic acquisition. Both of these services lean to the "scientific" aspects of a topic under study, in contrast to, for example, Google Scholar, which goes beyond strictly scientific academic journals to include other informative publications. Thus, the almost 400 articles in this study faced a rigorous test since Scopus and Web of Science are highly selective about which journals they evaluate. We limited our search to articles in English. To begin, we compiled all literature returned by searching for eight keywords as follows:

1. "Industrial symbiosis"
2. "By-product exchange"
3. "By-product synergy"
4. "Industrial ecosystem"
5. "Eco-industrial"
6. "Resource synergy"
7. "Recycling linkage"
8. "Recycling network"

These keywords, derived from the authors' existing sources on industrial symbiosis, are considered to be broad enough to capture the most significant literature in the field as it has evolved in different regions although surely some material is overlooked. The search covers articles published between 1995 and 2014. After screening for relevance, the 391 papers determined to pertain to industrial symbiosis were classified into one of the seven following categories: Foundations, Performance, Mechanism, Modeling, Structure, Case Study, and Proposal. To avoid multiple allocation of a paper into several categories, the main objective and content of the literature was carefully determined according to criteria described in Table 5.5 in the Appendix. The references for these 391 articles are available upon request.

4.2   Results and Analysis of Bibliometric Study

Figure 5.4 shows that the number of industrial symbiosis papers published in peer-reviewed journals has increased from one paper in 1995 (the first year of publication of a paper meeting our criteria) to 75 in 2014. In particular, 2007 is the year that shows a notable increase in the publication of papers, which is in line with the observation from Yu et al. (2014a), and, notably, the number of papers more than doubled between 2010 and 2014. It is not yet known whether the large increase of articles in 2014 is an outlier, but in general we expect this output to continue to rise at least in the short to medium term (2–5 years) given the trajectory.

Before 2004, the main topic of inquiry concerned conceptual aspects of industrial symbiosis. Case studies began to grow by 2004 with the most for any year being 11 case study articles in 2007. Later, more papers began to focus on analyzing performance and mechanisms of industrial symbiosis and proposing new ideas or strategies for its implementation. Modeling as a topic in industrial symbiosis emerged in 1998, but more than 90 % of modeling papers appeared after 2006. This implies that the focus of industrial symbiosis research has gradually shifted from introducing the concept and presenting specific case studies to delving more into performance and mechanisms of industrial symbiosis and presenting new ideas for industrial symbiosis.

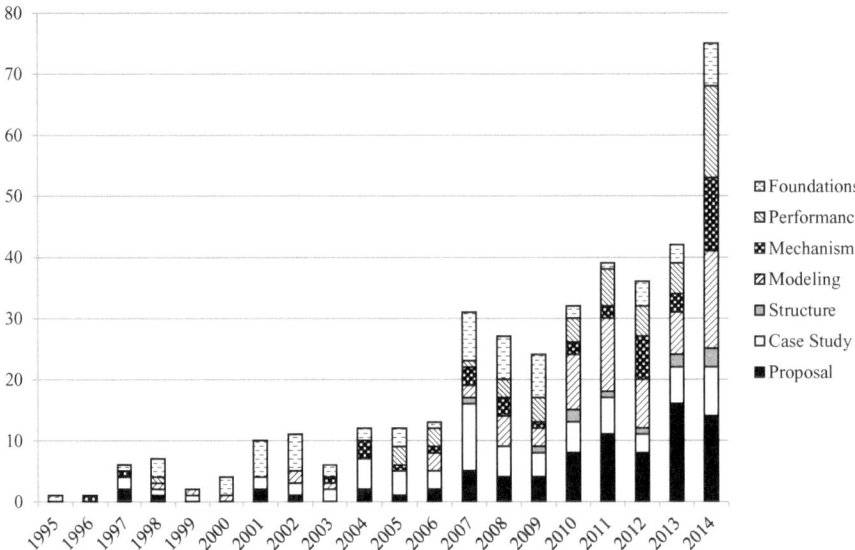

Fig. 5.4 The number of industrial symbiosis papers published between 1995 and 2014 divided into the seven thematic categories

According to Table 5.3, publications of industrial symbiosis papers are concentrated in a few journals, most of which also publish industrial ecology articles. Out of a total of 124 journals publishing industrial symbiosis articles, the top six journals published fifty percent of all the papers captured in the analysis. Figure 5.5 represents the number of industrial symbiosis papers published over time and differentiates whether these articles appeared in the top six journals or the others. Figure 5.5 indicates, in general, that the number of industrial symbiosis papers published in journals other than the top six has increased over time: from one paper in 1996 to 36 papers in 2014. This finding reveals that more industrial symbiosis-related research has emerged in a wider range of journals outside of the core, indicating that industrial symbiosis research is reaching a broader readership.

4.3 Discussion of Industrial Symbiosis Research

The evolution of industrial symbiosis research can be examined qualitatively by looking at common themes and questions addressed in the literature. One key question relates to the scale of industrial symbiosis given that geographic proximity is a well-recognized element that facilitates industrial symbiosis. Sterr and Ott (2004), for example, argued that the regional scale is favorable for industrial symbiosis because it is a scale that is broad enough to offer opportunities for economically viable material reuse among actors and, at the same time, small enough to allow communication and collaboration. Responding to these conceptual discussions

Table 5.3 Journals ranked according to the number of industrial symbiosis papers published

Name of journal	Number of IS papers published
Journal of Cleaner Production	90 (23.0 %)
Journal of Industrial Ecology	50 (12.8 %)
Progress in Industrial Ecology	21 (5.4 %)
Resources, Conservation and Recycling	14 (3.6 %)
Business Strategy and the Environment	11 (2.8 %)
Journal of Environmental Management	9 (2.3 %)
International Journal of Sustainable Development and World Ecology	7 (1.8 %)
Energy	6 (1.5 %)
Clean Technologies and Environmental Policy	5 (1.3 %)
Computers & Chemical Engineering	
Ecological Economics	
Industrial & Engineering Chemistry Research	
Shengtai Xuebao (Acta Ecologica Sinica)	
Environmental Science & Technology	4 (1.0 %)
Fresenius Environmental Bulletin	
Minerals Engineering	
Process Safety and Environmental Protection	
Regional Studies	

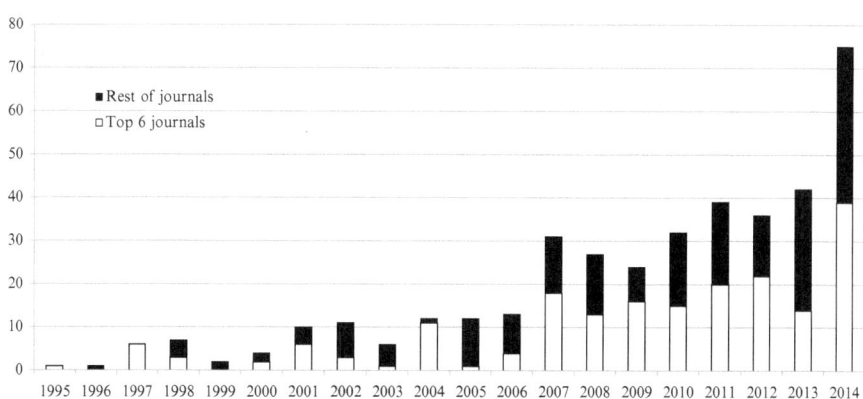

Fig. 5.5 The number of industrial symbiosis papers published over time in the top six journals versus all remaining journals

concerning scale, several studies provided empirical examination of the geographic scale of industrial symbiosis. Lyons (2007) found no preferable scale for recycling and remanufacturing in Texas and argued that recycling boundaries vary for different types of waste and depend on where demand occurs. Jensen et al. (2011) and Chen et al. (2012) presented similar findings about the relationship between reuse distance, the type of materials, and the location of demand. Data from the former

National Industrial Symbiosis Programme (NISP) in the UK showed that half of all resources were reused within 32.6 km (Jensen et al. 2011). According to the data from 88 recycling projects in 23 Japanese eco-towns, the average waste collection and product delivery distance ranged from 15 to 80 km (Chen et al. 2012).

Another key question about industrial symbiosis has evolved around the concept of self-organization and complex adaptive systems. While one stream of IS research has focused on how to replicate Kalundborg through deliberate planning (Potts Carr 1998; Roberts 2004; van Leeuwen et al. 2003), another stream of studies focused more on the organic nature of industrial symbiosis development. Based on historical appraisals, Desrochers (2004) argued that industrial symbiosis has existed and can exist primarily through market mechanisms instead of top-down planning. Chertow (2007) proposed an "uncovering" approach to industrial symbiosis, which stimulates the identification of existing precursors of symbiosis and nurtures them. Considering industrial symbiosis as a self-organizing phenomenon was then developed further by adopting the framework of complex systems science to understand industrial ecosystems as complex adaptive systems (Chertow and Ehrenfeld 2012). Along this line of understanding, tools from complex systems began to be applied to examine evolution and resilience of industrial ecosystems (Cao et al. 2009; Chopra and Khanna 2014; Romero and Ruiz 2014; Zheng et al. 2013; Zhu and Ruth 2013, 2014).

Early studies view industrial symbiosis mainly as a favorable outcome and focused on exploring ways to implement the most optimal form of industrial symbiosis from technological and economical perspectives. With increasing experiences with successes and failures, however, more studies have examined the role of social factors (Ashton 2008; Gibbs 2003; Hewes and Lyons 2008; Howard-Grenville and Paquin 2008; Jacobsen 2007). Some studies describe industrial symbiosis as a learning process and link it to innovation for sustainability at local and regional levels (Mirata and Emtairah 2005; Posch 2010; Ristola and Mirata 2007; Walter and Scholz 2006). Recently, industrial symbiosis was conceptualized as a dynamic process, which can offer new insights about the emergence, evolution, and dissolution of symbiotic relationships and broader institutional dynamics (Boons et al. 2011, 2014; Spekkink 2014).

Finally, measuring performance of industrial symbiosis has attracted much attention, particularly because economic and environmental benefits are what comprise the core industrial symbiosis approach. Some papers estimated net cost savings for different industrial symbiosis scenarios (Karlsson and Wolf 2008; Martin et al. 1998) or for existing industrial symbiosis networks in Kalundborg (Jacobsen 2006); Guayama, Puerto Rico (Chertow and Lombardi 2005); Oahu, Hawai'i (Chertow and Miyata 2011); and Kawasaki, Japan (Van Berkel et al. 2009a). Going beyond quantifying cost savings, Wen and Meng (2014) quantified changes in resource productivity through industrial symbiosis, and Park and Behera (2014) measured how symbiosis increases eco-efficiency. Park and Park (2014) showed how cost savings achieved through industrial symbiosis contributed to obtaining competitive advantage in the market.

Understanding the environmental performance of industrial symbiosis began with quantifying avoided landfilling or material/energy use reductions. While these are based on direct measurement of material/energy changes, more theoretical understanding of environmental performance came from applying thermodynamic indicators such as emergy or exergy (Geng et al. 2010, 2014; Valero et al. 2013; Wang et al. 2005; Yang et al. 2006). Responding to the issue of climate change, several studies focus on quantifying greenhouse gas or carbon reductions through industrial symbiosis (Dong et al. 2014; Hashimoto et al. 2010; Jung et al. 2012; Liu et al. 2012; Salmi and Wierink 2011). Application of life cycle assessment (LCA) tools to the area of industrial symbiosis was seen mainly after 2010 (Eckelman and Chertow 2013; Mattila et al. 2012; Sokka et al. 2011b; Yu et al. 2014c). Quantification efforts have developed along with the discussion of methodological aspects (Martin et al. 2013; Mattila et al. 2012; Sokka et al. 2008; Wolf and Karlsson 2008) and led to the further development of comprehensive indicator systems, particularly in China (Geng et al. 2009, 2012; Tian et al. 2014).

4.4   Section B: Industrial Symbiosis in Practice

Examining industrial symbiosis in practice for this section involved three primary activities: (1) a review of the project literature with database searches primarily in Google and LexisNexis and materials recommended by colleagues including the 2012 International Survey on Eco-innovation Parks published by the Swiss Federal Office for the Environment, the expanding activities of International Synergies, and the newsletters of the Eco-Industrial Development/Industrial Symbiosis (EID/IS) section of the International Society for Industrial Ecology, (2) the compilation of a list of industrial symbiosis-related events (Table 5.4), and (3) email correspondence with key practitioners and academics studying specific projects to reflect on questions posed:

1. What motivates industrial symbiosis in practice, including what is successful and what is less so?
2. What trends are you seeing and what mechanisms are being adopted for evolution and change?
3. What linkages/evidence do you see that might show the influence of academia on industrial symbiosis practice and industrial symbiosis practice on academia?

Our findings are outlined below.

Industrial symbiosis practitioners emphasize that to increase the success rate of industrial symbiosis, the needs and concerns of private businesses must be better understood and accounted for. Uncovering and identifying viable symbiosis opportunities are often not enough for success, as private firms may not take up the opportunity for many reasons such as uncertain risk, inflexibility owing to the ownership structure, and lack of trust. As an example, several textile dyeing companies in a Chinese industrial park refused low-cost steam from a government-owned coal plant, as they were reluctant to be tied to a giant. There are also issues with private businesses being hesitant to disclose internal information (Mangan 2015, January 19, Personal communication; Tian 2015, January 21, Personal communication).

Table 5.4 Key events related to industrial symbiosis, 1989–2014

Year	IS-related events
2014	US BCSD launches the Austin Materials Marketplace
	European Commission adopts the Communication "Towards a Circular Economy: A Zero Waste Programme for Europe"
	European Union's Horizon 2020 innovation and research funding plan calls on industrial symbiosis to deliver circular economy
	Finnish Industrial Symbiosis System launched
	Southern African Development Community (SADC) Knowledge Sharing Week on industrial symbiosis in Cape Town
	Green Industrial Symbiosis national program launched in Denmark
	NISP Canada launched
2013	European Industrial Symbiosis Association (EUR-ISA) launched
	European Resource Efficiency Platform recommends pan-European network of industrial symbiosis initiatives
	Chinese Association of Circular Economy established
	Western Cape Industrial Symbiosis Programme (WISP) pilot project, South Africa
2012	Indo-German Environment Partnership Programme fosters work on EIPs in India
2011	Industrial symbiosis embedded in European Resource Efficiency Flagship Initiative, part of the Europe 2020 Strategy
	European Commission adopts the Roadmap to a Resource Efficient Europe
	EU, Turkey, Romania, Bulgaria, and Moldova begin the Industrial Symbiosis Network for Environmental Protection and Sustainable Development in the Black Sea Basin (SymNet)
	Portugal's National Waste Management Plan introduces the concept of IS
2010	ECOMARK Initiative fosters development of EIPs in France, Greece, Italy, Slovenia, and Spain
2009	EU funds the project ZeroWIN (Towards Zero Waste in Industrial Networks) that terminated in 2014
2008	China passes the Law for the Promotion of the Circular Economy, set to begin in 2009
	Sweden launches SymbioCity project to promote symbiosis in urban systems
	France launches COMETHE project to design methodology and tools for the implementation of industrial ecology approaches on a business park scale (ended in 2011)
2006	China SEPA establishes national guidelines for EIPs
2005	The UK National Industrial Symbiosis Programme (NISP) launched with funding from Defra's Business Resource Efficiency and Waste Programme
	Korean National Eco-industrial Park Program begins
	China NDRC launches circular economy pilot demonstration projects
2004	*Progress in Industrial Ecology* begins publication
	Resource Optimization Initiative founded in India
	The annual Industrial Symbiosis Research Symposium initiated at Yale
2001	China SEPA (now MEP) approves the first national pilot EIP (Guangxi Guitang Group)
2000	The Business Council for Sustainable Development – United Kingdom (BCSD-UK) leads its first IS program in the Humber region, UK
	Waste & Resources Action Programme (WRAP) launched in the UK

(continued)

Table 5.4 (continued)

Year	IS-related events
1997	*Journal of Industrial Ecology* begins publication
	Eco-town program in Japan begins (ended in 2006)
1995	The Business Council for Sustainable Development – Gulf of Mexico (BCSD-GM) receives an EPA grant for by-product synergy
	South Korea passes the Act to Promote Environmental Friendly Industrial Structure (APEFIS)
1994	The US President's Council on Sustainable Development (US-PCSD) assigns demonstration EIPs
1993	*Journal of Cleaner Production* begins publication
1992	The predecessor to the US Business Council for Sustainable Development, which was the Business Council for Sustainable Development – Gulf of Mexico (BCSD-GM) is launched
1991	US National Academy of Sciences meeting on industrial ecology marks the first official academic meeting on IE
1990	The *Financial Times* reports on Kalundborg (November 14)
1989	The term "industrial symbiosis" is coined in Kalundborg

Practitioners consider it very important to find ways to move past known hurdles and to get buy-in from businesses, which is essential for success. Many practitioners noted the significance of company champions (Lowitt 2015, January 20, Personal communication with Director of Devens Enterprise Center; Mangan 2015, January 19, Personal communication) as well as the importance of using the language of business (costs, revenues, risk) to generate this buy-in (Laybourn 2015, January 26, Personal communication). Especially in many business situations, as one Nicaraguan business person was quoted as saying, "no company is looking for eco-technologies or eco-innovation. What they are looking for is a solution to a business problem" (Laybourn 2015, January 26, Personal communication).

Another effective and growing avenue for addressing these concerns is through platforms and tools offered by industry organizations or facilitator companies. Industry organizations such as the US BCSD, or facilitators such as NISP, allow businesses a safe and common platform for discussing synergies through symbiosis (Mangan 2015, January 19, Personal communication). A growing trend is the use of Internet-based technological tools (often offered by the aforementioned platforms) such as the Materials Marketplace by US BCSD, SYNERGie by International Synergies, or the Resource-eXchange-Platform as part of the ZeroWIN EU project to further promote coordination and exchanges.

Also emphasized by practitioners is the importance of supportive laws and policies, as well as enabling regulations to incentivize and promote industrial symbiosis. The slow uptake of industrial symbiosis in North America is perceived to be related to the lack of supportive national policies, in contrast to regions with strong supportive policies such as Europe and East Asia, which have seen growing numbers and scale of industrial symbiosis (Lowitt 2015, January 20, Personal communication with Director of Devens Enterprise Center; Massard 2015, January 28,

Personal communication; Shi 2015, January 18, Personal communication; Tian 2015, January 21, Personal communication).

A potential double-edged sword for industrial symbiosis noted by practitioners is that governments, organizations, and businesses have diverse motivations for pursuing and promoting zero-waste/by-product synergy/eco-development and are not necessarily strictly focused on industrial symbiosis (Massard 2015, January 28, Personal communication). A large motivation seems to be the growing cost of landfills. Few actors are overly concerned if their activities neatly fit an academic mold, as long as the cost of waste is reduced. While this is beneficial for the environment, it may not acknowledge or increase awareness of industrial symbiosis (Massard 2015, January 28, Personal communication). At the same time, the growing recognition of industrial symbiosis in various national and regional programs and policies, as illustrated in Table 5.4, is a countervailing positive trend for industrial symbiosis.

Generally, the relationship between practitioners and academics seems productive. There continue to be several business-interested groups, and academics and practitioners are both represented in the EID/IS section of the ISIE, which produces a newsletter and fosters other personal exchanges (Lowitt 2015, January 20, Personal communication with Director of Devens Enterprise Center). Practitioners appreciate academics for providing graduates who help identify symbiosis opportunities and seek to "determine what is working, what isn't, and why." Academics can also usefully address big questions such as "what is the function of global market competition for the industrial symbiosis system in a given area?" Still, there is much room for improvement. While the literature describes a wide diversity of industrial symbiosis cases, this diversity appears to be easily overlooked, meaning some scholarship is built on a limited empirical basis where even the most available anecdotes of resource sharing remain out of view (Howard-Grenville 2015, January 28, Personal communication; Spekkink 2015, January 21, Personal communication). Practitioners want more novel ideas beyond the existing canon of symbiosis such as combined heat and power (CHP) or gypsum reuse identified in the academic literature. Others feel that academics sometimes misunderstand what they see on the ground or may not sufficiently engage the work of practitioners in their research (Laybourn 2015, January 26, Personal communication; Mangan 2015, January 19, Personal communication; Tian 2015, January 21, Personal communication).

Going forward, it seems that industrial symbiosis will have to be more attuned to the wants and needs of private businesses to abate the trust and risk barriers identified earlier. Also, more attention needs to be paid to the individual and intra-firm level motivations and dynamics in choosing whether to engage industrial symbiosis practices (Walls and Paquin 2015). The role of facilitator companies such as International Synergies, Sofies, and many more regionally focused firms will continue to grow, as will the role of specialized waste companies (Massard 2015, January 28, Personal communication). While these facilitators can help identify more opportunities for industrial symbiosis and thus accomplish greater environmental gains, a side effect may be that the direct relationship between firms becomes weaker as the role of the facilitator grows. This will be an interesting issue for continued observation.

5 Conclusion: Industrial Symbiosis in a World of Difference

Industrial symbiosis construed as networks of organizations cooperatively sharing "wastes" has created irresistible imagery and high hopes for a time when virtually all water, energy, and materials will be used more than once, and not to do will have become societally unacceptable. On the one hand, industrial symbiosis research and practice have blossomed in the last 25 years and have established a significant and meaningful subfield of industrial ecology and a record of achievement both in policy and in the built environment. On the other hand, there is much more work ahead to prove the appropriateness and effectiveness of industrial symbiosis. The quest is to understand the level at which material, energy, and water reuse, spurred by the cooperative behavior inherent in industrial symbiosis, causes real reductions in total primary resource consumption, actual revenue-generating opportunities for businesses, net positive environmental externalities for communities, and enhanced people-to-people collaboration.

We see that industrial symbiosis has become and is becoming more and more diverse as it expands to additional regions of the world. There is, for example, a growing foothold in Turkey and Colombia and exploration in southern Africa, with facilitated or planned projects springing up, not necessarily in the traditional Kalundborg mold. This diversity is also reflected in the wide range of terminology (as illustrated by the use of eight keywords for the bibliometric search). Indeed, one of the most vexing puzzles of the last several years has been the difficulty of comparing across projects: how can we compare a huge, 5,000-company "Economic and Technology Development Zone" in China with Kalundborg or small-scale agriculturally based systems (Alfaro and Miller 2014; Klee 1999) or a biomass, biofuel, or biogas region developed in Norrköping and Linköping, Sweden (Martin and Eklund 2011)?

A new proposal by an international team the authors are working with has been trying to identify convergence based on the acknowledgement that there are many paths to industrial symbiosis. These researchers have sorted projects into a typology of seven industrial symbiosis dynamics that establish the differences but emphasize, also, that there are common building blocks even in seemingly disparate origins (Boons et al. 2015). Some colleagues are calling for a stronger linkage between industrial symbiosis and national and international efforts aimed at climate change, resource efficiency, and circular economy (Côté 2015, January 16, Personal Communication; Mangan 2015, January 19, Personal communication).

The industrial symbiosis community has a unique property with roots in practice aided by intellectual development, which is all the richer for the ability to bring this knowledge back to practice. In 1997, Volume 1 Number 1 of the *Journal of Industrial Ecology* had one of the early articles about industrial symbiosis coauthored by MIT's John Ehrenfeld and the graduate student he sent to explore Kalundborg, Nicholas Gertler. They recognized the opportunity that industrial symbiosis represents once the spirit of collaboration could be opened and embedded. They stated the goal we are still seeking that together with new institutional approaches, "more deep-seated cultural changes can provide a foundation from which symbioses and other forms of material exchange begin to actually move economies toward sustainability" (Ehrenfeld and Gertler 1997).

Acknowledgments This far-ranging chapter was greatly enhanced by the contributions of numerous colleagues. We would like to thank many thought leaders whose names are mentioned in the text and also some who helped with research and assembly, especially Shi Lei, Raymond Paquin, Weslynne Ashton, and Ray Côté. At Yale, we thank Reid Lifset, Tom Yang, and Lillian Childress.

Appendix

Table 5.5 Seven categories used for bibliometric analysis and criteria for classification

Category	Industrial symbiosis classification criteria followed by example paper(s)	
Foundations	*Address concepts or issues regarding industrial symbiosis at an abstract level*	
	Introduce ideas, concepts, potentials of IS	Lowe and Evans (1995)
	Emphasis on the "biological analogy"	Hardy and Graedel (2002)
	Establish a definition, framework, or theory	Chertow (2000) and Korhonen (2005)
	Conduct a comprehensive review of industrial symbiosis as a field of study	Yu et al. (2014a) and Zhang et al. (2014)
	Relate IS to a different field or concept	Parto (2000) and Gregson et al. (2012)
	Define and describe characteristics	Ashton (2009) and Schiller et al. (2014)
Performance	*Evaluate the performance* outcomes *of industrial symbiosis*	
	Economic performance	Jacobsen (2006) and Van Berkel et al. (2009a)
	Environmental performance	Mattila et al. (2010), Sokka et al. (2011a), Dong et al. (2013), and Eckelman and Chertow (2013)
	Thermodynamic performance	Wang et al. (2005), Yang et al. (2006), and Geng et al. (2014)
	Efficiency	Salmi (2007) and Park and Behera (2014)
	Implications of industrial symbiosis on regional cooperation and innovation	Mirata and Emtairah (2005) and Posch (2010)
	Develop evaluation methodologies	Sokka et al. (2008) and Mattila et al. (2012)
	Develop evaluation indicators	Geng et al. (2009) and Liu et al. (2014)

(continued)

Table 5.5 (continued)

Category	Industrial symbiosis classification criteria followed by example paper(s)	
Mechanism	Analyze underlying *processes* through which IS develops and related influencing factors such as institutional capacity, government policy, and social relations that play a significant role	Gibbs (2003), Mirata (2004), Baas and Huisingh (2008), Grant et al. (2010), and Domenech and Davies (2011)
Modeling	*Adopt and discuss various modeling schemes, for example, to optimize material flows or to understand evolution or changes in resilience for an industrial ecosystem*	
	Optimize water, energy, waste flows	Nobel and Allen (2000), Hipólito-Valencia et al. (2014), and Cimren et al. (2011)
	Adopt system dynamics	Bailey et al. (1999)
	Adopt agent-based modeling	Cao et al. (2009) and Kim et al. (2012)
	Model evolution of industrial symbiosis	Qin (2006), Huo and Chai (2008), and Zhu and Ruth (2014)
	Model resilience of industrial symbiosis	Chopra and Khanna (2014)
	Model stability of industrial symbiosis	Ng et al. (2014)
Structure	*Analyze structural elements of the industrial symbiosis network such as connectance, diversity, or scale*	Lyons (2007), Zhang et al. (2013), and Penn et al. (2014)
Case study	*Present a broad overview of specific industrial symbiosis initiatives/projects from various perspectives or examine multiple cases to conduct a review or make comparisons*	
	Case study at the level of firm	Zhu et al. (2007)
	Case study at the level of industry	Wolf and Petersson (2007) and Dong et al. (2013)
	Case study at the level of cluster	Park et al. (2008) and Taddeo et al. (2012)
	Case study at the level of city	Elabras Veiga and Magrini (2009)
	Case study at the level of region	Baas (2008) and Shi et al. (2010)
	Case study at the level of nation	Van Berkel et al. (2009b), Sakr et al. (2011), and Su et al. (2013)
Proposal	Propose a new idea or more specific plan or strategy for industrial symbiosis using a particular waste material, industry, or site	
	Proposal pertains to a specific plan or site	Alfaro and Miller (2014)
	Proposal pertains to a specific industry	Anh et al. (2011) and Martin and Eklund (2011)
	Proposal pertains to specific waste materials	Mirabella et al. (2014)
	Propose plans based on the evaluation of potential benefits or modeling results	Taskhiri et al. (2014)
	Propose a strategy to design, develop, and implement industrial symbiosis	Tsvetkova and Gustafsson (2012)

References

Alfaro, J., & Miller, S. (2014). Applying industrial symbiosis to smallholder farms: Modeling a case study in Liberia, West Africa. *Journal of Industrial Ecology, 18*(1), 145–154. doi:10.1111/jiec.12077.

Anh, P. T., My Dieu, T. T., Mol, A. P. J., Kroeze, C., & Bush, S. R. (2011). Towards eco-agro industrial clusters in aquatic production: The case of shrimp processing industry in Vietnam. *Journal of Cleaner Production, 19*(17–18), 2107–2118. doi:10.1016/j.jclepro.2011.06.002.

Ashton, W. (2008). Understanding the organization of industrial ecosystems: A social network approach. *Journal of Industrial Ecology, 12*(1), 34–51. doi:10.1111/j.1530-9290.2008.00002.x.

Ashton, W. S. (2009). The structure, function, and evolution of a regional industrial ecosystem. *Journal of Industrial Ecology, 13*(2), 228–246. doi:10.1111/j.1530-9290.2009.00111.x.

Baas, L. (2008). Industrial symbiosis in the Rotterdam Harbour and Industry Complex: Reflections on the interconnection of the techno-sphere with the social system. *Business Strategy and the Environment, 17*(5), 330–340. doi:10.1002/bse.624.

Baas, L. W., & Huisingh, D. (2008). The synergistic role of embeddedness and capabilities in industrial symbiosis: Illustration based upon 12 years of experiences in the Rotterdam Harbour and Industry Complex. *Progress in Industrial Ecology, 5*(5–6), 399–421.

Bailey, R., Bras, B., & Allen, J. K. (1999). Using robust concept exploration and systems dynamics models in the design of complex industrial ecosystems. *Engineering Optimization, 32*(1), 33–58.

Boons, F., Spekkink, W., & Mouzakitis, Y. (2011). The dynamics of industrial symbiosis: a proposal for a conceptual framework based upon a comprehensive literature review. *Journal of Cleaner Production, 19*(9–10), 905–911. doi:10.1016/j.jclepro.2011.01.003.

Boons, F., Spekkink, W., & Jiao, W. T. (2014). A process perspective on industrial symbiosis theory, methodology, and application. *Journal of Industrial Ecology, 18*(3), 341–355. doi:10.1111/jiec.12116.

Boons, F., Chertow, M., Park, J. Y., Spekkink, W., & Shi, H. (2015). *Industrial symbiosis dynamics and the problem of equivalence: Proposal for a comparative framework*. Forthcoming.

Cao, K., Feng, X., & Wan, H. (2009). Applying agent-based modeling to the evolution of eco-industrial systems. *Ecological Economics, 68*(11), 2868–2876. doi:10.1016/j.ecolecon.2009.06.009.

Chen, X., Fujita, T., Ohnishi, S., Fujii, M., & Geng, Y. (2012). The impact of scale, recycling boundary, and type of waste on symbiosis and recycling: An empirical study of Japanese eco-towns. *Journal of Industrial Ecology, 16*(1), 129–141. doi:10.1111/j.1530-9290.2011.00422.x.

Chertow, M. R. (2000). Industrial symbiosis: Literature and taxonomy. *Annual Review of Energy and the Environment, 25*, 313–337. doi:10.1146/annurev.energy.25.1.313.

Chertow, M. R. (2007). "Uncovering" industrial symbiosis. *Journal of Industrial Ecology, 11*(1), 11–30. doi:10.1162/jiec.2007.1110.

Chertow, M., & Ehrenfeld, J. (2012). Organizing self-organizing systems: Toward a theory of industrial symbiosis. *Journal of Industrial Ecology, 16*(1), 13–27. doi:10.1111/j.1530-9290.2011.00450.x.

Chertow, M. R., & Lombardi, D. R. (2005). Quantifying economic and environmental benefits of co-located firms. *Environmental Science & Technology, 39*(17), 6535–6541. doi:10.1021/es050050+.

Chertow, M., & Miyata, Y. (2011). Assessing collective firm behavior: Comparing industrial symbiosis with possible alternatives for individual companies in Oahu, HI. *Business Strategy and the Environment, 20*(4), 266–280. doi:10.1002/bse.694.

Chopra, S. S., & Khanna, V. (2014). Understanding resilience in industrial symbiosis networks: Insights from network analysis. *Journal of Environmental Management, 141*, 86–94. doi:10.1016/j.jenvman.2013.12.038.

Cimren, E., Fiksel, J., Posner, M. E., & Sikdar, K. (2011). Material flow optimization in by-product synergy networks. *Journal of Industrial Ecology, 15*(2), 315–332. doi:10.1111/j.1530-9290.2010.00310.x.

Cohen-Rosenthal, E., & Musnikow, J. (2003). *Eco-industrial strategies: Unleashing synergy between economic development and the environment.* Sheffield: Greenleaf.

Côté, R., & Hall, J. (1995). Industrial parks as ecosystems. *Journal of Cleaner Production, 3*(1–2), 41–46. doi:10.1016/0959-6526(95)00041-C.

Desrochers, P. (2000). Market processes and the closing of 'industrial loops': A historical reappraisal. *Journal of Industrial Ecology, 4*(1), 29–43.

Desrochers, P. (2001). Cities and industrial symbiosis: Some historical perspectives and policy implications. *Journal of Industrial Ecology, 5*(4), 29–44. doi:10.1162/10881980160084024.

Desrochers, P. (2004). Industrial symbiosis: The case for market coordination. *Journal of Cleaner Production, 12*(8–10), 1099–1110. doi:10.1016/j.jclepro.2004.02.008.

Domenech, T., & Davies, M. (2011). The role of embeddedness in industrial symbiosis networks: Phases in the evolution of industrial symbiosis networks. *Business Strategy and the Environment, 20*(5), 281–296. doi:10.1002/bse.695.

Dong, L., Zhang, H., Fujita, T., Ohnishi, S., Li, H. Q., Fujii, M., & Dong, H. J. (2013). Environmental and economic gains of industrial symbiosis for Chinese iron/steel industry: Kawasaki's experience and practice in Liuzhou and Jinan. *Journal of Cleaner Production, 59*, 226–238. doi:10.1016/j.jclepro.2013.06.048.

Dong, H., Ohnishi, S., Fujita, T., Geng, Y., Fujii, M., & Dong, L. (2014). Achieving carbon emission reduction through industrial & urban symbiosis: A case of Kawasaki. *Energy, 64*, 277–286. doi:10.1016/j.energy.2013.11.005.

Eckelman, M. J., & Chertow, M. R. (2013). Life cycle energy and environmental benefits of a US industrial symbiosis. *International Journal of Life Cycle Assessment, 18*(8), 1524–1532. doi:10.1007/s11367-013-0601-5.

Ehrenfeld, J., & Chertow, M. (2002). Industrial symbiosis: The legacy of kalundborg. In R. U. Ayres & L. W. Ayres (Eds.), *A handbook of industrial ecology.* Cheltenham: Edward Elgar.

Ehrenfeld, J., & Gertler, N. (1997). Industrial ecology in practice: The evolution of interdependence at Kalundborg. *Journal of Industrial Ecology, 1*(1), 67–79.

Elabras Veiga, L. B., & Magrini, A. (2009). Eco-industrial park development in Rio de Janeiro, Brazil: A tool for sustainable development. *Journal of Cleaner Production, 17*(7), 653–661. doi:10.1016/j.jclepro.2008.11.009.

Erkman, S. (1997). Industrial ecology: An historical view. *Journal of Cleaner Production, 5*(1–2), 1–10. doi:10.1016/S0959-6526(97)00003-6.

Frosch, R. A., & Gallopoulos, N. E. (1989). Strategies for manufacturing. *Scientific American, 261*, 144–152.

Geng, Y., & Côté, R. P. (2002). Scavengers and decomposers in an eco-industrial park. *International Journal of Sustainable Development and World Ecology, 9*(4), 333–340.

Geng, Y., Zhang, P., Côté, R. P., & Fujita, T. (2009). Assessment of the national eco-industrial park standard for promoting industrial symbiosis in China. *Journal of Industrial Ecology, 13*(1), 15–26. doi:10.1111/j.1530-9290.2008.00071.x.

Geng, Y., Zhang, P., Ulgiati, S., & Sarkis, J. (2010). Emergy analysis of an industrial park: The case of Dalian, China. *Science of the Total Environment, 408*(22), 5273–5283. doi:10.1016/j.scitotenv.2010.07.081.

Geng, Y., Fu, J., Sarkis, J., & Xue, B. (2012). Towards a national circular economy indicator system in China: An evaluation and critical analysis. *Journal of Cleaner Production, 23*(1), 216–224. doi:10.1016/j.jclepro.2011.07.005.

Geng, Y., Liu, Z. X., Xue, B., Dong, H. J., Fujita, T., & Chiu, A. (2014). Emergy-based assessment on industrial symbiosis: A case of Shenyang economic and technological development zone. *Environmental Science and Pollution Research, 21*(23), 13572–13587. doi:10.1007/s11356-014-3287-8.

Gertler, N. (1995). *Industrial ecosystems: Developing sustainable industrial structures*. Master of Science, Massachusetts Institute of Technology.

Gibbs, D. (2003). Trust and networking in inter-firm relations: The case of eco-industrial development. *Local Economy, 18*(3), 222–236. doi:10.1080/0269094032000114595.

Gibbs, D. (2009). Eco-industrial parks and industrial ecology: Strategic niche or mainstream development? In *The social embeddedness of industrial ecology* (pp. 73–102). Cheltenham: Edward Elgar.

Grant, G. B., Seager, T. P., Massard, G., & Nies, L. (2010). Information and communication technology for industrial symbiosis. *Journal of Industrial Ecology, 14*(5), 740–753. doi:10.1111/j.1530-9290.2010.00273.x.

Gregson, N., Crang, M., Ahamed, F. U., Akter, N., Ferdous, R., Foisal, S., & Hudson, R. (2012). Territorial agglomeration and industrial symbiosis: Sitakunda-Bhatiary, Bangladesh, as a secondary processing complex. *Economic Geography, 88*(1), 37–58. doi:10.1111/j.1944-8287.2011.01138.x.

Hardy, C., & Graedel, T. E. (2002). Industrial ecosystems as food webs. *Journal of Industrial Ecology, 6*(1), 29–38. doi:10.1162/108819802320971623.

Hashimoto, S., Fujita, T., Geng, Y., & Nagasawa, E. (2010). Realizing CO_2 emission reduction through industrial symbiosis: A cement production case study for Kawasaki. *Resources, Conservation and Recycling, 54*(10), 704–710. doi:10.1016/j.resconrec.2009.11.013.

Hewes, A., & Lyons, D. I. (2008). The humanistic side of eco-industrial parks: Champions and the role of trust. *Regional Studies, 42*(10), 1329–1342. doi:10.1080/00343400701654079.

Hipólito-Valencia, B. J., Rubio-Castro, E., Ponce-Ortega, J. M., Serna-González, M., Nápoles-Rivera, F., & El-Halwagi, M. M. (2014). Optimal design of inter-plant waste energy integration. *Applied Thermal Engineering, 62*(2), 633–652. doi:10.1016/j.applthermaleng.2013.10.015.

Howard-Grenville, J., & Paquin, R. (2008). Organizational dynamics in industrial ecosystems: Insights from organizational theory. In M. Ruth & B. Davidsdottir (Eds.), *Changing stocks, flows and behaviors in industrial ecosystems* (pp. 122–139). Cheltenham/Northampton: Edward Elgar.

Hu, J. (2012). *Making great efforts to promote ecological progress/civilization*. Report of Hu Jintao to the 18th national congress of the communist party of China. Retrieved April 5, 2015, from http://www.china.org.cn/chinese/18da/2012-11/19/content_27152706_9.htm

Huo, C. H., & Chai, L. H. (2008). Physical principles and simulations on the structural evolution of eco-industrial systems. *Journal of Cleaner Production, 16*(18), 1995–2005. doi:10.1016/j.jclepro.2008.02.013.

Jacobsen, N. B. (2006). Industrial symbiosis in Kalundborg, Denmark: A quantitative assessment of economic and environmental aspects. *Journal of Industrial Ecology, 10*(1–2), 239–255. doi:10.1162/108819806775545411.

Jacobsen, N. B. (2007). Do social factors really matter when companies engage in industrial symbiosis? *Progress in Industrial Ecology, 4*(6), 440–462. doi:10.1504/PIE.2007.016353.

Jensen, P. D., Basson, L., Hellawell, E. E., Bailey, M. R., & Leach, M. (2011). Quantifying 'geographic proximity': Experiences from the United Kingdom's national industrial symbiosis programme. *Resources, Conservation and Recycling, 55*(7), 703–712. doi:10.1016/j.resconrec.2011.02.003.

Jiao, W. T., & Boons, F. (2014). Toward a research agenda for policy intervention and facilitation to enhance industrial symbiosis based on a comprehensive literature review. *Journal of Cleaner Production, 67*, 14–25. doi:10.1016/j.jclepro.2013.12.050.

Jung, S., An, K. J., Dodbiba, G., & Fujita, T. (2012). Regional energy-related carbon emission characteristics and potential mitigation in eco-industrial parks in South Korea: Logarithmic mean Divisia index analysis based on the Kaya identity. *Energy, 46*(1), 231–241. doi:10.1016/j.energy.2012.08.028.

Kalundborg Symbiosis. (2014). *Kalundborg symbiosis 40th anniversary: Grafisk forum A/S*. Denmark.

Karlsson, M., & Wolf, A. (2008). Using an optimization model to evaluate the economic benefits of industrial symbiosis in the forest industry. *Journal of Cleaner Production, 16*(14), 1536–1544. doi:10.1016/j.jclepro.2007.08.017.

Kim, H., Ryu, J. H., & Lee, I. B. (2012). Development of an agent-based modeling methodology for an industrial byproduct exchange network design. *Industrial and Engineering Chemistry Research, 51*(33), 10860–10868. doi:10.1021/ie201915e.

Klee, R. (1999, February). Zero waste system in paradise. *BioCycle, 40*, 66–67.

Korhonen, J. (2005). Industrial ecology for sustainable development: Six controversies in theory building. *Environmental Values, 14*(1), 83–112. doi:10.3197/0963271053306096.

Liu, L. X., Zhang, B., Bi, J., Wei, Q., & Pan, H. (2012). The greenhouse gas mitigation of industrial parks in China: A case study of Suzhou Industrial Park. *Energy Policy, 46*, 301–307. doi:10.1016/j.enpol.2012.03.064.

Liu, J. G., Lü, B., Zhang, N., & Shi, Y. (2014). Definition and evaluation indicators of ecological industrial park's complex eco-efficiency. *Shengtai Xuebao/Acta Ecologica Sinica, 34*(1), 136–141. doi:10.5846/stxb201212071764.

Lowe, E. A., & Evans, L. K. (1995). Industrial ecology and industrial ecosystems. *Journal of Cleaner Production, 3*(1–2), 47–53. doi:10.1016/0959-6526(95)00045-G.

Lyons, D. I. (2007). A spatial analysis of loop closing among recycling, remanufacturing, and waste treatment firms in Texas. *Journal of Industrial Ecology, 11*(1), 43–54. doi:10.1162/jiec.2007.1029.

Martin, M., & Eklund, M. (2011). Improving the environmental performance of biofuels with industrial symbiosis. *Biomass and Bioenergy, 35*(5), 1747–1755. doi:10.1016/j.biombioe.2011.01.016.

Martin, S. A., Cushman, R. A., Weitz, K. A., Sharma, A., & Lindrooth, R. C. (1998). Applying industrial ecology to industrial parks: An economic and environmental analysis. *Economic Development Quarterly, 12*(3), 218–237.

Martin, M., Svensson, N., & Eklund, M. (2013). Who gets the benefits? An approach for assessing the environmental performance of industrial symbiosis. *Journal of Cleaner Production*. doi:10.1016/j.jclepro.2013.06.024.

Mattila, T. J., Pakarinen, S., & Sokka, L. (2010). Quantifying the total environmental impacts of an industrial symbiosis-a comparison of process-, hybrid and input–output life cycle assessment. *Environmental Science and Technology, 44*(11), 4309–4314. doi:10.1021/es902673m.

Mattila, T., Lehtoranta, S., Sokka, L., Melanen, M., & Nissinen, A. (2012). Methodological aspects of applying life cycle assessment to industrial symbioses. *Journal of Industrial Ecology, 16*(1), 51–60. doi:10.1111/j.1530-9290.2011.00443.x.

Mirabella, N., Castellani, V., & Sala, S. (2014). Current options for the valorization of food manufacturing waste: A review. *Journal of Cleaner Production, 65*, 28–41. doi:10.1016/j.jclepro.2013.10.051.

Mirata, M. (2004). Experiences from early stages of a national industrial symbiosis programme in the UK: Determinants and coordination challenges. *Journal of Cleaner Production, 12*(8–10), 967–983. doi:10.1016/j.jclepro.2004.02.031.

Mirata, M., & Emtairah, T. (2005). Industrial symbiosis networks and the contribution to environmental innovation: The case of the Landskrona industrial symbiosis programme. *Journal of Cleaner Production, 13*(10–11), 993–1002. doi:10.1016/j.jclepro.2004.12.010.

Morioka, T., Hanaki, K., & Moriguchi, Y. (2011). *Establishing a resource-circulating society in Asia: Challenges and opportunities* (T. Morioka, K. Hanaki, & Y. Moriguchi, Eds.). United Nations University Press. Tokyo, Japan.

Ng, R. T. L., Wan, Y. K., Ng, D. K. S., & Tan, R. R. (2014) Stability analysis of symbiotic bioenergy parks. In: *17th conference on process integration, modelling and optimisation for energy saving and pollution reduction*, PRES 2014 (Vol. 39, pp. 859–864). Italian Association of Chemical Engineering – AIDIC.

Nobel, C. E., & Allen, D. T. (2000). Using Geographic Information Systems (GIS) in industrial water reuse modelling. *Process Safety and Environmental Protection, 78*(4), 295–303.

Panyathanakun, V., Tantayanon, S., Tingsabhat, C., & Charmondusit, K. (2013). Development of eco-industrial estates in Thailand: Initiatives in the northern region community-based eco-industrial estate. *Journal of Cleaner Production, 51*, 71–79. doi:10.1016/j.jclepro.2012.09.033.

Park, H. S., & Behera, S. K. (2014). Methodological aspects of applying eco-efficiency indicators to industrial symbiosis networks. *Journal of Cleaner Production, 64*, 478–485. doi:10.1016/j.jclepro.2013.08.032.

Park, J. Y., & Park, H. S. (2014). Securing a competitive advantage through industrial symbiosis development the case of steam networking practices in Ulsan. *Journal of Industrial Ecology, 18*(5), 677–683. doi:10.1111/jiec.12158.

Park, H. S., Rene, E. R., Choi, S. M., & Chiu, A. S. F. (2008). Strategies for sustainable development of industrial park in Ulsan, South Korea-from spontaneous evolution to systematic expansion of industrial symbiosis. *Journal of Environmental Management, 87*(1), 1–13. doi:10.1016/j.jenvman.2006.12.045.

Parto, S. (2000). Industrial ecology and regionalization of economic governance: An opportunity to 'localize' sustainability? *Business Strategy and the Environment, 9*(5), 339–350.

Penn, A. S., Jensen, P. D., Woodward, A., Basson, L., Schiller, F., & Druckman, A. (2014). Sketching a network portrait of the Humber region. *Complexity, 19*(6), 54–72. doi:10.1002/cplx.21519.

Pi, Q., & Wang, K. (2004). *Out of the island: Overview of Chinese ETDAs*. SDX Joint Publishing Company, China.

Posch, A. (2010). Industrial recycling networks as starting points for broader sustainability-oriented cooperation? *Journal of Industrial Ecology, 14*(2), 242–257. doi:10.1111/j.1530-9290.2010.00231.x.

Potts Carr, A. J. (1998). Choctaw Eco-Industrial Park: An ecological approach to industrial land-use planning and design. *Landscape and Urban Planning, 42*(2–4), 239–257. doi:10.1016/S0169-2046(98)00090-5.

Qin, S. T. (2006). The research of modeling of Eco-Industrial Park based on evolution game. *Dynamics of Continuous Discrete and Impulsive Systems-Series a-Mathematical Analysis, 13*, 1322–1329.

Reuter, M., Heiskanen, K., Boin, U., Van Schaik, A., Verhoef, E., Yang, Y., & Gerorgalli, G. (2005). *The metrics of material and metal ecology: Harmonizing the resource, technology, and environmental cycles*. Amsterdam/London: Elsevier.

Ristola, P., & Mirata, M. (2007). Industrial symbiosis for more sustainable, localised industrial systems. *Progress in Industrial Ecology, 4*(3–4), 184–204. doi:10.1504/PIE.2007.015186.

Roberts, B. H. (2004). The application of industrial ecology principles and planning guidelines for the development of eco-industrial parks: An Australian case study. *Journal of Cleaner Production, 12*(8–10), 997–1010. doi:10.1016/j.jclepro.2004.02.037.

Romero, E., & Ruiz, M. C. (2014). Proposal of an agent-based analytical model to convert industrial areas in industrial eco-systems. *Science of the Total Environment, 468*, 394–405. doi:10.1016/j.scitotenv.2013.08.049.

Sagar, A. D., & Frosch, R. A. (1997). A perspective on industrial ecology and its application to a metals-industry ecosystem. *Journal of Cleaner Production, 5*(1–2), 39–45.

Sakr, D., Baas, L., El-Haggar, S., & Huisingh, D. (2011). Critical success and limiting factors for eco-industrial parks: Global trends and Egyptian context. *Journal of Cleaner Production, 19*(11), 1158–1169. doi:10.1016/j.jclepro.2011.01.001.

Salmi, O. (2007). Eco-efficiency and industrial symbiosis – A counterfactual analysis of a mining community. *Journal of Cleaner Production, 15*(17), 1696–1705. doi:10.1016/j.jclepro.2006.08.012.

Salmi, O., & Wierink, M. (2011). Effects of waste recovery on carbon footprint: A case study of the Gulf of Bothnia steel and zinc industries. *Journal of Cleaner Production, 19*(16), 1857–1864. doi:10.1016/j.jclepro.2011.04.014.

Schiller, F., Penn, A., Druckman, A., Basson, L., & Royston, K. (2014). Exploring space, exploiting opportunities the case for analyzing space in industrial ecology. *Journal of Industrial Ecology, 18*(6), 792–798. doi:10.1111/jiec.12140.

Schwarz, E. J., & Steininger, K. W. (1997). Implementing nature's lesson: The industrial recycling network enhancing regional development. *Journal of Cleaner Production, 5*(1–2), 47–56.

Shi, H., & Shi, L. (2014). Identifying emerging motif in growing networks. *Plos One, 9*(6), e99634. doi:10.1371/journal.pone.0099634.

Shi, L., & Yu, B. (2014). Eco-industrial parks from strategic niches to development mainstream: The cases of China. *Sustainability (Switzerland), 6*(9), 6325–6331. doi:10.3390/su6096325.

Shi, H., Chertow, M., & Song, Y. Y. (2010). Developing country experience with eco-industrial parks: A case study of the Tianjin economic-technological development area in China. *Journal of Cleaner Production, 18*(3), 191–199. doi:10.1016/j.jclepro.2009.10.002.

Sokka, L., Melanen, M., & Nissinen, A. (2008). How can the sustainability of industrial symbioses be measured? *Progress in Industrial Ecology, 5*(5–6), 518–535.

Sokka, L., Lehtoranta, S., Nissinen, A., & Melanen, M. (2011). Analyzing the environmental benefits of industrial symbiosis life cycle assessment applied to a Finnish forest industry complex. *Journal of Industrial Ecology, 15*(1), 137–155. doi:10.1111/j.1530-9290.2010.00276.x.

Spekkink, W. (2014). Building capacity for sustainable regional industrial systems: An event sequence analysis of developments in the Sloe Area and Canal Zone. *Journal of Cleaner Production.* doi:10.1016/j.jclepro.2014.08.028.

Sterr, T., & Ott, T. (2004). The industrial region as a promising unit for eco-industrial development – Reflections, practical experience and establishment of innovative instruments to support industrial ecology. *Journal of Cleaner Production, 12*(8–10), 947–965. doi:10.1016/j.jclepro.2004.02.029.

Su, B., Heshmati, A., Geng, Y., & Yu, X. M. (2013). A review of the circular economy in China: Moving from rhetoric to implementation. *Journal of Cleaner Production, 42*, 215–227. doi:10.1016/j.jclepro.2012.11.020.

Taddeo, R., Simboli, A., & Morgante, A. (2012). Implementing eco-industrial parks in existing clusters. Findings from a historical Italian chemical site. *Journal of Cleaner Production, 33*, 22–29. doi:10.1016/j.jclepro.2012.05.011.

Taskhiri, M. S., Behera, S. K., Tan, R. R., & Park, H. S. (2014). Fuzzy optimization of a waste-to-energy network system in an eco-industrial park. *Journal of Material Cycles and Waste Management.* doi:10.1007/s10163-014-0259-5.

Tian, J., Liu, W., Lai, B., Li, X., & Chen, L. (2014). Study of the performance of eco-industrial park development in China. *Journal of Cleaner Production, 64*, 486–494. doi:10.1016/j.jclepro.2013.08.005.

Tsvetkova, A., & Gustafsson, M. (2012). Business models for industrial ecosystems: A modular approach. *Journal of Cleaner Production, 29–30*, 246–254. doi:10.1016/j.jclepro.2012.01.017.

Valero, A., Usón, S., Torres, C., Valero, A., Agudelo, A., & Costa, J. (2013). Thermoeconomic tools for the analysis of eco-industrial parks. *Energy, 62*, 62–72. doi:10.1016/j.energy.2013.07.014.

Van Berkel, R., Fujita, T., Hashimoto, S., & Fujii, M. (2009a). Quantitative assessment of urban and industrial symbiosis in Kawasaki, Japan. *Environmental Science and Technology, 43*(5), 1271–1281. doi:10.1021/es803319r.

Van Berkel, R., Fujita, T., Hashimoto, S., & Geng, Y. (2009b). Industrial and urban symbiosis in Japan: Analysis of the Eco-Town program 1997–2006. *Journal of Environmental Management, 90*(3), 1544–1556. doi:10.1016/j.jenvman.2008.11.010.

Van Ha, N. T., Ananth, A. P., Visvanathan, C., & Anbumozhi, V. (2009). Techno policy aspects and socio-economic impacts of eco-industrial networking in the fishery sector: Experiences from an Giang Province, Vietnam. *Journal of Cleaner Production, 17*(14), 1272–1280. doi:10.1016/j.jclepro.2009.03.014.

van Leeuwen, M. G., Vermeulen, W. J. V., & Glasbergen, P. (2003). Planning eco-industrial parks: An analysis of Dutch planning methods. *Business Strategy and the Environment, 12*(3), 147–162. doi:10.1002/bse.355.

Walls, J. L., & Paquin, R. (2015). Organizational perspectives of industrial symbiosis. *A Review and Synthesis Organization & Environment, 28*(1), 32–53. doi:10.1177/1086026615575333.

Walter, A. I., & Scholz, R. W. (2006). Sustainable innovation networks: An empirical study on interorganisational networks in industrial ecology. *Progress in Industrial Ecology, 3*(5), 431–450. doi:10.1504/PIE.2006.012270.

Wang, L. M., Zhang, J. T., & Ni, W. D. (2005). Emergy evaluation of Eco-Industrial Park with power plant. *Ecological Modelling, 189*(1–2), 233–240. doi:10.1016/j. ecolmodel.2005.02.005.

Wen, Z., & Meng, X. (2014). Quantitative assessment of industrial symbiosis for the promotion of circular economy: A case study of the printed circuit boards industry in China's Suzhou New District. *Journal of Cleaner Production, 3*, 1–9. doi:10.1016/j.jclepro.2014.03.041.

Wolf, A., & Karlsson, M. (2008). Evaluating the environmental benefits of industrial symbiosis: Discussion and demonstration of a new approach. *Progress in Industrial Ecology, 5*(5–6), 502–517. doi:10.1504/PIE.2008.023413.

Wolf, A., & Petersson, K. (2007). Industrial symbiosis in the Swedish forest industry. *Progress in Industrial Ecology, 4*(5), 348–362. doi:10.1504/PIE.2007.015616.

Yang, L., Hu, S., Chen, D., & Zhang, D. (2006). Exergy analysis on eco-industrial systems. *Science in China, Series B Chemistry, 49*(3), 281–288. doi:10.1007/s11426-006-0281-0.

Yu, C., Davis, C., & Dijkema, G. P. J. (2014a). Understanding the evolution of industrial symbiosis research: A bibliometric and network analysis (1997–2012). *Journal of Industrial Ecology, 18*(2), 280–293. doi:10.1111/jiec.12073.

Yu, C., Dijkema, G. P., & De Jong, M. (2014b). What makes eco-transformation of industrial parks take off in China? *Journal of Industrial Ecology.* doi:10.1111/jiec.12185.

Yu, F., Han, F., & Cui, Z. (2014c). Assessment of life cycle environmental benefits of an industrial symbiosis cluster in China. *Environmental Science and Pollution Research.* doi:10.1007/ s11356-014-3712-z.

Zhang, Y., Zheng, H., Chen, B., & Yang, N. (2013). Social network analysis and network connectedness analysis for industrial symbiotic systems: Model development and case study. *Frontiers of Earth Science, 7*(2), 169–181. doi:10.1007/s11707-012-0349-4.

Zhang, Y., Zheng, H., Chen, B., Su, M., & Liu, G. (2014). A review of industrial symbiosis research: Theory and methodology. *Frontiers of Earth Science.* doi:10.1007/ s11707-014-0445-8.

Zheng, K. F., Jia, S. L., & Wang, H. S. (2013). *Evolution of industrial ecosystem with government's intervention: Integration of evolutionary game model into multi-agent simulation.* Paper presented at the 19th international conference on industrial engineering and engineering management, Changsha.

Zhu, J. M., & Ruth, M. (2013). Exploring the resilience of industrial ecosystems. *Journal of Environmental Management, 122*, 65–75. doi:10.1016/j.jenvman.2013.02.052.

Zhu, J. M., & Ruth, M. (2014). The development of regional collaboration for resource efficiency: A network perspective on industrial symbiosis. *Computers, Environment and Urban Systems, 44*, 37–46. doi:10.1016/j.compenvurbsys.2013.11.001.

Zhu, Q., Lowe, E. A., Wei, Y. A., & Barnes, D. (2007). Industrial symbiosis in China: A case study of the Guitang Group. *Journal of Industrial Ecology, 11*(1), 31–42. doi:10.1162/jiec.2007.929.

Chapter 6
A Socio-economic Metabolism Approach to Sustainable Development and Climate Change Mitigation

Timothy M. Baynes and Daniel B. Müller

Abstract Humanity faces three large challenges over the coming decades: urbanisation and industrialisation in developing countries at unprecedented levels; concurrently, we need to mitigate against dangerous climate change and we need to consider finite global boundaries regarding resource depletion.

Responses to these challenges as well as models that inform strategies are fragmented. The current mainstream framework for measuring and modelling climate change mitigation focuses on the flows of energy and emissions and is insufficient for simultaneously addressing the material and infrastructure needs of development. The models' inability to adequately represent the multiple interactions between infrastructure stocks, materials, energy and emissions results in notable limitations. They are inadequate: (1) to identify physically realistic (mass balance consistent) mitigation pathways, (2) to anticipate potentially relevant co-benefits and risks and thus (3) to identify the most effective strategies for linking targets for climate change mitigation with goals for sustainable development, including poverty eradication, infrastructure investment and mitigation of resource depletion.

This chapter demonstrates that a metabolic approach has the potential to address urbanisation and infrastructure development and energy use and climate change, as well as resource use, and therefore to provide a framework for integrating climate change mitigation and sustainable development from a physical perspective. Metabolic approaches can represent the cross-sector coupling between material and energy use and waste (emissions) and also stocks in the anthroposphere (including fixed assets, public and private infrastructure). Stocks moderate the supply of services such as shelter, communication, mobility, health and safety and employment opportunities.

The development of anthropogenic stocks defines boundary conditions for industrial activity over time. By 2050 there will be an additional three billion urban

T.M. Baynes (✉)
CSIRO Land and Water, PO Box 52, North Ryde, NSW 1670, Australia
e-mail: Tim.Baynes@csiro.au

D.B. Müller
NTNU: Norges Teknisk-naturvitenskapelige Universitet, Trondheim, Norway

© The Author(s) 2016
R. Clift, A. Druckman (eds.), *Taking Stock of Industrial Ecology*,
DOI 10.1007/978-3-319-20571-7_6

dwellers, almost all of them in developing countries. If they are to receive the level of services converging on those currently experienced in developed nations, this will entail a massive investment in infrastructure and substantial quantities of steel, concrete and aluminium (materials that account for nearly half of industrial emissions). This scenario is confronted by the legacy of existing infrastructure and the limit of a cumulative carbon budget within which we could restrain global temperature rise to <2 °C.

A metabolic framework incorporating stock dynamics can make an explicit connection between the timing of infrastructure growth or replacement and the material and energy needs of that investment. Moreover, it provides guidance on the technical and systemic options for climate mitigation concurrent with a future of intense urban development and industrialisation.

Keywords Climate change • Cross-sector coupling • Embodied energy and emissions • Flows • Infrastructure • Metabolic framework • MFA • Socio-economic metabolism • Stocks

1 Background

There is strong consensus among scientists that climate change is upon us and that mitigation action is both worthwhile and urgent (UNFCCC 2011; IPCC 2014a). At the same time, there is widespread recognition that the poverty and inequality in the developing world is unsustainable and there are internationally agreed goals to rectify this (UN 2012). Climate change research has defined the problem: through measuring and modelling the flows of CO_2 and other greenhouse gases (GHGs), monitoring extreme weather events, acidification of oceans and other observations, the causes and consequences of climate change have been identified. We can attribute global climate change to a host of different economic activities with some degree of spatial detail (e.g. Hertwich and Peters 2009; Peters 2010). Current mainstream models for climate change mitigation (CCM) frame the problem predominantly as one of the energy systems and one that is located where energy or emissions are produced or where energy is finally consumed. They emphasise energy and emissions directly or indirectly associated with activity in different sectors of society (including land use change) – see Fig. 6.1 – but they omit (1) the linkages between energy use sectors through nonenergy resource flows, (2) the drivers of resource use (e.g. from infrastructure development) and (3) the secondary resource availability (e.g. from infrastructure retirement). Thus, the current mainstream CCM models omit the material boundary conditions of the global system and opportunities for energy and emissions saving through recycling and reuse of materials.

When we refer to solutions, we ask: 'what can we do about climate change?' What are the technical and behavioural responses to the challenge? Analysis of energy and emissions flows is essential but insufficient to address these questions

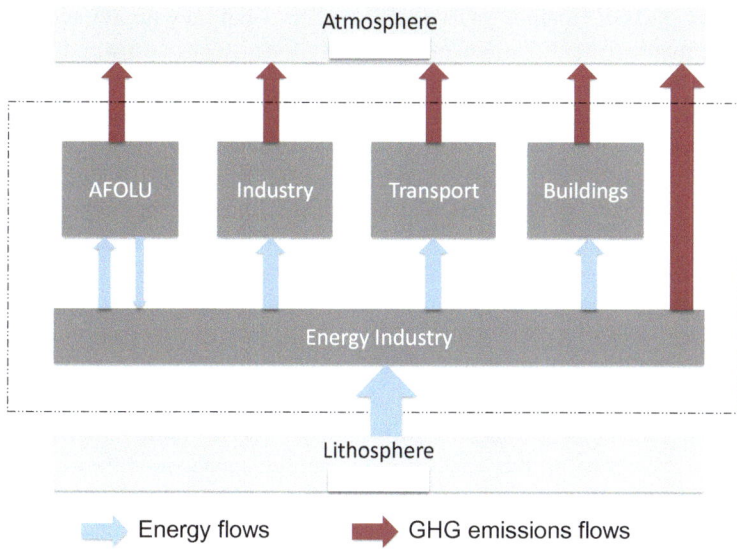

Fig. 6.1 Summary of the standard approach for emissions accounting, e.g. 2006 IPCC Guidelines for National Greenhouse Gas Inventories (IPPC 2006)

because it has only a poor representation of the capital stocks and infrastructure that are the mediators of many mitigation initiatives.

Drivers of emissions are seen as consumption and production, connected by primarily linear flows of materials and energy. What is important for CCM interventions and what is usually not well covered is the underlying activity, which is often associated with infrastructure growth and maintenance. For example, we attach emissions to aggregate steel production and steel consumption but frequently *not* to what the steel is used for, or for how long. If we seek to substitute for, or reduce, steel in-use ('lightweighting'), prolong its lifetime in-use or increase the intensity of use for the product containing steel, these are questions about the stocks in-use (Pauliuk and Müller 2014). Infrastructures are currently a blind spot in the framing of climate change mitigation, and they are also an essential link to sustainable development.

The sectors of buildings, transport and industry account for more than 40 % of direct GHG emissions and nearly 65 % of direct and indirect energy-related emissions (IPCC 2014a), but to enact mitigation in these sectors requires knowledge of the stocks in-use, their age structure, current and potential future efficiency and the material resources needed to create them. The same information is also needed in assessing the requirements (and speed) of future development. It is the services from capital stocks and infrastructure that are essential to improving and maintaining quality of life, whether they be productive assets that are linked to employment or roads that enable mobility or schools, hospitals and other public infrastructures that facilitate education, health and social prosperity.

Some integrated assessment models do include the population dynamics of stocks and their characteristic efficiency – for example, the TIMES (Loulou et al.

2005) and IMAGE (Stehfest et al. 2014) models – but they do not account rigorously for materials and the dynamics of flows from stocks at the end of life. The existing assessment and modelling frameworks are not lacking in scope – they include many sectors of society in detail – but often in-use stock dynamics and cross-sectoral linkages are absent (e.g. Allwood et al. (2010)). Moreover, studies of material flows and models of aggregate stocks (Davis et al. 2010) rarely couple the material flow to a capital stock that has both a direct energy efficiency characteristic in operation and an indirect energy and emissions requirement in its own construction. These issues are also addressed in Chap. 8.

In this chapter, we discuss a socio-economic metabolic framework that can represent the interlinked nature of sustainable development and climate change mitigation from a physical perspective. It incorporates stocks and flows of infrastructures, materials, energy and emissions as well as their multiple linkages through processes that are treated using mass and energy balances. This allows for a representation of feedbacks and delays in these material and energy connections through recycling and maintenance. The metabolic framework is not intended to replace the energy and emissions framework, but rather expands it in order to reconcile CCM with sustainable development, to understand the side effects of CCM (co-benefits and risks) and thereby identify effective mitigation pathways.

2 A Socio-economic Metabolism Framework

The concept of socio-economic metabolism is relatively young in the literature. Here we interpret the term in an inclusive sense that is synonymous with social (Fischer-Kowalski 1998; Fischer-Kowalski and Weisz 1999), industrial (Ayres 1989) and anthropogenic metabolism (Baccini and Brunner 1991).

Modelling methods that incorporate anthropogenic stocks and flows can trace their lineage back to the early works of Forrester (1958), but examples where social metabolism and material and energy flow accounting are united with dynamic stock analysis are rare (Baccini and Bader 1996; Müller et al. 2004; Lennox et al. 2005; Müller 2006; Baynes et al. 2009; Müller et al. 2013; Pauliuk and Müller 2014). Even within that handful, few talk of the services from stocks that are germane to our discussion on sustainable development.

There are a number of proponents of this integrated thinking and there is a natural application in urban systems. The 'Social-Ecological-Infrastructural Systems Framework' of Ramaswami et al. (2012) begins with questions of urban sustainability and expands to the same scope as the socio-economic metabolism framework. They specifically include services from internal and 'trans-boundary' infrastructures, but the key difference from the socio-economic metabolism framework lies in the dynamic treatment of stocks in-use in the latter. In the following sections, we discuss key features of the socio-economic metabolism framework (hereafter the 'metabolic framework' or approach) – see Fig. 6.2.

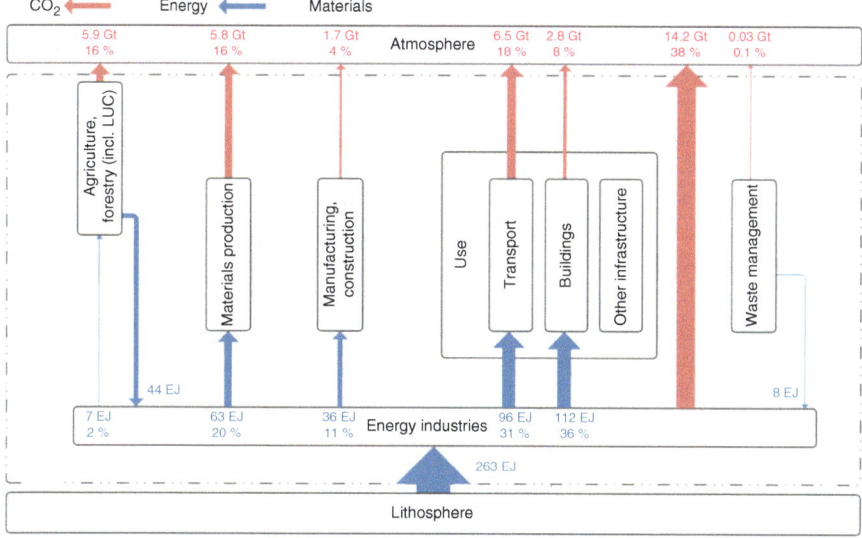

Fig. 6.2 Socio-economic metabolism framework including energy and emissions flows (as in IPCC Guidelines) and material flows. Stocks of materials and energy can occur in all segments of the diagram though in-use stocks lie within the boundary around the economic system

2.1 Energy

Energy and emissions flows, as calculated in the standard formulations of energy and emissions accounting (OECD/IEA and Eurostat 2005), are still essential components of the socio-economic metabolism framework. The quantity of energy required is a reliable indicator of aggregate economic activity and affluence. The way in which energy is used in society and sectors of the economy is a measure of socio-economic status and structure.

The major role of energy accounting is one of monitoring the problem: recording how much of what type of energy has been used and in which sector of the economy. This may enable assessment of excessive energy use or of overdependence on a particular fuel or an emissions-intensive means of providing energy. A socio-economic metabolism framework extends this accounting framework to have a more explanatory function by linking energy flows with in-use stocks and material cycles across sectors. From these explicit connections, we may proceed to simulations to develop possible solutions to global challenges: exploring the questions of where to effectively intervene in the system to achieve lesser impacts, greater efficiencies or other beneficial outcomes for society.

For example, about one-third of the world's population still relies on the use of animal power and non-commercial fuels. More than 1.4 billion people have no access to electricity (Global Energy Assessment 2012). If we are to redress this situation in the future, there will likely be a need for both an increase in the flow of energy and a change to the distribution of its use. This, of course, would need to be monitored, but it also begets the question of *how* this change will occur, and we are immediately drawn into the need to include the energy producing and consuming stocks in the future analysis, their interconnection and the materials that those stocks require for operation (and, in the complete picture, the energy efficiency of those stocks and the energy embodied in their constituent materials).[1]

An energy analysis is basically linear (thermodynamics prevents energy from being truly recycled in feedback loops). The energy and emissions flow emphasis of most integrated assessment models has lead analysts to view opportunities for climate change mitigation as if sectors operated separately. Energy flows into a sector; it is used in production and transformed into waste heat; and local or upstream GHG emissions are co-produced. Yet industrial materials are used in transport and building stocks, and building stocks are part of the capital stock of industry. Clearly, emission mitigation schemes in different sectors are interrelated.

The flow of materials shown in Fig. 6.2 reveals connections between sectors and shows how energy-using sectors are dependent. Furthermore, it represents physical feedbacks – material recycling – that highlight the importance of circular relations in production. Lastly, the metabolic picture reveals many more potential intervention points, within and across sectors, than in the linear conception of energy and emissions accounting in Fig. 6.1.

2.2 Materials

To record energy flows independent of materials is to mute half of the story of social metabolism. Material flows, as in the economy-wide material flow accounting EWMFA (EUROSTAT 2009) and as feedback flows in recycled material from stocks, are an essential component of the metabolic framework (see Chap. 8), and it is important to track material flows along with energy use. Materials production is an important energy user, and the production of energy technologies depends on many critical materials.

While a large portion of energy is consumed directly in providing services like thermal comfort, lighting, communication or entertainment, industry accounts for more than 32 % of global final energy use[2] (115EJ in 2005). Production of just five materials – cement, iron and steel, chemicals, pulp and paper and aluminium – accounts for more than half of industrial energy use (Global Energy Assessment 2012).

While still acknowledging the importance of energy flows in climate change mitigation, we must also recognise that energy is inextricably linked to material flows and in-use stocks. For instance, the consumption of hot clean water involves

---

[1] These issues are addressed from a 'Global South' perspective in Chap. 12.
[2] This includes final electricity consumption in arc furnaces and smelters that ultimately requires a great deal more upstream primary energy.

the treatment, pumping and heating of the water, and, effectively, we consume the energy and material simultaneously whether it is in a hot shower or a hot cup of coffee. Similarly, the energy use and emissions associated with industry are linked to the movement and transformation of materials: energy may be used in a factory but the output is the material product.

It has been noted already that there is more embodied energy in household consumption of goods and services than directly consumed energy (Lenzen et al. 2004). Furthermore, as housing standards improve and operational energy use declines, there is an increasing importance of embedded emissions in the materials of dwelling construction (Giesekam et al. 2014).

Lastly, and relevant to the next section, materials exhibit dynamic path dependency. Material flows enter and stay in the system as stocks and leave, or are recycled, at a future date. Most energy embedded in products is not recoverable, but an interesting and useful feature of material stocks in-use is their latent ability to contribute to material and energy saving through recycling in the future.

2.3   The Importance of Representing Stocks

Operational efficiency and embodied energy are not simply properties of the aggregated material stock but of individual items making up infrastructure and other artefacts (appliances, durable goods, etc.). These stocks contain a great deal of material, but they also have discrete lifetimes and often interact across sectors to supply services to society. This key point has been made by Fischer-Kowalski and Weisz (1999): 'physical infrastructure (buildings, machines, artefacts in-use and livestock)' is 'core to our understanding of the society-nature interrelation.' In-use stocks represent large monetary investments and determine the long-term dynamics of social metabolism. Through their long lifetime in-use, they are responsible for lock-ins of lifestyles and emission pathways.

Urban in-use stocks also relate to the density and accessibility of urban spaces and the capacity and utility of urban systems such as public transport or water supply systems. A study of global cities over 10 years (Angel et al. 2010) revealed that recent trends are away from denser cities, and the implications in both the industrialised and developing world are for cities to take up yet more space and for more infrastructure to be needed to fill and connect that space.

Stocks and flows play different roles over different time scales. Over the short-term (less than 5 years), physical and economic flows can change with the vicissitudes of markets, income (GDP), prices and events. Over the long-term, deeper structural changes related to population dynamics, urbanisation and infrastructure development, long-run economic policy, cumulative savings, resource depletion and institutional arrangements are more connected to the development of stocks. Thus, long-term change is recorded in the quantity and quality of in-use stocks, and existing stocks and systems of stocks influence the long-term future. This is valid from both the metabolic and the wealth and income perspectives (Piketty 2013).

Stocks and flows are certainly interrelated. Flows are the 'material income' to a system, and any net addition to stocks (NAS) is limited through the conservation of mass by direct material input (DMI) minus domestic processed outputs (DPO) minus exports. Yet, the use of stocks by society determines the resource flows needed to operate, maintain or expand the physical stock. The services from stocks of infrastructure, machines and durable goods are closest to the interests of society; the socio-economic metabolic framework is driven first by stocks. Flows of resource inputs and waste outputs are driven by the need for services provided by stocks rather than the demand for the flows themselves (Müller et al. 2004; Pauliuk and Müller 2014). For example, travel by automobile requires a flow of energy, but it is the stock (the automobile) that enables the conversion of energy into the service of mobility determines the amount and type of fuel used and resulting emissions.

Stocks in-use record the cumulative resource flows – materials and energy – embedded in the infrastructure and artefacts of the socio-economic system. Through their role in production, the age and efficiency of stocks are key factors in the operational consumption of resource flows in the economy and related GHG emissions. At the same time, the services from stocks are key for social and material development.

3   Problem Shifting

An important issue in responding to global challenges is the danger of 'problem shifting': the potential for an intervention that alleviates one challenge to exacerbate the response to another. Without an integrated approach that encompasses stocks and flows of materials and energy, such a trade-off is underestimated or even invisible. What follows is an example from Müller et al. (2013) using an indicator for the carbon footprint of infrastructure stocks, the 'carbon replacement value' (CRV) of the stocks, which is defined as the carbon emissions required to replace an existing infrastructure stock or to build a new infrastructure stock using currently available technologies. Müller et al. (2013) based their CRV on upstream carbon emissions starting with primary production, CRV_P. This indicator is used to estimate the carbon emissions caused by developing nations if they were to converge on the level of service provided in industrialised nations. The CRV_P for scenarios of infrastructure development was calculated using a metabolic approach that incorporates stock dynamics.

3.1   Sustainable Development and the Carbon Budget

Emissions from the operation of infrastructure are generally considered to be the main concern in the standard energy and emissions approach to GHG accounting: the models used, such as Davis et al. (2010), represent infrastructure stocks as energy users and GHG emitters but omit the energy and emissions embodied in the stocks. However, for the developing world to converge on the quality of life enjoyed in the industrialised world by 2050, there will need to be a significant increase in the material and monetary quantity of infrastructure stocks. If we are to use current energy sources and technology to construct them, this must lead to a large carbon impost.

To estimate the GHG emissions from the materials needed in such a development scenario, Müller et al. (2013) used data on the key materials of steel, aluminium and cement (other materials having either less associated emissions or less importance in infrastructure stocks). They found that current CRV_P is similar for most industrialised countries at a level of about 50 t CO_2 per capita. Assuming a population growth from currently 6.8 to 9.3 billion, the direct material requirement for infrastructure and other assets needed to maintain or improve human welfare would involve an indirect carbon footprint $CRV_P = 350$ Gt CO_2 (see Fig. 6.3). The cumulative emissions during the 2000–2050 time period cannot exceed 1000–1440 Gt CO_2, if we are to have a 75 % or 50 % probability of limiting warming to less than 2 °C, respectively (Meinshausen et al. 2009). About 420 Gt of this amount has already been emitted between 2000 and 2011 (IPCC2014a) which leaves an emissions budget of approximately 600–1000 Gt CO_2 for the period from 2012 to 2050. Just the emissions embedded in the stock yet to be built therefore constitute 35–60 % of the remaining carbon budget, provided developing countries invest in built environment stocks similar to industrialised countries and use currently available technology. This leaves precious little in the carbon budget for using the stock and emissions beyond 2050.

There is a premise to these calculations that should be acknowledged that achieving Western-style infrastructure stock is a desirable endpoint of sustainable development and that obtaining the same level of services from infrastructure and in-use stocks involves the same intensity of resource use as seen currently in the developed world. The former assumption is certainly debateable in terms of environmental sustainability, and the latter is not necessarily the case as, quite apart from probable technical improvements, it is possible to realise a better quality of life without the need for a high-income, high impact society. As a model for this, there is a group of countries in the so-called Goldemberg corner that have relatively high income and long average life expectancy with low-carbon lifestyles (Steinberger et al. 2012).

The salient point, however, is the significant trade-off between the aims of sustainable development and climate change mitigation. The very poor access to basic infrastructure in developing nations is untenable. If we are to alleviate this situation, then industrial development and urbanisation in these countries will dominate growth in infrastructure construction for several decades; this will increase global material demand and thereby produce GHG emissions.

Problem shifting is not limited to the issues of developing nations. While attempting to depress the carbon intensity of production and consumption, we face increasing material demands, sometimes for critical materials. Hence, it is important in any scenario analysis of climate change mitigation to include materials and anticipated large-scale in-use stocks of materials.

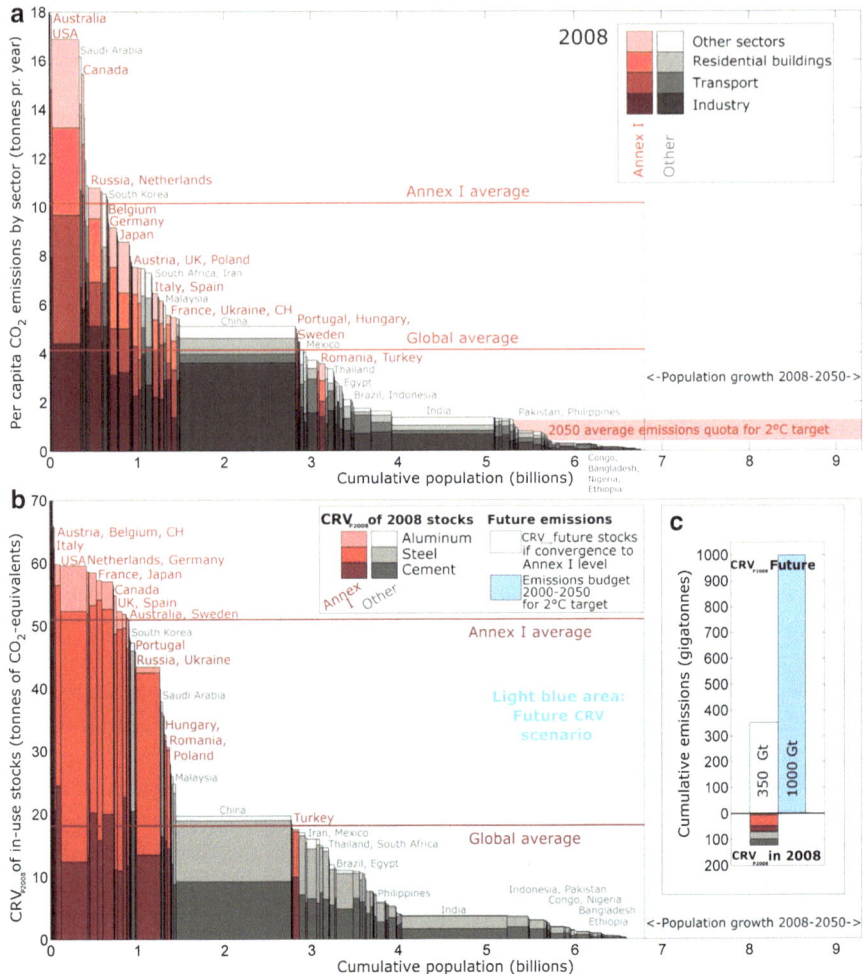

Fig. 6.3 (**a**) Total fuel-related per-capita CO$_2$ emissions by country (*red and grey bars*) compared to the global per-capita emission level in 2050 to remain within the 2 °C target with a 50–75 % probability (*red horizontal bar*); (**b**) CRV$_P$ at 2008 per capita of existing stocks by country (*red and grey*) and of as-yet unbuilt stocks if developing countries converge on the current average Annex I level (*light blue*); (**c**) comparison with emission budget for the period 2000–2050 to reach the 2 °C target with a 75 % probability. Of this emission budget (1000 Gt CO$_2$), approximately 420 Gt CO$_2$ was already emitted during the period from 2000 to 2011 (Graphic reproduced from Müller et al. 2013)

One problem is that many renewable energy technologies are reliant on critical minerals, for example, neodymium in wind turbines and indium in thin film solar cells. Scenarios based on widespread implementation of wind, solar and other alternative means of generating electricity shift the problem of decarbonising the energy sector to one of material criticality. Material criticality is not to be confused with material scarcity. Material criticality may be measured using two orthogonal dimensions: the quantity or importance of a material in an activity and the risk of supply interruption. Criticality can be represented in a matrix format using these measures, as shown in Fig. 6.4. This analysis identifies a few mineral elements that are at high risk for supply disruption in the United States, including the rare earth elements neodymium and dysprosium used extensively in permanent magnets that enable a number of lightweight electronics and 'green' energy technology.

Despite their name, 'rare earth metals' are not as limited in supply as, for comparison, platinum or palladium. Criticality can be a complex function of factors such as geographical distribution of reserves and stability of government in the nation owning those reserves (Dawson et al. 2014; Roelich et al. 2014). Although rare earth elements are often found in other metal ores, e.g. zinc, 95 % of the world's current rare earth metal supply comes from China and 72 % of the known geological reserves of dysprosium are also in China. Criticality is sensitive to such a monopoly and one uncertainty is the speed at which new mines can be opened outside of China, but another part of the supply issue is the limited opportunities for recycling before 2050 due to the long lifetimes of end-use products. These issues are not limited to critical materials; see, for example, Kushnir and Sandén (2012) on lithium supply. Roelich et al. (2014) examined electricity system transitions in the United Kingdom in this light, with a focus on neodymium. While supply disruption was anticipated to decrease by almost 30 % by 2050, the criticality of low-carbon electricity production increases ninefold because of an increasing demand for neodymium across a range of technologies.

Economists could argue that more demand will raise prices and thereby make it economic to exploit more reserves. Whether or not this is valid, and whether or not production is sufficiently responsive to changes in demand, the challenge remains: over the next human generation, large infrastructure investment decisions will be made in developing nations, and unless they have affordable greener options, they will revisit an industrial history in a way that the available carbon budget does not permit.

Fig. 6.4 Medium-term (5–15 years) relation of supply risk to the importance in clean energy technologies (From Fig. 8.2 in the US Department of Energy *Critical Materials Strategy* (USDOE 2011))

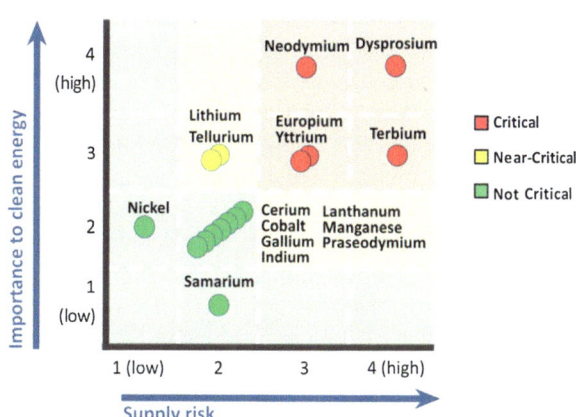

4   Effective Policymaking: The Case of the Aluminium Sector

The preceding sections expanded on the challenges of climate change mitigation and the physical aspects of sustainable development. Effective policy that enacts responsible management of energy, emissions and materials needs to consider:

- Increased future demand for services from infrastructure and other fixed capital – such as carbon capture and storage (CCS), water treatment plant, roads and renewable energy technology – nearly a billion new dwellings globally
- The boundaries imposed by emissions reduction targets
- Material resource use and availability

In contrast to some adaptive strategies that aim for resilience through flexibility and reversibility in investments and even reducing the lifetime of investments (Hallegatte 2009), climate change mitigation is about commitment to long-term change: setting in place the economic, institutional and physical structures to enable a sustained transition to a low-carbon future. The first priority is *effective* interventions to limit climate change (UNFCCC 2011), followed by the question of whether a response is efficient in terms of cost or resources required.

What are the options for effective climate change mitigation policy and how does the metabolic framework generate answers or enable assessment? We use the global aluminium sector to illustrate a range of policy actions addressing technical and behavioural change. The energy intensity of producing new aluminium makes it a major contributor to GHG emissions, and there is also the need for aluminium in the future infrastructure stocks of both the industrialised and developing world.

Stabilising global average temperature at 2 °C above pre-industrial levels by 2050 has been translated into a general reduction of global GHG emissions of 50–85 % below levels in 2000 (IPCC 2007). Reducing the emissions from the aluminium sector by 50 % would entail a reduction in emissions *intensity* of nearly 85 % because of the expected threefold increase in global demand for aluminium by 2050 (IEA 2009).

Under these targets, Liu et al. (2013) analysed mitigation options for the aluminium industry through estimating demand in current and future in-use stocks. Their dynamic stock-driven model captured global flows of aluminium from reserves to post-consumer scrap (shown in Fig. 6.5) and calculated direct and indirect (energy-related) emissions arising from each process (not shown). A 50 % reduction in emissions compared to 2000 levels at 2050 was found to be only feasible with a combination of optimistic assumptions about rates of recycling, uptake of new technology, including CCS, and low levels of aluminium in stocks needed per person (200 kg/person or roughly double the current global average). The latter assumption implies a significant contraction in access to aluminium stocks per capita in developed countries.

If developing nations were to attain the 200–600 kg/person allocation of aluminium in stocks currently observed in developed nations, the aluminium industry would not be able to contribute proportionally to the 2 °C target. These results indi-

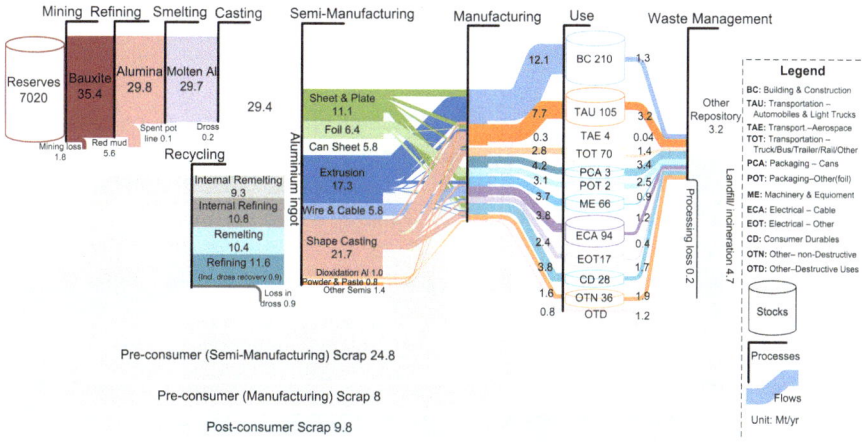

Fig. 6.5 The global aluminium material cycle (Graphic reproduced from Liu et al. 2013)

cate the magnitude of the problem at the largest scale and underline the importance of coupling mitigation strategies with material efficiency strategies. The following sections expand on a selection of strategies, again relating to aluminium stocks and their use in society.

4.1 Reducing Resource Use in the Product

Currently, global aluminium recycling is dominated by pre-consumer scrap (32.8 Mt). This is useful for reducing energy demand per unit of production by substituting for energy-intensive virgin aluminium. However, for every ton of aluminium finally consumed, approximately half a ton goes through various production processes, consuming energy and producing emissions but without ever forming a final product. These yield losses can be as high as 90 % in aircraft manufacture, and there is certainly room for improvement in reducing losses in production and using less material by design in the end product (e.g. 'lightweighting'). Allwood and Cullen (2012) have a number of other practical suggestions about reducing material and energy wasted in production in general, including diverting scrap 'blanks' to making smaller components prior to recycling and reusing components rather than recycling, e.g. steel I beams in construction can readily be recovered from demolition for direct use in new building construction.

Post-consumer aluminium scrap recycling has the potential to significantly lower total energy use and emissions, reducing energy intensity by 90 % (IEA 2009), but at present post-consumer scrap (9.8 Mt) is available mainly in the form of used beverage cans and end-of-life vehicles with 45 % of post-consumer aluminium going to waste or other repositories.

4.2   Changing the Demand for Stocks in Providing Services

Ultimately, in-use stocks provide a service and the question here is: can we use fewer stocks by using them more to obtain the same level of service? Answering this has less to do with the consumption of a particular material, like aluminium, and more to do with the characteristics and lifetime of the stock in-use. We may also seek to change behaviours in the use of stocks.

We look at the example of the stock of US automobiles in which aluminium is increasingly a material component (Ducker Worldwide 2008). Even with optimistic improvements in the efficiency of aluminium production, embodied energy in automotive parts is likely to increase (Cheah et al. 2009), but there are options for changing the way auto stocks are used.

For example, collective consumption in the form of car sharing reduces the need for people to own vehicles in exchange for the inconvenience of not having a private vehicle on demand. An analysis of the US car sharing market found members of carshare schemes reduced their number of cars per household, and on aggregate, every carshare vehicle took between 9 and 13 private vehicles off the road (Martin et al. 2010). Pauliuk and Müller (2014) also found that reducing vehicle ownership by a third, reducing vehicle use by 20 % and shifting to smaller vehicles could reduce emissions in the Chinese passenger vehicle fleet by more than 40 % at 2050 (an effect that would be greater than doubling fuel efficiency in the same period).

Another way to provide services from less stock is to have longer lived stocks and infrastructure that are more intensely used. This is counter to the current practice of planned obsolescence but is compatible with business models using adaptable design (Allwood and Cullen 2012) and more radical approaches (Chap. 8). It is sometimes argued that newer technology introduces greater operational efficiencies and so faster stock turnover is an effective mitigation strategy. However, this needs to be considered in conjunction with the emissions embodied in the stocks. Kagawa et al. (2011) demonstrated that, even based on technology from the 1990s and even when considering the longevity of less efficient vehicles, extending vehicle lifetimes in Japan has a net benefit in terms of reducing GHG emissions. Further examples are discussed in Chap. 8.

4.3   Timing

The current impacts of climate change and the prospect of yet higher global temperatures and more extreme weather should already instil urgency in policymakers to institute climate adaptation and mitigation measures. There are also long-term benefits of early action that interact with large-scale urbanisation and infrastructure growth. For example, implementing more stringent building codes now that anticipate and avoid the impacts of future climate events can have a net economic benefit over the long term (Baynes et al. 2013; Wang et al. 2015). This raises the importance of timing in effective policymaking.

Delaying mitigation action can have long-lived consequences. Retaining or augmenting carbon- or energy-intensive technology stocks locks in their emissions intensity for at least the life of those assets (IPCC 2014a). Where stocks operate in a system, e.g. roads and private transport, the inertia against change can be even greater. In many sectors, the technology to achieve cuts in emissions is already available (Pacala and Socolow 2004; IPCC 2014b), which undermines the contrary view that it may be better to wait for yet more efficient technology and hold off investment until that is available.

Returning to the example of the aluminium industry, Liu et al. (2013) explored four options to reduce resource use in products: scrap collection, minimising losses, efficiency improvement and decoupling emissions from electricity supply. They also considered three high-level scenarios of reducing the demand for stocks by assuming different saturation levels for global in-use aluminium per capita. They found that the effectiveness of mitigation options depended heavily on the stock dynamics and the timing of stock creation and scrap availability. Among the options they simulated for reducing emissions in aluminium production was CCS in decarbonising electricity. Introducing efficiency measures and CCS early (before 2030) had a greater effect because emissions related to aluminium in new building, transport and communication infrastructure were reduced. Conversely, later in the century, maximising scrap collection had an increasing benefit as earlier cohorts of stock came to the end of their lifetime.

5 The Socio-economic Metabolism Framework and Wealth

Building up the infrastructure capacity, as an essential component of wealth, also involves energy and emissions. A dynamic metabolic analysis properly represents both physical wealth in in-use stocks and the material and energy flows needed to create and maintain those stocks. Economic or physical flow measures may represent growth and development, but they are insufficient to understand long-term change and the lasting effect of wealth creation. On the basis of environmental and economic flows alone, developing nations appear highly materialised compared with industrialised countries and have lower productivity, but a more balanced assessment, taking into account physical stocks and infrastructure, can provide information on their relative socio-economic situation. In reporting and modelling sustainable development, physical in-use stocks are an essential complement to the current information on flows in the system of environmental and economic accounting.

There are conceptual parallels between the economic and environmental accounting systems even if their valuation and methodology differ. In the UN system of national accounting (European Commission et al. 2009), economists measure income as separate from wealth. Although the income measure of GDP is commonly misperceived as a measure of wealth, there is a separate ledger of assets and liabilities from which 'net worth' is calculated. This is entirely a calculation on values of capital stocks that has not been fully translated into the environmental extensions of national accounts. There is a substantial literature on the valuation of natural and anthropogenic capital, but there is a strong tendency to use a common denominator of monetary value or utility. Important physical information is then lost. For example, there is a world of difference between having a small amount of high quality, recently built road, and a much larger quantity of ageing, highly depreciated road that will soon be due for repair or replacement; the way capital value is determined means these two situations could be equivalent in net worth but they are very different in the services they provide and in their long-term material and energy requirements.

Concentrating on the flow measure of GDP is rightly criticised as a false measure of human well-being, which might be better defined by the accumulation of capacity to provide employment, food, education, safety and security, an approach explored in Chap. 8. That capacity corresponds more with the quantity and quality of infrastructure, productive and non-productive capital stocks. Stocks are not merely the accounting residuals of the net difference in bulk material flows. Through the services they provide, stocks are indicators of physical wealth, and our ability (through flows) to maintain and sustain that wealth is equally an important dimension of reporting and modelling the physical aspect of human well-being.

Just as seeking greater wealth by increasing GDP can produce perverse outcomes (Costanza et al. 2014), appraising environmental performance through flow measures alone is insufficient. For example, the metric of domestic material consumption (DMC) is used widely as a macroeconomic indicator of material requirements. Developing countries will always have a relatively high DMC until they attain sufficient infrastructure and capital stock to satisfy the physical demands of their socio-economic aspirations. Evidence for this comes from growth in DMC of developing nations in the Asia-Pacific region (UNEP and CSIRO 2013); for example, the DMC per capita of China has increased by 640 % over the last 40 years.

Most developed countries have achieved relative material decoupling (lower DMC/GDP) over the last 30 years (Giljum et al. 2014). Müller et al. (2006) and Wiedmann et al. (2013) have both suggested that industrialised nations have lower DMC because they have already established their major infrastructure and their population has grown more slowly than developing countries or has even saturated. However, the interpretation may not be so simple. Matthews et al. (2000) calculated NAS as the residual between input and output flows for a sample of developed nations with established infrastructure and concluded that the NAS is still 8–12 tons/capita per year. The authors attributed this to a combination of factors including lower occupancy, urban expansion and affluence.

In this chapter, we have used concepts from industrial ecology to frame problems of the long-term future in terms of both stocks and flows and show how solutions are substantially influenced by the creation of physical wealth in stocks and how stocks can mitigate (or exacerbate) impacts and deliver services to society. The socio-economic metabolism framework is intended to represent the interaction of stocks across sectors and enable a more complete integrated assessment of policy options. As in many areas of industrial ecology, data availability is a hurdle, but the socio-economic metabolism framework discussed in this chapter provides a structured way for interpreting statistics on physical stocks; automated data collection and informatics provide the opportunity to capture physical transactions alongside monetary exchanges.

References

Allwood, J. M., & Cullen, J. M. (2012). *Sustainable materials: With both eyes open*. Cambridge: UIT Cambridge.

Allwood, J. M., Cullen, J. M., & Milford, R. L. (2010). Options for achieving a 50 % cut in industrial carbon emissions by 2050. *Environmental Science & Technology, 44*, 1888–1894.

Angel, S., Parent, J., Civco, D., Blei, A., & Potere, D. (2010). *A planet of cities: Urban land cover estimates and projections for all countries, 2000–2050* (Lincoln Institute of Land Policy Working Paper). New York: Lincoln Institute of Land Policy.

Ayres, R. U. (1989). Industrial metabolism. In J. H. Ausubel & H. E. Sladovich (Eds.), *Technology and environment* (pp. 23–49). Washington, DC: National Academy of Engineering.

Baccini, P., & Bader, H. P. (1996). *Regionaler Stoffhaushalt. Erfassung, Bewertung und Steuerung*. Heidelberg: Spektrum Akademischer Verlag.

Baccini, P., & Brunner, P. (1991). *Metabolism of the anthroposphere*. Berlin: Springer.

Baynes, T. M., West, J., & Turner, G. M. (2009). Design approach frameworks, regional metabolism and scenarios for sustainability. In M. Ruth & B. Davidsdottir (Eds.), *Dynamics of regions and networks in industrial ecosystems*. Northampton: Edward Elgar.

Baynes, T., West, J., McFallan, S., Wang, C., Khoo, Y., Herr, A., Langston, A., Beaty, M., Li, Y., & Lau, K. (2013). Net benefit assessment of illustrative climate adaptation policy for built assets. In J. Piantadosi, R. S. Anderssen, & J. B. (Eds.), *MODSIM 2013: 20th international congress on modelling and simulation, Adelaide, Australia, 1–6 December* (pp. 2207–2213). Adelaide: Modelling and Simulation Society of Australia and New Zealand.

Cheah, L., Heywood, J., & Kirchain, R. (2009). Aluminum stock and flows in U.S. passenger vehicles and implications for energy use. *Journal of Industrial Ecology, 13*, 718–734.

Costanza, R., Kubiszewski, I., Giovannini, E., Lovins, H., McGlade, J., Pickett, K. E., Ragnarsdóttir, K. V., Roberts, D., De Vogli, R., & Wilkinson, R. (2014). Development: Time to leave GDP behind. *Nature, 505*, 283–285.

Davis, S. J., Caldeira, K., & Matthews, H. D. (2010). Future CO2 emissions and climate change from existing energy infrastructure. *Science, 329*, 1330–1333.

Dawson, D. A., Purnell, P., Roelich, K., Busch, J., & Steinberger, J. K. (2014). Low carbon technology performance vs infrastructure vulnerability: Analysis through the local and global properties space. *Environmental Science & Technology, 48*, 12970–12977.

Ducker Worldwide. (2008). *Aluminum Association Auto and Light Truck Group, 2009 update on aluminum content in North American light vehicles, Phase I*. Troy: Ducker Worldwide.

European Commission, International Monetary Fund, Organisation for Economic Co-operation and Development, & United Nations & World Bank. (2009). *System of national accounts 2008*. New York: United Nations Statistics Division.

EUROSTAT. (2009). *Economy-wide material flow accounts: Compilation guidelines for reporting to the 2009 Eurostat questionnaire*. Luxembourg: Office for Official Publications of the European Communities.

Fischer-Kowalski, M. (1998). Society's metabolism. *Journal of Industrial Ecology, 2*, 61–78.

Fischer-Kowalski, M., & Weisz, H. (1999). Society as hybrid between material and symbolic realms: Toward a theoretical framework of society-nature interaction. *Advances in Human Ecology, 8*, 215–252.

Forrester, J. W. (1958). Industrial dynamics: A major breakthrough for decision makers. *Harvard Business Review, 36*, 37–66.

Giesekam, J., Barrett, J., Taylor, P., & Owen, A. (2014). The greenhouse gas emissions and mitigation options for materials used in UK construction. *Energy and Buildings, 78*, 202–214.

Giljum, S., Dittrich, M., Lieber, M., & Lutter, S. (2014). Global patterns of material flows and their socio-economic and environmental implications: A MFA study on all countries world-wide from 1980 to 2009. *Resources, 3*, 319–339.

Global Energy Assessment. (2012). *Global energy assessment – Toward a sustainable future*. Cambridge/New York/Laxenburg: Cambridge University Press and the International Institute for Applied Systems Analysis.

Hallegatte, S. (2009). Strategies to adapt to an uncertain climate change. *Global Environmental Change, 19*, 240–247.

Hertwich, E. G., & Peters, G. P. (2009). Carbon footprint of nations: A global, trade-linked analysis. *Environmental Science & Technology, 43*, 6414–6420.

IEA. (2009). *IEA energy technology transitions for industry: Strategies for the next industrial revolution*. Paris: International Energy Agency.

IPPC. (2006). *2006 IPCC guidelines for national greenhouse gas inventories* (Vols. 1–5). Retrieved April 10, 2015, from http://www.ipcc-nggip.iges.or.jp/public/2006gl/index.html

IPCC. (Ed.) (2007). Contribution of Working Group III to the fourth assessment report of the Intergovernmental Panel on Climate Change. Cambridge/New York: Cambridge University Press.

IPCC. (2014a). AR5 summary for policymakers. In C. B. Field, V. R. Barros, D. J. Dokken, K. J. Mach, M. D. Mastrandrea, T. E. Bilir, M. Chatterjee, K. L. Ebi, Y. O. Estrada, R. C. Genova, B. Girma, E. S. Kissel, A. N. Levy, S. MacCracken, P. R. Mastrandrea, & L. L. White (Eds.), *Climate change 2014: Impacts, adaptation, and vulnerability. Part A: Global and sectoral aspects. Contribution of Working Group II to the fifth assessment report of the Intergovernmental Panel on Climate Change* (pp. 1–32). Cambridge/New York: Cambridge University Press.

IPCC. (2014b). Climate change 2014: Mitigation of climate change. In O. Edenhofer, R. Pichs-Madruga, Y. Sokona, E. Farahani, S. Kadner, K. Seyboth, A. Adler, I. Baum, S. Brunner, P. Eickemeier, B. Kriemann, J. Savolainen, S. Schlömer, C. von Stechow, T. Zwickel, & J. C. Minx (Eds.), *Contribution of Working Group III to the fifth assessment report of the Intergovernmental Panel on Climate Change*. Cambridge/New York: Cambridge University Press.

Kagawa, S., Nansai, K., Kondo, Y., Hubacek, K., Suh, S., Minx, J., Kudoh, Y., Tasaki, T., & Nakamura, S. (2011). Role of motor vehicle lifetime extension in climate change policy. *Environmental Science & Technology, 45*, 1184–1191.

Kushnir, D., & Sandén, B. A. (2012). The time dimension and lithium resource constraints for electric vehicles. *Resources Policy, 37*, 93–103.

Lennox, J. A., Turner, G., Hoffman, R. B., & McInnis, B. C. (2005). Modelling Australian basic industries in the Australian stocks and flows framework. *Journal of Industrial Ecology, 8*, 101–120.

Lenzen, M., Dey, C., & Foran, B. (2004). Energy requirements of Sydney households. *Ecological Economics, 49*, 375–399.

Liu, G., Bangs, C. E., & Müller, D. B. (2013). Stock dynamics and emission pathways of the global aluminium cycle. *Nature Climate Change, 3*, 338–342.

Loulou, R., Remne, U., Kanudia, A., Lehtila, A., & Goldstein, G. (2005). *Documentation for the TIMES model*. Paris: International Energy Network, ETSAP.

Martin, E., Shaheen, S. A., & Lidicker, J. (2010). Impact of carsharing on household vehicle holdings: Results from North American shared-use vehicle survey. Transportation research record. *Journal of the Transportation Research Board of the National Academies*, Washington, DC, *2143*, 150–158.

Matthews, E., Amann, C., Bringezu, S., Hüttler, W., Ottke, C., Rodenburg, E., Rogich, D., Schandl, H., Van, E., & Weisz, H. (2000). *The weight of nations-material outflows from industrial economies*. Washington, DC: World Resources Institute.

Meinshausen, M., Meinshausen, N., Hare, W., Raper, S. C. B., Frieler, K., Knutti, R., Frame, D. J., & Allen, M. R. (2009). Greenhouse-gas emission targets for limiting global warming to 2°C. *Nature, 458*, 1158–1162.

Müller, D. B. (2006). Stock dynamics for forecasting material flows—Case study for housing in The Netherlands. *Ecological Economics, 59*, 142–156.

Müller, D. B., Bader, H.-P., & Baccini, P. (2004). Long-term coordination of timber production and consumption using a dynamic material and energy flow analysis. *Journal of Industrial Ecology, 8*, 65–88.

Müller, D. B., Wang, T., Duval, B., & Graedel, T. E. (2006). Exploring the engine of anthropogenic iron cycles. *Proceedings of the National Academy of Science, 103*, 16111–16116.

Müller, D. B., Liu, G., Løvik, A. N., Modaresi, R., Pauliuk, S., Steinhoff, F. S., & Brattebø, H. (2013). Carbon emissions of infrastructure development. *Environmental Science & Technology, 47*, 11739–11746.

OECD/IEA & Eurostat. (2005). *Energy statistics manual*. Paris: OECD/IEA/Eurostat.

Pacala, S., & Socolow, R. (2004). Stabilization wedges: Solving the climate problem for the next 50 years with current technologies. *Science, 305*, 968–972.

Pauliuk, S., & Müller, D. B. (2014). The role of in-use stocks in the social metabolism and in climate change mitigation. *Global Environmental Change, 24*, 132–142.

Peters, G. P. (2010). Carbon footprints and embodied carbon at multiple scales. *Current Opinion in Environmental Sustainability, 2*, 245–250.

Piketty, T. (2013). *Capital in the twenty first century*. Harvard: Harvard University Press.

Ramaswami, A., Weible, C., Main, D., Heikkila, T., Siddiki, S., Duvall, A., Pattison, A., & Bernard, M. (2012). A social-ecological-infrastructural systems framework for interdisciplinary study of sustainable city systems. *Journal of Industrial Ecology, 16*, 801–813.

Roelich, K., Dawson, D. A., Purnell, P., Knoeri, C., Revell, R., Busch, J., & Steinberger, J. K. (2014). Assessing the dynamic material criticality of infrastructure transitions: A case of low carbon electricity. *Applied Energy, 123*, 378–386.

Stehfest, E., van Vuuren, D., Kram, T., & Bouwman, L. (Eds.). (2014). *Integrated assessment of global environmental change with IMAGE 3.0: Model description and policy applications*. The Hague: PBL Netherlands Environmental Assessment Agency.

Steinberger, J. K., Timmons, R. J., Peters, G. P., & Baiocchi, G. (2012). Pathways of human development and carbon emissions embodied in trade. *Nature Climate Change, 2*, 81–85.

UN. (2012). *United Nations General Assembly Resolution a/RES/66/288 – The future we want*. New York: United Nations.

UNEP & CSIRO. (2013). *Recent trends in material flows and resource productivity in Asia and the Pacific*. Bangkok: United Nations Environment Program.

UNFCCC. (2011). *The Cancun agreements: Outcome of the work of the Ad Hoc working group on long-term cooperative action under the convention (CP16 CMP6)*. New York: United Nations.

USDOE. (2011). *U.S. Department of Energy Critical Materials Strategy*. Washington, DC: U.S. Department of Energy.

Wang, C. -H., Baynes, T., McFallan, S., West, J., Khoo, Y. B., Wang, X., Quezada, G., Mazouz, S., Herr, A., Beaty, R. M., Langston, A., Li, Y., Wai Lau, K., Hatfield-Dodds, S., Stafford-Smith, M., & Waring, A. (2015). Rising tides: Adaptation policy alternatives for coastal residential buildings in Australia. *Structure and Infrastructure Engineering*, published online: 16 March. doi:10.1080/15732479.2015.1020500.

Wiedmann, T. O., Schandl, H., Lenzen, M., Moran, D., Suh, S., West, J., & Kanemoto, K. (2013). The material footprint of nations. *Proceedings of the National Academy of Sciences, 112*(20), 6271–6276.

Chapter 7
Stocks and Flows in the Performance Economy

Walter R. Stahel and Roland Clift

Abstract The performance economy is a concept which goes beyond most interpretations of a "circular economy": the focus is on the maintenance and exploitation of stock (mainly manufactured capital) rather than linear or circular flows of materials or energy. The performance economy represents a full shift to servicisation, with revenue obtained from providing services rather than selling goods. While the form of industrial economy which has dominated the industrialised countries since the industrial revolution is arguably appropriate to overcome scarcities in a developing economy, the performance model is applicable in economies close to saturation, when the quantities of new goods entering use are similar to the quantities of goods being scrapped at the end of life.

Key elements of the performance economy are re-use and re-manufacturing, to maintain the quality of stock and extend its service life by reducing material intensity, i.e. the material flow required to create and maintain the stock. Because material flows represent costs which reduce the revenue from service provision, business models inherent in the performance economy support the macro-level objective of extending service life and thereby minimising material intensity. Product life in the performance economy is limited by technological improvements in the efficiency of manufactured capital rather than by damage, wear or fashion.

Re-use and re-manufacturing tend to be more labour-intensive and less capital-intensive than virgin material production or primary manufacturing. This enables re-use and remanufacturing to be economically viable at smaller scales. It also enables these activities to substitute labour for energy, reversing the trend which has characterised industrial economies and offering ways to alleviate current environmental, economic and global challenges; i.e. to make the economy more sustainable. However, there are significant barriers to adoption of the performance economy model, partly because economic and business models generally focus on flows (GDP or added value) rather than prioritising the quality, value and use of stock. Promoting the performance model may require a complete re-think of public policy,

W.R. Stahel (✉)
The Product-Life Institute, Geneva, Switzerland
e-mail: wrstahel2014@gmail.com

R. Clift
Centre for Environmental Strategy, University of Surrey, Guildford, UK
e-mail: r.clift@surrey.ac.uk

© The Author(s) 2016
R. Clift, A. Druckman (eds.), *Taking Stock of Industrial Ecology*,
DOI 10.1007/978-3-319-20571-7_7

away from subsiding to taxing use of non-renewable resources and away from taxing the use of renewable resources, of which labour is possibly the most important. Recent analyses of the social costs of unemployment and potential social benefits of a more resource efficient performance economy provide some of the evidence supporting a shift from flow to stock management.

Keywords Circular economy • Flows • Goods as services • Manufactured capital • Molecules as services • Performance economy • Re-use • Recycling • Remanufacturing • Reprocessing • Service life • Stocks

1 Introduction

Economic activities, particularly trade, are usually characterised in terms of flows of goods, energy and services. Many areas of industrial ecology focus on flows, and concepts such as the "circular economy" are framed in terms of recycling and re-use flows. However, wealth itself is a stock, not a flow: the wealth of societies is based on their stocks of goods and capital. Arguably, the quality of life in a developed society depends more on the quantity and quality (Q&Q) of its stock than on the flows through the economy. Stocks represent accumulated flows. Therefore, the flows needed to develop and maintain capital stocks are important, particularly in developing economies; the relationship between flows and stocks is explored in Chap. 6. But this chapter explores some of the insights revealed by focussing on managing and using the stocks themselves rather than flows. This is one of the central principles in the concept of the "performance economy" (Stahel 2010).

Various authors (e.g. Forum for the Future 2015) have proposed categorisations of the stocks available to a society. Most categorisations propose five forms of capital; for example, in the context of this chapter:

- Natural capital
- Cultural capital
- Human capital
- Manufactured capital
- Financial capital.

Natural capital can be treated as a Global Commons, using a stock management approach, such as sustainable fishing and forestry. If a flow maximisation approach is applied instead, natural capital will inevitably be degraded, repeating the "Tragedy of the Commons"; obvious examples are over-fishing, loss of biodiversity and accumulation of foreign substances such as toxins, salts and plastic debris.

Maintaining quality with its associated tensions is also a widespread concern for *Cultural capital*; the UNESCO world heritage programme is regularly faced with the dilemma of balancing protecting sites against their economic exploitation.

Cultural capital includes immaterial stocks such as music and local traditions; these can be shared but commercialised sharing may lead to loss of quality through dilution.

Human capital is intrinsically linked with intangible *acquired capital; i.e. skills and capabilities*. Human capital is the only resource whose quality can be improved through education and training. Developing and maintaining human capital as a renewable resource is one of the issues explored in this chapter.

The Role of Incentives in Flow Versus Stock Management – The Case of Reforestation

If the wealth of societies is based on the quantity *and* quality (Q&Q) of capital or stocks, a delicate balance is necessary to maintain stock quantity without compromising stock quality. Growth can be defined as increase in Q&Q of stock rather than flow. To achieve this, policy measures need to shift from a flow to a stock approach: witness reforestation programmes in countries like Nepal, which are directed and financed by Western organisations.

Local people are paid to grow seedlings, plant them on barren slopes, water them regularly and protect them by fences - a typical production-driven flow approach. Local people respond in a flow-focused way by minimising their work load, which means reforesting slopes near villages rather than the slopes presenting the biggest hazards (the stock approach). As villagers have conflicting priorities, such as feeding their goats and cattle and working as tourist guides, they will often neglect the maintenance of the fences, and goats will eat the young trees. A new reforestation programme will follow, financed by well-meaning sponsors.

In a stock management approach, the villagers are instead paid a modest fee for each tree on the village property. The driver for reforestation is then villagers looking for a higher donor income by increasing their stock of trees, which includes protecting young trees from being eaten by animals as much as planting new trees. The barrier is that donors, including the World Bank, generally do not pay for stock management but only for flow. The same applies to preventive measures versus post-disaster repairs.

The focus in this chapter is on *manufactured capital*: i.e. material goods and fixed assets. This includes infrastructure (for energy distribution, communications, transport, and water and other services), buildings (industrial, commercial, institutional and domestic), equipment (both "productive" and appliances used by "consumers") and durable consumer goods (including, for example, furniture and garments). Focussing on the stock brings out the importance of something which is often overlooked: durability or service life of manufactured capital. This is explored later in this chapter.

2   The Circular Economy – "Loop", "Lake" and "Performance" Models

Distinguishing between capital stock and flows opens up useful perspectives on the idea of a "circular economy" which aspires to replace the linear "once through" *industrial economy* which has dominated business in the industrialised countries since the industrial revolution. In the linear economy, the focus is on management of throughput flows. On the macro-economic level, the performance of the industrial economy is judged by measuring the sum of all flows (GDP); on the micro-economic level, by calculating the value added to the flows. Its optimisation stops at the point of sale where the responsibility for operating and disposing of the goods is passed to the buyer.

Arguably, the industrial economy is the best strategy to increase stock and expand economic activity to overcome scarcities of food, housing, infrastructure and/or equipment, as they exist in many developing countries. However, in markets near saturation, so that the number of new goods is similar to the number of scrapped goods, the relevance of the paradigm of economic growth has been questioned (e.g. Jackson 2009); in the absence of quantum leap innovation, the circular economy is a more viable business model than the industrial economy with regard to environmental, economic and social factors. Even where technology quantum leaps do take place through rapid innovation, the circular economy will complement the industrial economy.

The *circular economy* is based on value preservation, not value added. The basic elements are shown schematically in Fig. 7.1. *The Reuse Loop* includes second-hand markets (from garage sales and flea markets to eBay) as well as commercial and private reuse of goods (e.g. refilling of beverage containers, reuse and resale of garments). These activities are usually carried out locally. *Loop 1*, labelled *Remanufacturing*, includes repair, remanufacturing and "upgrading" to meet new technological standards or to meet new fashion expectations (Smith and Keoleian 2004). Remanufacturing may be a local activity (e.g. refurbishing of domestic appliances or cars) or may be carried out via regional service centres. *Loop 2* represents *recycling* in which, rather than repairing or re-using manufactured goods and components, the product is reprocessed to recover secondary materials for return to the same use. *Reprocessing* may be a regional activity or may be part of a global supply system. Reprocessing includes operations such as recycling of paper and plastics, re-refining of fluids such as lubrication oils (Clift 2001) and, where practical, depolymerisation of polymers (Clift 1997). Some end-of-life goods and materials may go to other uses, such as export for re-use in other locations or "cascading" into lower specification applications (*downcycling*) including energy recovery, or may leak from the economic system as *waste*.

Different interpretations of the circular economy place different emphasis on the elements in Fig. 7.1. At its simplest (but also least profitable and materially efficient), the circular economy takes the form of the *Loop* economy, focussed on recycling (loop 2 in Fig. 7.1) in which the material of the product is not lost as waste but

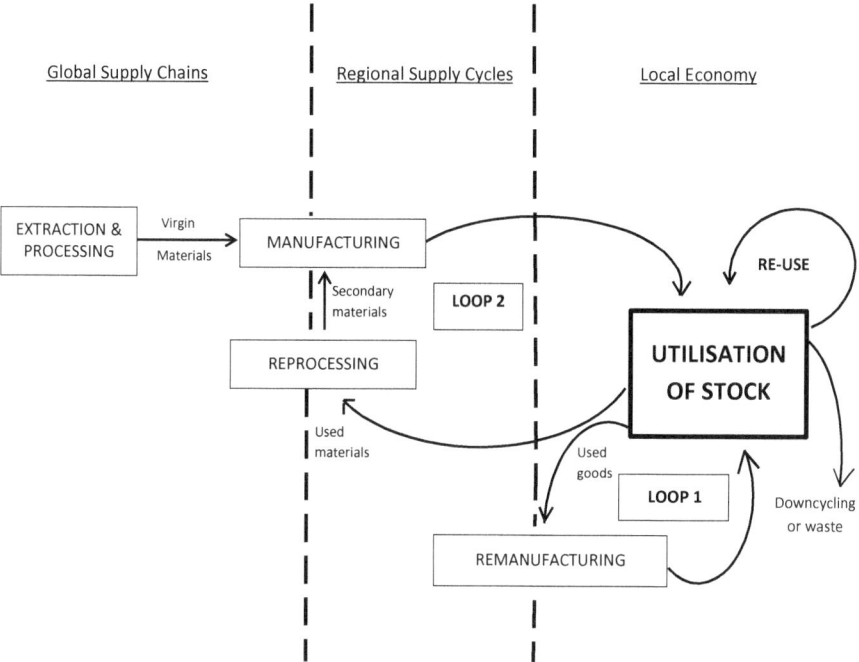

Fig. 7.1 The basic loops of a circular economy (Adapted from Stahel and Reday-Mulvey 1976)

is reprocessed for return to the same use. This is the interpretation embodied in China's "Circular Economy" laws, for example. In the Loop Economy, ownership changes with each loop, each owner aiming to achieve the highest profit when passing on material goods.

The "Loop" or "cradle-to-cradle" concept is still framed in terms of flows and therefore overlooks ways to optimise the physical and economic performance of the economy based on optimising the use of the stock. An example of good stock management is provided by the 2008 EU Waste Directive, which appropriately calls for the reuse and service-life extension of goods as priorities, putting waste prevention before waste management. The *Lake* economy uses the same loops as the Loop economy, but with a focus on value preservation through stewardship and without changes in ownership: The main economic actors in such an economy are here termed "fleet managers": they operate a fleet of similar goods, such as the vehicles of haulage or transport companies, and maintain their components (e.g. engines and tyres in the case of road vehicles). These owners may have their goods repaired or remanufactured by independent service companies. The focus is on managing the stock rather than the flows: "fleet managers" maintain ownership of their stock and therefore profit from the economic advantages of re-use and remanufacturing strategies.

If economic actors are selling performance, through business models such as "goods as services" or "molecules as services", which means earning revenue and profits from stocks instead of flows, they shift from the Lake Economy, with its focus on maintaining the stock, to the *Performance* Economy which focuses on maximising the value obtained from using the stock. The Performance Economy demands an internalisation of the costs of waste and of risk over the full service-life of goods, which in turn are substantial financial incentives to include waste prevention and loss prevention at all stages in the product cycle from design to decommissioning.

The Performance Economy is primarily driven by competitiveness, the Lake Economy by long-term operation and maintenance cost optimisation and the Loop Economy by (environmental) legislation. Some of the business implications of the shift to the Performance model are explored by Stahel (2010) and in Sect. 4 of this chapter.

The stock perspective is routine for infrastructure and buildings but less familiar for other forms of manufactured capital. The shift from a flow to a stock perspective is enabled when economic actors (companies, consumers and public entities) assume longer-term ownership or stewardship of, for example, fleets of vehicles or goods, changing their business approach from the bigger-better-faster-safer model of an industrial flow economy to the functional view of goods of a Lake Economy. The motivation for commercial actors is usually to reduce operating costs (for example, retreading truck tyres by haulage companies, remanufacturing diesel engines), whereas public actors (armed forces and public administrations like railways, NASA) seek to reduce long-term system costs (mothballing of warships, "cemeteries" of aircraft for access to spare parts). For consumers, the motivation may be a personal relationship with goods (the "teddy bear" effect which leads individuals to keep personal souvenirs such as watches or pens, or family heritage objects such as paintings or vintage cars). By retaining the ownership of the goods and their embodied resources, fleet managers gain a resource security with regard to both future availability of resources and commodity prices. Expected scarcity of some critical materials therefore provides another driver to take the stock perspective.

If loops involve professional services, transaction costs occur, adding to the costs and often influencing the choice of the next owner: for example, sales such as buildings, domestic premises and artworks require fees to individuals or specialist dealers. However, some OEMs take back their own goods, disassemble them and reuse components as service parts, a strategy pursued by many IT manufacturers, or remanufacture and remarket them in exchange for faulty products. Such service exchange systems are used by some European car manufacturers: damaged car engines and gearboxes which cannot be repaired locally are returned to the OEM in exchange for an OEM-remanufactured product[1]; Sony Computer Entertainment Europe offers a remanufactured exchanged product when customers return a faulty product for repair, in order to reduce the time a customer is without the product. Further illustrative examples are discussed in later sections.

---

[1] VW annually remanufactures 50,000 engines and the same number of gearboxes in a dedicated plant located in Kassel.

3 Remanufacturing, Reprocessing and Product Life

3.1 Material Intensity and Product-Service Intensity

To understand the importance of the quality and durability of stock, it is convenient to use a formulation introduced by the IPCC in the 5th Assessment Report (2014).[2] For a sector producing materials and products for stocks that deliver quantifiable services, the energy use (e_p) and associated GHG emissions (g_p) over a specified accounting period (for example, GJ per year and tonnes CO_2e per year) can be broken down in a form of Kaya (1990) relationship as:

$$e_p = (e/p) \times (p/S) \times (S/d) \times d \tag{7.1}$$

$$g_p = (g/e) \times e_p = (g/e) \times (e/p) \times (p/S) \times (S/d) \times d \tag{7.2}$$

where e represents the energy input to manufacturing and processing, e_p is the energy used specifically to produce the flow p of materials and products (e.g. tonnes per year) needed to maintain the stock S of the relevant manufactured capital (e.g. tonnes) and d is the quantity of service delivered in the time period through use of that capital (e.g. passenger-km per year for the personal transport sector).

These expressions are conceptual, but they reveal the significance of the different terms:

(g/e) is the *emission intensity* of the sector expressed as a ratio of GHG emissions to energy used. The emissions arise largely from energy use (directly from combusting fossil fuels, and indirectly through purchasing electricity and steam) and therefore depend most critically on the emission intensity of the background energy system of the economy where the goods in question are made. However, emissions also arise from industrial chemical reactions; in particular, producing cement, chemicals and non-ferrous metals leads to release of significant "process emissions" regardless of background energy sources.

(e/p) is the *energy intensity of production* . Approximately three quarters of industrial energy use worldwide is required to create materials from ores, oil or biomass, with the remaining quarter used in the downstream manufacturing and construction sectors that convert materials to products (IPCC 2014). In some cases, particularly for metals, e/p can be reduced by production from reused components or recycled material (see below) and can be further reduced by exchange of waste heat and/or by-products between sectors through industrial symbiosis.

(p/S) is the *material intensity* of the sector: the material flow required to create and maintain the stock.

---

[2] A form of this equation is given in IPCC (2014: 746) but the explanation and interpretation given here differs from that in IPCC (2014). In the notation used here, upper case letters denote stocks and lower case denote flows.

(*S/d*) is the *product-service intensity*; i.e. the quantity of stock required to deliver the required service.

Equation 7.1 has a number of important implications, underpinning the specific business strategies explored in Sect. 4. The emission intensity term, (*g/e*), depends primarily on the fuel mix used in the extractive, processing and manufacturing operations and the background economies in which they are located. The energy intensity of production, (*e/p*), which is the usual target of industrial improvement and innovation, is also obviously important. These terms underlie concerns over the "carbon leakage" resulting from the potential migration of primary manufacturing to economies with high carbon intensity and possibly lax environmental standards (see for example Clift et al. 2013) but the scope for reducing (*e/p*) is limited by technological and thermodynamic constraints (Allwood et al. 2012). We therefore concentrate here on the other terms in Eq. 7.1 which are key for the circular and performance economies: material intensity, (*p/S*), and product-service intensity, (*S/d*). These parameters are measures of the efficiency and quality of the stock: low values indicate good system performance. Reducing the product-service intensity means using capital goods more intensively, for example through "pooled" use of vehicles or appliances. Material intensity can be reduced by design measures such as "lightweighting" and, importantly but less obviously, is also dependent on product life, re-use, remanufacturing and recycling.

3.2 Remanufacturing and Reprocessing

To reveal the significance of material and product-service intensities and product life, we explore the flows needed to maintain the stock of manufactured capital, S, as shown in Fig. 7.2. The "Re-use" loop (see Fig. 7.1) is treated here as an activity needed to keep the stock in use; it is therefore not shown in Fig. 7.2 because inputs to and losses from re-use are included in i_S (see below). Of the flow of goods into stock, p (as in Eqs. 7.1 and 7.2), a fraction r_1 is remanufactured and the balance is newly manufactured goods incorporating a fraction r_2 of recycled material. The input of primary material to the product system is therefore $(1-r_2)(1-r_1)p$. Unsurprisingly, increase in r_1 or r_2 reduces the need for primary material.

More interesting insights emerge from examining other relationships within the service system, illustrated by Fig. 7.2. The outflow from stock at the end of its (first) service life is denoted by q. A fraction f_1 is routed to remanufacturing, a fraction f_2 to recycling and the balance leaves the product system as downcycled goods or materials or as waste. Material losses, degradation and/or contamination in remanufacturing and reprocessing are thermodynamically inevitable and also occur in the associated logistics, so that $r_1p < f_1q$ and $r_2(1-r_1)p < f_2q$. The relatively high population density in urban areas makes the logistics of collection easier and therefore supports the development of a circular economy (see, for example van Berkel et al. 2009; Kennedy et al. 2011). The main losses and degradation usually occur in

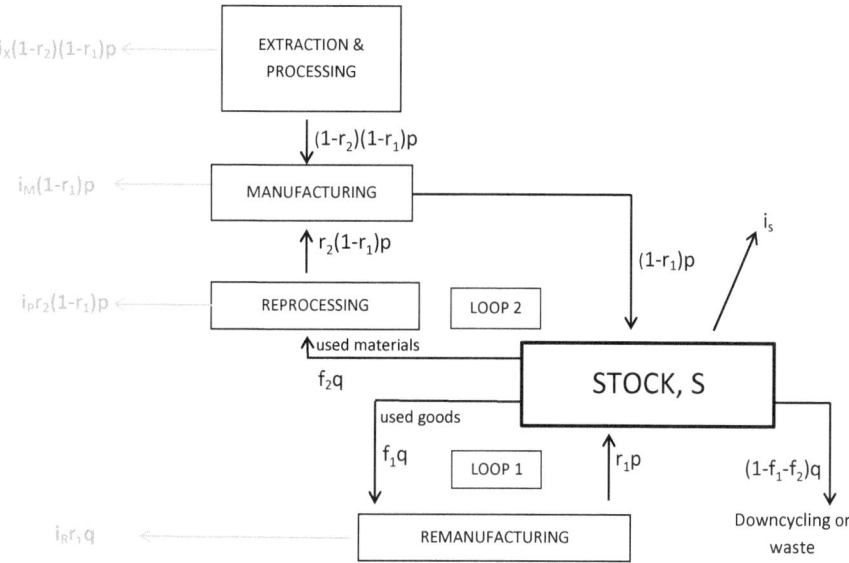

Fig. 7.2 Remanufacturing and recycling to maintain manufactured stock (adapted from Clift et al. 2009 and Clift and Allwood 2011). Interventions per unit of output from operations: i_R remanufacturing, i_M manufacturing, i_P reprocessing/recycling, i_X extraction and primary processing

reprocessing. It follows that the economics of the circular economy depend on using the shorter, more local and less dissipative loops (i.e. re-use and remanufacturing) and keeping materials separate to minimise contamination.

As well as the material flows within the system, Fig. 7.2 shows the interventions: i.e. the exchanges across the system boundary. These interventions may be emissions from the system (for example of greenhouse gases or waste material) but the analysis also applies to inputs (energy or labour, for example) or costs. They are associated with the operations shown in Fig. 7.2, expressed in proportion to the quantity of material produced. Interventions, i_S, are also associated with use of the stock to deliver the service. For the whole service system, the total interventions, i_T, are:

$$i_T = [i_R r_1 + (1 - r_1)\{i_M + r_2 i_P + (1 - r_2)i_X\}]p + i_S \qquad (7.3)$$

We focus initially on the first term in Eq. 7.3, i.e. the interventions associated with the material flows. If the objective is to reduce i_T, the single most important action is to reduce the material throughput, p; this is explored below. For some materials, the energy use, emissions, material degradation and/or costs of reprocessing may be prohibitive (see, for example, Allenby 1999; Allwood 2014). Reprocessing is then undesirable on cost or environmental grounds, representing a limit to the circular economy in any of its forms. More commonly, and particularly for manufactured goods, the energy use and emissions are least for remanufacturing and greatest for primary material production; i.e.

$$i_R < i_P < i_X \qquad (7.4)$$

while labour input is in the reverse rank order. This underlines the business models for the performance economy explored in Sect. 4: whilst reprocessing is commonly (but not universally) beneficial compared to primary production, remanufacturing is to be preferred over reprocessing. In addition to energy and resource efficiency, remanufacturing has the benefit of substituting labour input for energy input.

3.3 Product Life

It was noted above that the circular economy model is most appropriate for a mature economy, with markets near saturation, where the stock of manufactured capital has already been built up. Under these circumstances the change of stock over time is relatively small, so that $dS/dt \approx 0$ and $p \approx q$. This approximation also applies to personal property, for example clothing (where the stock is limited by storage space) and furniture (limited by living area). The throughput of material to maintain stock with long life is then:

$$p = q = S / T \qquad (7.5)$$

where T is the average service life of the stock. As noted above, the priority is to reduce p. From Eqs. 7.3, 7.4, and 7.5, the options for reducing energy use and associated emissions are, in priority order:

1. Extend service life, T, to reduce throughput, p
2. Intensify use to reduce stock needed, S
3. Increase the proportion of post-use products remanufactured, f_1
4. Increase the proportion of post-use products reprocessed, f_2
5. Reduce the inputs required for manufacturing, i_M
6. Reduce the inputs required for reprocessing, i_P
7. Reduce the inputs required for remanufacturing, i_R
8. Reduce the inputs required for primary material production, i_X.

This rank order underlines why focussing on stock leads to the priorities which underpin the performance economy model (see Sect. 4 below) but which are overlooked in many policy measures. It also leads to useful insights into sustainable consumption concerning the importance of durable quality goods rather than disposable purchases (Clift et al. 2013).

For products with a short service life, such as beverage containers, the priority is on the efficiency of loop 2; i.e. on reprocessing/recycling rather than remanufacturing. Figure 7.3 shows a simple recycling loop. The fraction of material returned to

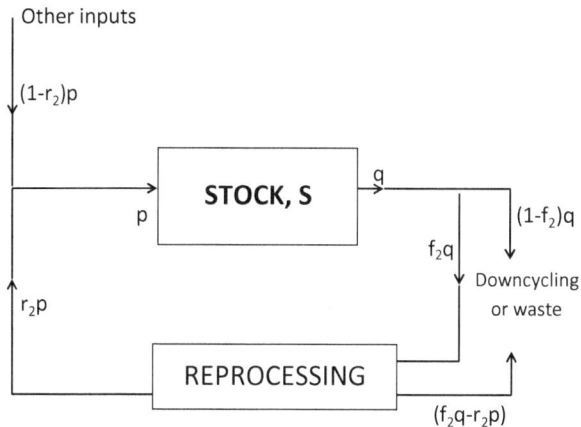

Fig. 7.3 Simple recycling loop

use after each loop is r_2. After n loops, the fraction remaining in use is r_2^n, which reduces rapidly even when r_2 is relatively high. The average life of the material leaving use in terms of number of uses, $\tilde{n}$, is:[3]

$$\tilde{n} = 1/(1-r_2) \tag{7.6}$$

For paper, recycling damages the fibres so that, to keep $\tilde{n}$ from becoming too high, r_2 is limited and a minimum introduction of new material is needed to maintain the paper properties (Hart et al. 2005). Even for such a valuable material as nickel, used in many alloys but also in items such as fashion jewellery, only 55 % returns around the loop (i.e. $r_2 = 0.55$) so that 30 % remains after two loops and only 17 % after three. In this particular case, about 2/3 of the losses arise from dissipative use $((1-f_2)q$ in Fig. 7.3) and the remaining 1/3 from losses during reprocessing, mainly of alloy bars used to reinforce concrete structures (Bihoux, quoted in Levy and Aurez (2014)). For beverage cans, which typically have a life of 3 weeks from canning plant to disposal, even with a high recycling rate of 75 % half the stock is lost after 6 months in use and effectively all is lost after a year. Even with a highly optimistic recycling rate of 90 %, the metal stock is effectively lost in less than 2 years of use. By contrast, and emphasising the importance of re-use to extend service life (T), refillable glass bottles are typically refilled 27 times before reprocessing, so that the first recycling occurs after about 1 year and a half (Stahel 2010).

In addition to the interventions required to produce and maintain the material stock, Eq. 7.3 includes those needed to operate it:

$$i_S = (i/d) \times d \tag{7.7}$$

---

[3] For a more complete analysis, specific to paper recycling allowing for inputs from other sources, see Göttsching and Pakarinen 2000: 394 and Hart et al. 2005.

where (i/d) is the input or emission per unit of service delivered; e.g. the energy or labour intensity of the service, providing a further measure of the efficiency or quality of the stock. Technological improvements usually mean that (i/d) decreases over time. The balance between the interventions needed to operate and to replace the stock leads to a systematic approach to defining optimal service life (Kim et al. 2003; Keoleian 2013). Combining Eqs. 7.1 and 7.7, the ratio of operating energy (e_S) to production energy (e_p) is:

$$e_S / e_p = (e_S / d) / [(e_p / p) \times (p / S) \times (S / d)] \qquad (7.8)$$

Operating energy is thus more significant, and therefore the optimal service life is shorter, when the material intensity (p/S) and product-service intensity (S/d) are low; i.e. when the manufactured capital is well designed and efficiently used.

4   Economic and Social Implications

The principal of stock management is caring (stewardship) to maintain the quantity and quality of stock. This applies to most stocks, including natural, human and manufactured capital, and is radically different from the bigger-better-faster-safer (fashion) thinking underpinning the industrial economy. The throughput (flow) optimisation of production in global supply chains is replaced by asset (stock) management in the circular economy; the economic concept of value added is replaced by the objective of value preservation. What is of interest to investors is the fact that the return on investment (ROI) of a remanufacturing plant is usually many times that of a plant manufacturing the same goods from scratch, due to lower capital cost. On the other hand, the operating costs, notably labour, are typically much higher. These differences have a number of important implications.

4.1   Business Models in the Performance Economy

The essence of the performance economy lies in producing, selling and managing performance over time (Stahel 2010). Stock management lies at the heart of the business model because each flow (repair or stock loss) represents a cost. The three essential components and actors in the performance economy are shown schematically in Fig. 7.4:

1. Retained ownership of goods and their embodied resources by a manufacturer or fleet manager; this supports objectives (1) and (2) of the priority list developed in Sect. 3;
2. The skills and powers of an original equipment manufacturer (OEM), to support objectives (3) to (8);

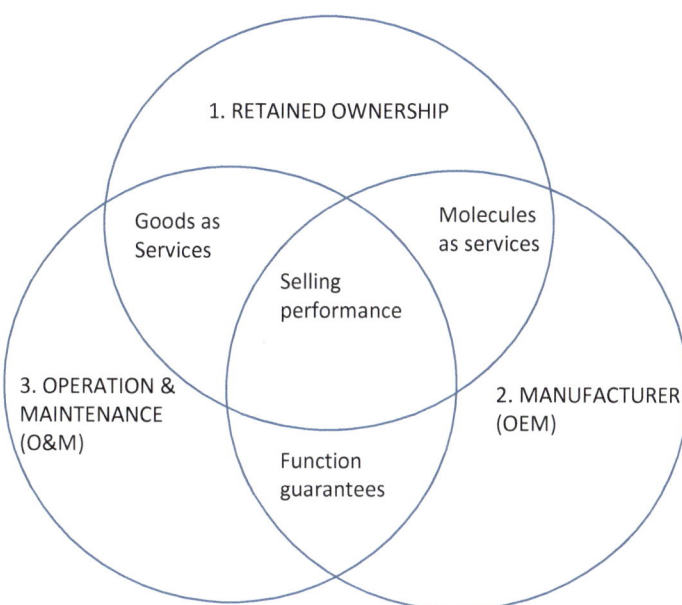

Fig. 7.4 The business models of the performance economy

3. The skills of an economic actor responsible for the operation and maintenance (O & M) of a fleet of goods, to support objectives (3) and (7).

Successful operation in the performance economy incorporates all three components. Selling performance (or "servicisation") entails internalising the costs of risk and waste over the full service life of the manufactured capital. As a result, different ways have emerged to combine the roles of the three different types of actor to increase stock life, quality and performance and reduce transaction costs. They are illustrated by the specific examples shown in Fig. 7.4 and itemised in the text box. Manufacturers exercising the O & M of their goods through service contracts give their customers a function guarantee; this provides reassurance of quality and encourages users to retain the stock. This also encourages modular design to facilitate upgrading rather than complete replacement; for example, some lift manufacturers adapt existing elevators by replacing single doors with modern double door sets; devices with electric motors can be equipped with electronic speed control to improve energy efficiency; office equipment companies (e.g. Xerox) use modular system design with standardised components across different product lines. Retaining ownership encourages management of end-of-life goods. Performance monitoring of stock in use and preventive maintenance to guarantee uninterrupted performance are essential, where possible using maintenance strategies which minimise or eliminate the need to stock spares.

As indicated in Fig. 7.1, the performance economy entails intelligent decentralisation, with generally more localisation of economic activity than in the industrial economy. This provides further opportunities for the development of industrial symbioses (see Chap. 5). Re-use is inherently local. The geographic scale of Loop 1 activities (remanufacturing) is determined by the mobility of goods, the extent of standardisation of goods and the batch size necessary to reach a competitive remanufacturing volume. For immobile goods, such as infrastructure and buildings, service-life extension activities require mobile labour and mobile workshops. For stand-alone goods, such as engines of tractors, buses, ambulances and vintage vehicles, local remanufacturing workshops may be optimal because the remanufacturing costs are secondary to the clients' wish to continue operating the vehicle. In textile leasing (hotel or hospital textiles, which have to be washed or sterilised daily), the optimal transport distance is about 100 km (Stahel 1995: 249); franchising can therefore be a better business model than centralised treatment. The geographic scale of Loop 2 activities (reprocessing) is determined by the technology used; many metallurgical processes, associated with high capital cost, need to be operated at large scale to be competitive with primary production. However, the dominance of labour and logistic costs, rather than capital, in re-use and remanufacturing reduce the economies of scale and are therefore consistent with smaller scale, decentralised operations.

Examples of Business Activities in the Performance Economy (Figure 7.4)

(1) RETAINED OWNERSHIP + (2) OEM + (3) O&M

Selling Performance

- tyre use by the mile (e.g. Michelin)
- power by the hour (e.g. Rolls-Royce turbines)
- illumination: "pay per lux" (e.g. Philips)
- office equipment: "pay per copy" (e.g. Xerox)

(1) RETAINED OWNERSHIP + (2) OEM

Molecules as Services

- chemical leasing: "rent a molecule" (e.g. lubricants, cleaning solvents)
- mining: nation state grants "licence to operate" but retains ownership of the output (WEF 2015)

(2) OEM + (3) O&M

Function Guarantee

- commercial and service equipment (e.g. freezers, elevators)
- chemical management systems
- integrated crop management

(continued)

(3) O&M + (1) RETAINED OWNERSHIP

Goods as Services

– transport (e.g. shipping, buses, containers)
– "wet" leasing of aircraft: aircraft plus crew plus fuel
– real estate (e.g. hotels, time-shares)
– short-term equipment rental (e.g. vehicles, tools)

4.2   Employment

What was probably the first research report to articulate the idea of a circular economy set out to identify the potential for substituting manpower for energy (Stahel and Reday 1976). It found that on a macro-economic level, three quarters of energy is used in mining activities and basic material production, while one quarter is used in manufacturing goods from basic material; the same estimates are reported by IPCC (2014). For manpower, the proportions are reversed: three quarters of the labour input into finished products is in the manufacturing of goods, with only one quarter in resource exploitation and primary production. Service-life extension focuses on activities which are kin to manufacturing before automation and are the most labour intensive. As noted in Sects. 2 and 3, end-of-life goods must be collected using non-destructive and non-diluting collection processes (rather than compactor lorries); the collected goods must be disassembled into components and materials (instead of being shredded); the components and materials must be screened to separate parts to preserve value (using skills and experience) so that components can be repaired or remanufactured and materials can be sorted to be recycled in the purest form and kept separate to minimise contamination. Each of these steps is more labour-intensive and resource saving than in the linear make-use-waste approach.

Macro-economic analyses of the impact of service life extension are scarce, although a recent study by Skanberg for the Club of Rome (2015) has used input/output analysis to explore the effect on the Swedish economy of increased resource efficiency. At the micro-economic level, numerous case studies exist (see text box) on the manpower intensity of reuse, repair and remanufacturing. All confirm the conclusion that service-life extension activities substitute manpower for energy and materials, when compared to manufacturing equivalent new goods. This is exemplified by Fig. 7.5, which shows the results of an analysis of different sectors (Product-Life Institute 2015) with regards to value per weight at the point of sale (€/kg) and labour input per weight (mh/kg; i.e. the labour components of i_M and i_R in the terms of Fig. 7.2) associated with manufacturing or remanufacturing: the key activities of the performance economy are comparable to life-sciences and nanotechnologies, and a far distance from manufacturing (Product-Life Institute 2015).

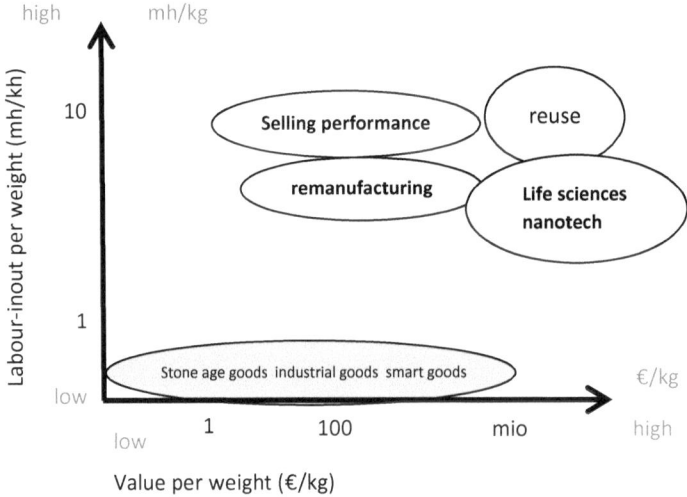

Fig. 7.5 Comparison of different sectors by value per weight (€/kg) and labour input per weight (mh/kg) (Source: Product-Life Institute 2015)

Examples of the Manpower Intensity of Re-use, Repair and Remanufacture (Stahel 2010)

An analysis of the running costs of a 30-year-old automobile (Toyota Corona Mk II of 1969) reveals that the share of labour costs as a percentage of accumulated total costs increases from 18 % after 10 years to 34 % after 20 years to reach 48 % after 30 years (Buhrow et al. 1999).

The remanufacture of a Jaguar XJ6 engine necessitated 120 h of labour and new parts of a total weight of 20 kg; the labour-input per weight ratio of 6 man-hour/kg material input is 240 times that of manufacturing a new car engine (pp. 198/9).

The Eiffel Tower in Paris is completely repainted every 7 years; 25 painters spend more than 12 months to apply 60 tonnes of paint on 250,000 m² of iron elements; the labour input per weight ratio is 0.7 man-hours per kg (p. 191).

The remanufacture of the gearbox of a Lorry needs 59 h of manpower if parts are individually repaired, at a cost of CHF 10,400; or 23.5 h at a cost of CHF 16,000 if damaged parts are replaced by new ones. (p. 220).

Medieval cathedrals have teams of masons spending a lifetime repairing the stonework to check decay and prevent the eventual collapse of the landmark buildings. A similar workshop exists for the Golden Gate Bridge in San Francisco. (p. 208).

The biggest potential for creating jobs and developing skills lies in repairing and remanufacturing standardised goods, such as infrastructure, buildings and stand-alone mobile and immobile goods, preferably in local workshops using and developing the human capital represented by skilled labour. They thus contribute to preserving the "industrial commons" – an industrial network of customers, skilled labour and subcontractors which is an essential knowledge base for innovation and the emergence of new industries (Pisano and Shih 2012). Some companies active in repair, maintenance and remanufacturing activities acquire sufficient knowledge to move upstream into manufacturing innovative new products in their field of activity (see box).

> **The Hidden Innovation Potential of Operation and Maintenance Services**
> Some companies active in repair, maintenance and remanufacturing activities acquire sufficient knowledge to move upstream into manufacturing innovative new products in their field of activity. Witness Israeli Aircraft Industries (IAI), which started as maintenance service provider before assembling fighter planes and today is a leading manufacturer of drones (pilot-less remotely controlled aircraft); and Stadler Rail in Switzerland, moving from a local maintenance service provider for rolling stock (railways) to become a major European manufacturer of regional trains.

4.3 Fiscal Policy

By reducing dependence on primary resources and energy and providing skilled employment, the performance economy has the potential to alleviate many current economic, environmental and social problems, i.e. to contribute to improved sustainability. However, promoting and developing this economic model requires a comprehensive re-examination of public policies, moving away from the fragmented policies of most public administrations. The "whole system" approach of industrial ecology, which provides the background to the idea of the performance economy, reveals the inconsistencies between policies to preserve stocks (natural and cultural capital, for example), to tax flows (levies like Value Added Tax) and stocks (human and financial capital) and to subsidise consumption of stock (such as fossil fuels).

These inconsistencies are exemplified by the 2012 "cash for clunkers" policy followed in 22 countries, which had the ambitions of stimulating national economies by increasing the demand for new cars (i.e. the flow of new goods) and simultaneously reducing GHG emissions from vehicles in use. The policy ignored the GHG emissions resulting from recycling old cars (i.e. the 'clunkers') and manufacturing their replacements. In many countries, the cars bought were imports; in some cases, "clunkers" were not destroyed but exported and sold abroad (Ashok et al. 2012; Antoniades 2013). In a circular economy perspective, the same payments could have been offered to remanufacture/retrofit stock, e.g. to remanufacture the car engines and convert them to compressed natural gas. As this job is best done in local workshops, it would have created local jobs. It would have avoided the GHG emissions from recycling the clunkers and manufacturing their replacements, and substantially reduced emissions in use.

Most current policies promoting a circular economy focus on environmental performance or resource efficiency, exemplified by:

- OECD Guidance (2015b): due diligence guidance for responsible supply chains of minerals from conflict-affected and high-risk areas ("Guidance");
- Rules of the U.S. Securities and Exchange Commission on Disclosure of Payments by Resource Extraction Issuers (US SEC 2012);
- UN (2013) Mercury agreement, designed to reduce emissions and ensure an efficient utilisation of stocks of resources;
- EU (2008) waste directive to foster waste prevention through reuse and service-life extension of goods (see Chap. 14).

To promote sustainable development, a new approach should be more holistic, defining policies which are simple, convincing and cross-cutting with the objective of preserving stock. The full societal cost of not working has generally been underestimated, but a recent OECD study has shown a strong link between mental health and employment (OECD 2015a). Both health and acquired capital of unused workers (e.g. sub-employed or unemployed people and non-active retirees) can deteriorate rapidly and even lead to a high risk of developing mental problems (Coulmas 2012). On the other hand, just as for poorly maintained manufactured capital, overuse of human capital can also lead to a loss of the stock through, for instance, burn-out. In general, to promote sustainable development, policies are needed which promote economic sectors that intelligently use and also preserve all forms of stocks or capitals. As these activities are more labour and considerably less resource intensive than manufacturing, sustainable taxation emerges as a key lever to promote change to a low-carbon resource-efficient society (Stahel 2013). Examples of a different approach to taxation include:

- Do not tax renewable resources, noting that human labour is renewable, but exclusively tax non-renewable resources, wastes and emissions;
- Do not levy value added tax (VAT) on the value preservation of stock (such as reuse and service-life extension activities);
- Give carbon credits to carbon emission prevention (smart stock management) at the same rate as to carbon emission reductions (cleaner flow).

It must be emphasised that the proposal here is for a shift in the tax base rather than an increase in tax levels. A fiscal policy of sustainable taxation could make many subsidy policies redundant; taxing non-renewable resources instead of labour would give clear incentives to economic actors to shift from flow to stock business models. In addition, it would make all stock management activities (looking after people's health, looking after natural and cultural capital) more competitive. In the USA, eleven States do not tax labour (human capital) but flow of non-renewable resources (for example, the construction industry in Florida, the oil and gas industry in Texas). In Canada, the move in British Columbia towards taxing GHG emissions (B.C. 2013) appears to be having effects which are both environmentally and economically beneficial (Elgie and Clay 2013). Not levying value-added tax (VAT) on value preservation activities would give goods in the circular economy a substantial cost advantage over new goods (around 20 % in most EU countries), again giving economic actors a clear incentive to change from flow to stock management.

5   Industrial Ecology and the Performance Economy

The idea of the performance economy has developed separately from developments in industrial ecology (Stahel 2010). One of the main objectives of this chapter is to show how thinking on the performance economy embodies the idea of an industrial ecosystem articulated by Frosch and Gallopoulos (1989), so that the performance economy is underpinned by and applies industrial ecology concepts and tools including life cycle management, accounting of material flows and stocks, resource efficiency, urban metabolism, servicisation and dematerialisation. Localisation of re-use, remanufacturing and some reprocessing can open up new opportunities for symbiosis between industrial activities. Although the drivers for the performance economy are primarily economic (or could be primarily economic under an appropriate fiscal regime), the model has the potential to alleviate the same environmental, economic and social challenges which industrial ecology seeks to address.

The performance economy represents industrial ecology in action.

Notation

d	Quantity of service delivered in specified time period (e.g. passenger-km per year for personal transport)
e	Energy use (e.g. GJ/year)
e_p	Energy input to a sector producing material products (e.g. GJ/year)
e_S	Energy input to using manufactured stock (e.g. GJ/year)
f_1	Fraction of q routed to remanufacturing
f_2	Fraction of q routed to recycling/reprocessing
g	GHG emissions from energy sector (e.g. tonnes CO_{2e} per year)
g_p	GHG emissions associated with a sector producing material products (e.g. tonnes CO_{2e} per year)
i	Interventions (i.e. exchanges across the system boundary) associated with operations and activities, per unit of output; i.e. inputs (e.g. energy or labour or financial costs) or outputs (e.g. emissions or wastes)
p	Flow of materials or products into use (e.g. tonnes/year)
q	Outflow of materials or products from use phase (e.g. tonnes/year)

r$_1$ Fraction remanufactured goods in flow p
r$_2$ Fraction recycled material in newly manufactured goods
S Stock of manufactured capital (e.g. tonnes)

Subscripts Indicate Operation or Activity

M Manufacturing
P Reprocessing
R Remanufacturing
S Use of stock to deliver services
T Total
X Extraction and primary processing

References

Allenby, B. R. (1999). *Industrial ecology: Policy framework and implementation.* Upper Saddle River: Prentice Hall.

Allwood, J. M. (2014). Squaring the circular economy: The role of recycling within a hierarchy of material management strategies. In E. Worrell & M. A. Reuter (Eds.), *Handbook of recycling.* Waltham: Elsevier.

Allwood, J. M., Cullen, J. M., Carruth, M. A., Cooper, D. R., McBrien, M., Milford, R. L., Moynihan, M. C., & Patel, A. C. H. (2012). *Sustainable materials with both eyes open.* Cambridge: UIT.

Antoniades, A. (2013). Whoops—'Cash for Clunkers' Actually Hurt the Environment. Yahoo News. Retrieved March. 15, 2015 from: http://news.yahoo.com/why-cash-clunkers-hurt-environment-more-helped-024848694.html.

Ashok, K., Pfeifer, G., & Witte, S. (2012). *The incidence of cash for clunkers: An analysis of the 2009 car scrappage scheme in Germany.* Working Paper No. 68, University of Zurich, Department of Economics, Working Paper Series.

B.C. (2013). *Carbon tax report and plan.* June budget update 2013/4 to 2015/6 (pp. 66–68). Tax Measure, British Columbia Legislature, Victoria.

Buhrow, J., Bierter, W., & Stahel W. R. (1999, November). Lebenskostenanalysen von drei Fahrzeugen über 30/17 Jahre. Studie zu den Auswirkungen einer Nutzungsdauerverlängerung von Gütern auf die Kosten für Arbeit, Ersatzteile und Beschaffung. Retrieved March 15, 2015, from http://product-life.org/de/archive/case-studies/langzeit-kostenanalyse-von-fahrzeugen-pkw-und-lkw

Clift, R. (1997). Clean technology – The idea and the practice. *Journal of Chemical Technology and Biotechnology, 68,* 347–350.

Clift, R. (2001). Clean technology and industrial ecology. In R. M. Harrison (Ed.), *Pollution – Causes, effects and control* (pp. 411–444). London: Royal Society of Chemistry.

Clift, R., & Allwood, J. (2011, March). Rethinking the economy. *The Chemical Engineer,* pp. 30–31.

Clift, R., Basson, L., & Cobbledick, D. (2009, September). Accounting for carbon. *The Chemical Engineer,* pp. 35–37.

Clift, R., Sim, S., & Sinclair, P. (2013). Sustainable consumption and production: Quality, luxury and supply chain equity. In I. S. Jawahir, Y. Huang, & S. K. Sikdar (Eds.), *Treatise on sustainability science and engineering* (pp. 291–309). Dordrecht: Springer.

Club of Rome. (2015). *Societal benefits of a more resource-efficient circular economy—The case of Sweden – What could be achieved until the year 2030?* To be published.

Coulmas, F. (2012). *Lernen, nur nicht aufhören zu lernen.* Neue Zürcher Zeitung. Retrieved March 15, 2015, from http://www.nzz.ch/aktuell/feuilleton/uebersicht/lernen-nur-nicht-aufhoeren-zu-lernen-1.16973270

Elgie, S., & Clay, J. A. (2013). *BC's carbon tax shift after five years: Results.* Ottawa: University of Ottawa.

EU. (2008). Directive 2008/98/EC on waste (Waste Framework Directive). Retrieved March 15, 2015, from http://ec.europa.eu/environment/waste/framework

Forum for the Future. (2015). *The five capitals.* Retrieved March 15, 2015, from http://www.forumforthefuture.org/project/five-capitals/overview

Frosch, R. A., & Gallopoulos, N. E. (1989). Strategies for manufacturing. *Scientific American, 261*(3), 144–152.

Göttsching, L., & Pakarinen, H. (Eds.). (2000). *Recycled fiber and deinking. Papermaking science and technology series, Book no. 7.* Helsinki: Finnish Paper Engineers Association/TAPPI, Fapet Oy.

Hart, A., Clift, R., Riddlestone, S., & Buntin, J. (2005). Use of life cycle assessment to develop industrial ecologies – A case study: Graphics paper. *TransIChemE, Process Safety and Environmental Protection, 83*(B4), 359–363.

IPCC (Intergovernmental Panel on Climate Change). (2014). 5th assessment report, Working Group III, Chapter 10, United Nations, p. 746.

Jackson, T. (2009). *Prosperity without growth: Economics for a finite planet.* London: Earthscan/Routledge.

Kaya, Y. (1990). *Impact of carbon dioxide emission control on GNP growth: Interpretation of proposed scenarios.* IPCC Energy and Industry Subgroup. Response Strategies Working Group, Paris.

Kennedy, C. A., Pincetl, S., & Bunje, P. (2011). The study of urban metabolism and its application to urban planning and design. *Environmental Pollution, 159,* 1965–1973.

Keoleian, G. A. (2013). Life cycle optimization methods for enhancing the sustainability of design and policy decisions. In I. S. Jawahir, Y. Huang, & S. K. Sikdar (Eds.), *Treatise on sustainability science and engineering* (pp. 3–17). Dordrecht: Springer.

Kim, H. C., Keoleian, G. A., Grande, D. E., & Bean, J. C. (2003). Life cycle optimization of automobile replacement: Model & application. *Environmental Science & Technology, 37,* 5407–5413.

Levy, J. C., & Aurez, V. (2014). *L'économie circulaire: un désir ardent des territoires.* Paris: Transition écologique, Presses des Ponts.

OECD. (2015a). *Fit mind, fit job: From evidence to practice in mental health and work.* Paris: OECD.

OECD. (2015b). *Due diligence guidance for responsible supply chains of minerals from conflict-affected and high-risk areas.* Retrieved March 15, 2015, from http://www.oecd.org/corporate/mne/mining.htm

Pisano, G. P., & Shih, W. C. (2012). *Producing prosperity: Why America needs a manufacturing renaissance.* Boston: Harvard Business Review Press.

Product-Life Institute. (2015). *Performance sustainability rating of sectors.* Retrieved March 15, 2015, from http://product-life.org/en/major-publications/performance-economy

Smith, V. M., & Keoleian, G. A. (2004). The value of remanufactured engines: Lifecycle environmental and economic perspectives. *Journal of Industrial Ecology, 8,* 193–222.

Stahel, W. R. (1995). *Handbuch Abfall 1, Allgemeine Kreislauf- und Rückstandswirtschaft—intelligente Produktionsweisen und Nutzungskonzepte.* Landesanstalt für Umweltschutz Baden-Württemberg.

Stahel, W. R. (2010). *The performance economy* (2nd ed.). Basingstoke: Palgrave MacMillan.

Stahel, W. R. (2013). Policy for material efficiency – Sustainable taxation as a departure from the throwaway society. *Philosophical Transactions of the Royal Society A, 371*(1986), 20110567.

Stahel, W. R., & Reday-Mulvey, G. (1976). *The potential for substituting manpower for energy: A report to the European Commission*. Subsequently published as *Jobs for tomorrow*. New York: Vantage Press (1981).

UN (2013). *Minamata convention on mercury*. Retrieved March 15, 2015, from http://www.cbc.ca/news/world/un-treaty-to-limit-mercury-emissions-signed-by-140-countries-1.1405350

US SEC (2012). *New rules and an amendment to a new form pursuant to Section 1504 of the Dodd-Frank Wall Street Reform and Consumer Protection Act relating to disclosure of payments by resource extraction issuers*. Retrieved March 15, 2015, from http://www.sec.gov/rules/final/2012/34-67717.pdf

Van Berkel, R., Fujita, T., Hashimoto, S., & Fujii, M. (2009). Quantitative assessment of urban and industrial symbiosis in Kawasaki, Japan. *Environmental Science and Technology, 43*, 1271–1281.

WEF. (2015). *Licence to mine, making resource wealth work for those who need it most*. World Economic Forum, *to be published*.

Chapter 8
Impacts Embodied in Global Trade Flows

Thomas Wiedmann

Abstract The steep and unprecedented growth of globalisation and trade over the last few decades has led to accelerated economic activity with mixed outcomes. Continued economic growth and alleviation of poverty in many countries has been accompanied with an overall increase and shifting of environmental pressures between countries. Industrial ecology research has contributed decisively to the knowledge around impacts in trade. This chapter summarises the latest empirical findings on global change instigated by trade, discusses new methodological developments and reflects on the sustainability of globalised production and consumption. Significant proportions of up to 64 % of total environmental, social and economic impacts can be linked to international trade. Impacts embodied in trade have grown much more rapidly than their total global counterparts. Policies aimed at increasing the sustainability of production and consumption need to go beyond domestic regulation and seek international cooperation to target production practices for exports worldwide.

Keywords Trade-embodied impacts • Consumption-based accounting • Environmental footprint • Global resource use • Multi-region input-output analysis • Sustainability of trade

1 Introduction

The steep and unprecedented growth of globalisation and trade over the last few decades has led to accelerated economic activity with mixed outcomes. Continued economic growth and alleviation of poverty in many countries has been accompanied with an overall increase and shifting of environmental pressures between countries. Industrial ecology research has contributed decisively to the knowledge around

T. Wiedmann (✉)
Sustainability Assessment Program (SAP), School of Civil and Environmental Engineering, University of New South Wales Australia, Sydney, NSW 2052, Australia

ISA, School of Physics A28, The University of Sydney, Sydney, NSW 2006, Australia
e-mail: t.wiedmann@unsw.edu.au

© The Author(s) 2016
R. Clift, A. Druckman (eds.), *Taking Stock of Industrial Ecology*,
DOI 10.1007/978-3-319-20571-7_8

159

impacts in trade. This chapter summarises the latest empirical findings on global change instigated by trade, discusses new methodological developments and reflects on the sustainability of globalised production and consumption. Significant proportions of up to 64 % of total environmental, social and economic impacts can be linked to international trade. Impacts embodied in trade have grown much more rapidly than their total global counterparts. Policies aimed at increasing the sustainability of production and consumption need to go beyond domestic regulation and seek international cooperation to target production practices for exports worldwide.

International trade is not a new phenomenon. People have exchanged goods and services since prehistoric and ancient times. One prominent example of early trade links between countries and continents is the Silk Roads, a network of trading routes established between Asia and Europe during the Han Dynasty in China (206 BC – 220 AD) (Liu 2010). The trade in Chinese silk and many other goods extended over 6,000 km and was very lucrative. It boosted the economic development of China and its Middle Asian and European trading partners and became so important that it was protected militarily by fortified watch towers. The Great Wall was extended to ensure the protection of the trade route. The Silk Roads' importance during ancient times and up to its golden age during the early middle age was confirmed in 2014, when parts of the network were declared a UNESCO World Heritage Site.[1]

How does this compare to trade in modern times? It is certainly true to say that international trade accelerates economic development – nowadays as it did thousands of years ago. What is different – due to economic globalisation and technological advances, especially in the last 20 years – is the unprecedented scale, speed and complexity of trade movements and transactions.

Over the last few decades, international trade has grown much more rapidly compared to other indicators of development such as, for example, GDP (gross domestic product), population or CO_2 emissions (Kanemoto and Murray 2013 and Fig. 8.1). The value of exports of goods and services is almost 300 times larger today than it was in 1950 (35 times larger by volume; WTO 2013). On average, exports make up 30 % of a country's GDP (World Bank 2015). The value added along global production chains (outside the country of completion) has steadily increased since 1995, only briefly interrupted in 2008 due to the global financial crisis (Los et al. 2015; Timmer et al. 2014). This trend is seen as a clear sign that production has shifted from the regional to the global scale. The expansion of international trade has changed production and consumption patterns almost everywhere, with wide-ranging implications for economies, societies and the environment.

Undoubtedly, globalisation and trade have helped to alleviate poverty and social hardship in many countries. According to the World Resources Institute, over the last 20 years 'Real incomes in low- and middle-income countries have doubled and poverty rates have halved. Two billion people have gained access to improved drinking water. Maternal mortality has dropped by nearly half, and the share of those who

---

[1] Retrieved February 23, 2015 from http://whc.unesco.org/en/list/1442

Fig. 8.1 World merchandise exports (by value and volume) and gross domestic product 1950–2012 (index 2005 = 100) (Data from WTO 2013)

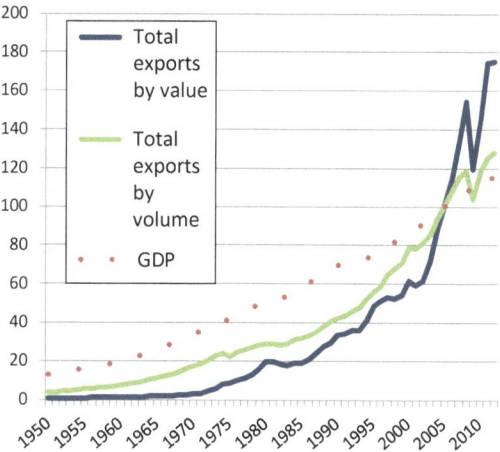

are malnourished has fallen by a third' (WRI 2014). At the same time, pressures on the natural environment have increased tremendously: 'Every minute of every day we have been losing the equivalent of 50 soccer fields of forest. Over one billion people already face water scarcity, and this may triple by 2025. Climate change is costing $700 billion per year, with the greatest impact on the poor' (WRI 2014).

The sheer amount of goods shipped around the world is also unprecedented. Ten billion tonnes (10 gigatons, Gt) of materials and products were shipped between countries in 2005 (Dittrich and Bringezu 2010). And this figure includes only the direct physical trade, i.e. actual shipment of materials and goods. As will be shown later in this chapter, raw materials are also extracted and processed in order to enable exports, even though they never leave the country. Adding these indirect material flows or 'raw material equivalents' to the actual physical trade resulted in a total amount of 29 Gt of materials associated with trade flows between countries in 2008 (Wiedmann et al. 2015).

In addition to growing in scale, trade has become more complex and fragmented. The production process of many products occurs in small stages in different countries, interlinked through complex global supply chain networks. Supply chains have become longer, more fragmented and more complex. World merchandise exports of intermediate and final products were almost identical in 1993 (7–8 % of world GDP), but exports of intermediate exports have grown faster since and were 15 % of GDP in 2012, whilst exports of final goods only reached 11 % of GDP.

Longer and more fragmented supply chains also mean that places of production and consumption are more separated and that it becomes more difficult to establish the link between environmental impacts exerted by the production process and the final destination of the product. In other words, the 'cradle-to-gate' life cycle becomes longer, more convoluted and more difficult to assess. Increasingly sophisticated global models have had to be developed to evaluate impacts embodied in global supply chains (Tukker and Dietzenbacher 2013; Wiedmann 2009; Wiedmann et al. 2007, 2011).

Most of the growth in trade value, volume and complexity occurred in the last couple of decades only. According to Richard Baldwin, possibly the most influential change over the last 20 years was the international movement of firm-specific know-how (Baldwin 2013). Changes in both technology and legislation have made it easier for multinational companies to exchange knowledge and coordinate internal processes, enabling them to quickly respond to changing demands and ramp up production capacities in varying locations. Some large multinational companies, such as Apple, Exxon Mobil, Royal Dutch Shell or IBM, have market values that are comparable or exceed the GDP of countries such as Belgium, Switzerland, Sweden, Norway or Saudi Arabia. Therefore, with respect to international production and consumption, national borders might not be as influential as they seem.

2 Impacts of Trade: New Insights from Recent Research

2.1 Taking a Consumption-Based Perspective: What Are Impacts Embodied in Trade?

Two words in this question require further explanation: 'impacts' and 'embodied'. The term 'impact' is used here in a very wide sense, comprising both pressure and impact indicators as defined by the causal DPSIR framework (Driving Forces-Pressures-State-Impacts-Responses) that describes interactions between society and the environment.[2] Environmental pressures include the use of resources, such as land, water or materials as well as the emissions of greenhouse gases (GHG) or pollutants. In the stricter definition provided by Life Cycle Assessment (LCA) standards (ISO 2006; Hellweg et al. 2014), environmental impacts represent the (actual or potential) damage exerted by pressures, e.g. global warming, toxicity or biodiversity loss. Especially in the context of international trade, the expression environmental 'burden' or 'load' has been used as well as 'burden shifting' (e.g. Giljum and Eisenmenger 2008; Schütz et al. 2004; Zhang et al. 2013) or 'displacement of pressures' (e.g. Steen-Olsen et al. 2012) to describe the change of location where environmental pressures or impacts occur when resources from other countries are used indirectly through trade. For social and economic indicators, the distinction between pressures and impacts is less well defined. For the sake of simplicity, the term 'impact' has been used for all indicators in this chapter (see also Table 8.1).

The word 'embodied' describes indirect impacts that can be 'attributed to', are 'associated with' or are 'embedded in' activities that are not directly linked to the impacts. In the context of trade, consuming a product in one country can lead to impacts in many other countries, depending on where the production and supply chain processes occur that are required to produce the final consumer product. All

---

[2] Retrieved February 23, 2015 from http://root-devel.ew.eea.europa.eu/ia2dec/knowledge_base/Frameworks/doc101182

these supply chain impacts are said to be 'embodied' in the product, even if there is no direct physical connection. This may be exemplified best in the context of water use, where the term 'virtual' has been used widely (e.g. Chen and Chen 2013; Dalin et al. 2012; Orlowsky et al. 2014). The virtual water is not actually physically embodied in a traded product – yet the term 'embodied' is widely used in the literature to describe indirect impacts. Another expression introduced by Lenzen et al. 2012 is the word 'implicated' which was used by the authors to indicate a connection between consumption in one country and threat to species in other countries, even though it would be difficult to prove a direct causal relationship between the two (a point also made with respect to CO_2 emissions embodied in trade, see Jakob and Marschinski 2013). The term 'implicated' is again used in Alsamawi et al. 2014b to indicate the inequality associated with the trade of commodities between nations.

As an overarching model of evaluating the embodied impacts of consumption, the concept of environmental footprints has been used widely (Hoekstra and Wiedmann 2014). Applied at the country level, a nation's total footprint is calculated as follows:

$$\textit{Territorial} \text{ impacts}$$
$$+ \text{ impacts embodied in } \textit{imports}$$
$$- \text{ impacts embodied in } \textit{exports}$$
$$= \text{national } \textit{footprint}$$

The footprint takes a consumption perspective, in most cases equivalent to a 'cradle-to-shelf' perspective in LCA. Evaluating footprints has therefore also been referred to as consumption-based accounting (CBA), in particular in the context of accounting for national GHG emissions and resource use (Barrett et al. 2013; Kander et al. 2015; Peters 2008). Countries can use CBA to measure both their impact as well as their dependence on foreign economies and environments. It is well known that impacts have increasingly been shifted abroad (Table 8.1). The consumption view provided by national footprints offers consumer information and policy options for the mitigation of emissions and resource use that are complementary to measures based on territorial accounting (Andrew et al. 2013; Barrett et al. 2013). Both perspectives,[3] the production (territorial) and the consumption perspective, provide important insights into the sources and drivers of impacts, and both have their pros and cons. The production perspective is easier to implement, refers to environmental pressures at the source and is widely accepted as an accounting method for national GHG emissions (UNFCCC). However, it does not account for burden shifting or carbon leakage, both of which can occur if domestic production is moved abroad. CBA, on the other hand, adds back embodied impacts in imports to the national balance sheet and correctly accounts for impacts of total national

---

[3] A third perspective, named income-based (or downstream) responsibility, was introduced by Marques et al. (2012). This allows for calculating carbon emissions occurring abroad associated with the trade from which a region or country derives its income (also called 'enabled emissions') (Marques et al. 2013).

consumption. CBA is more difficult to measure and implement though, and impacts occurring in foreign jurisdictions are hard if not impossible to influence or control (Jakob et al. 2014). Furthermore, CBA provides no incentive for countries to produce clean exports (since impacts embodied in exports are subtracted). It has been suggested recently to address this drawback by using the world-average carbon intensity for exporting industries, rather than the domestic average, when calculating export-related emissions (Kander et al. 2015). Doing so rewards countries that produce export commodities that are cleaner than their counterparts on the world market.

2.2 Recent Research on Environmental, Social and Economic Impacts Embodied in International Trade

2.2.1 Scope and Scale of Embodied Impacts

Numerous studies have been conducted in the last few years to shed light on the question how trade influences the use and distribution of natural, social and economic capital. Table 8.1 summarises some high-level results, in particular the fraction of total global impact that can be attributed to international trade as well as the major bilateral embodied trade flows. Note that these values depend on the number of countries or regions used in the various calculation models. As a general rule, the finer the spatial resolution of the model, the higher the international trade flows, and the lower the intra-regional trade movements. Where possible, individual countries were identified as main traders in Table 8.1.

At least a fifth and up to 64 % of global environmental impacts can be linked to trade (for all references refer to Table 8.1). Greenhouse gas emissions are the best-studied indicator. About one quarter of all global CO_2 emissions are linked to the production of goods and services that are exported and used to satisfy demand in countries other than the country where the emissions occur. One study suggests that the fraction of CO_2 embodied in trade could be as high as a third of global emissions. And if the trade of fossil fuels is taken into account, then the amount of 'dislocated' CO_2 emissions from the point of extraction to the point of final consumption is 37 % or more than 10 Gt of CO_2. According to Meng et al. (2015), the median export share of a country's territorial emissions was 29 % in 2007, and emissions embodied in imports made up almost half of the carbon footprints of countries (median 49 %). The largest bilateral flows of embodied CO_2 emissions with well over 1 Gt of CO_2 are from China to the USA. This finding is not surprising given the large volumes of exports from China and imports to the USA and the fact that China's production system is very carbon intensive (Minx et al. 2011). The EU is also a large importer of GHG emissions from Asia (0.8 Gt CO_2e). When accounting for international CO_2 emissions embodied in investments (instead of total final demand), China also emerges as the main exporter of investment-embodied emissions and Western Europe and North America as the main importers.

Table 8.1 Global studies quantifying environmental, social and economic impacts embodied in international trade

Impact	Fraction of total global impact embodied in trade (absolute amount, year)	Largest exporter (i), largest importer (ii), largest bilateral trade flow (iii), gross flows, not net flows[a]	Method (name of database/model)	References
CO_2 embodied in traded products	(a) **23** % (6.2 Gt CO_2, 2004)	(a) (i) China (1.43 Gt CO_2) (ii) USA (1.22 Gt CO_2) (iii) From China to USA (395 Mt CO_2)	(a) MRIO analysis (GTAP)	(a) Davis and Caldeira (2010)
	(b) **23** % (6.4 Gt CO_2, 2004)	(b) (i) China (1.24 Gt CO_2, 2004) (ii) USA (1.22 Gt CO_2)	(b) MRIO analysis (GTAP)	(b) Davis et al. (2011)
	(c) **22** % (1.7 Gt C = 6.1 Gt CO_2, 2004)	(c) n.p.[b]	(c) Synthesis of MRIO-based studies	(c) Peters et al. (2012)
	(d) **25** % (7.5 Gt CO_2, 2006)	(d) (iii) From Canada to USA (195 Mt CO_2)	(d) EEBT[c] analysis with life cycle inventory factors for carbon intensity of products	(d) Sato (2014)
	(e) **n.p.** (6.9 GtCO_2, 2007)	(e) n.p.	(e) MRIO analysis (GTAP)	(e) Andrew et al. (2013)
	(f) **33** % (8.3 Gt CO_2, 2007)	(f) n.p.	(f) MRIO analysis (WIOD)	(f) Xu and Dietzenbacher (2014)
	(g) *26 % (7.8 Gt CO_2 in 2008)*	(g) (iii) From China to USA (207 Mt CO_2, average 1998–2008)	(g) MRIO analysis (GTAP) using MRIO (global supply chains) and EEBT (domestic supply chains) balances	(g) Peters et al. (2011)
CO_2 emissions embodied in investments	**n.p.**	(i) Greater China (2.3 Gt CO_2, 2004) (ii) Western Europe (3.6 Gt CO_2, 2004)	Global Interregional Social Accounting Matrix (GTAP)	Bergmann (2013)

(continued)

Table 8.1 (continued)

Impact	Fraction of total global impact embodied in trade (absolute amount, year)	Largest exporter (i), largest importer (ii), largest bilateral trade flow (iii), gross flows, not net flows[a]	Method (name of database/model)	References
CO_2 emissions from traded fossil fuels	(a) **37** % (10.2 Gt CO_2, 2004) (b) **n.p.** (10.8 Gt CO_2 in 2007)	(a) (i) Russia (1.47 Gt CO_2, 2004) (ii) USA (2.08 Gt CO_2) (b) n.p.	(a) MRIO analysis (GTAP) b) MRIO analysis (GTAP)	(a) Davis et al. (2011) b) Andrew et al. (2013)
GHG emissions (CO_2, CH_4, N_2O)	(a) **23** % (8.7 Gt CO_2e, 2007) (b) **27** % (10.4 Gt CO_2e, 2008)	(a) (iii) From Asia to EU (0.79 Gt CO_2e) (b) (i) China (2.9 Gt CO_2e) (ii) USA (1.8 Gt CO_2e)	(a) MRIO analysis (EXIOBASE) (b) MRIO analysis (WIOD)	(a) Tukker et al. (2014) (b) Arto et al. (2012)
Water	(a) **26** % (2,320 Gm³, 1996–2005) (b) **24** % (1900 Gm³, 2000) (c) **30** % (2004) (d) **22** % (2651 Gm³, 2008)	(a) (i) USA (314 Gm³/y) (ii) USA (234 Gm³/y) (iii) From USA to Mexico (b) (i) USA (180 Gm³) (ii) USA (300 Gm³) (iii) From USA to Mexico (34.2 Gm³) (c) (i) China (204 Gm³) (ii) USA (178 Gm³) (d) (i) China (472 Gm³) (ii) USA (427 Gm³)	(a) Water Footprint Network method (Hoekstra et al. 2011) (b) MRIO analysis (Eora) (c) MRIO (GTAP) (d) MRIO analysis (WIOD)	(a) (Hoekstra and Mekonnen (2012); (b) Lenzen et al. (2013) (c) Chen and Chen (2013) (d) Arto et al. (2012)
Scarce water	(**32** % (480 Gm³, 2000)	(i) India (30 Gm³), (ii) USA (45 Gm³) (iii) From Pakistan to USA (7.9 Gm³)	MRIO analysis (Eora)	Lenzen et al. (2013)

Land	(a) **24 %** (1800 Mgha, 2004) (biologically productive land area)	(a)(i) China (218 Mgha)	(a) MRIO analysis (GTAP)	(a) Weinzettel et al. (2013)
		(ii) USA (326 Mgha)		
		(iii) From China to USA (59 mgha)		
	(b) **n.p.**	(b) (i) Russia (258 Mha)	(b) MRIO analysis (GTAP)	(b) Yu et al. (2013)
		(ii) USA (198 Mha)		
		(iii) From Russia to China (64 Mha)		
	(c) **23 %** (1660 Mha, 2008)	(c) (i) China (160 Mha)	(c) MRIO analysis (WIOD)	(c) Arto et al. (2012)
		(ii) USA (260 Mha)		
Cropland	**20 %** (271 Mha, 2008)	(i) USA (37 Mha, 2009)	Analysis of bilateral trade data (FAOSTAT)	Kastner et al. (2014a)
		(ii) China (34 Mha, 2009) (MRIO analysis suggests that China is a major exporter Kastner et al., (2014b))		
		(iii) North America to East Asia (18 Mha)		
Threatened species	**30 %** (7500 species threats, 2009)	(i) Indonesia (238 species threats)	MRIO analysis (Eora)	Lenzen et al. (2012)
		(ii) USA (1262)		
		(iii) Papua New Guinea to Japan (91)		
Energy	**35 %** (n.p., 2007)	(i) Russia (23 PJ)	MRIO analysis (EXIOBASE)	Simas et al. (2015)
		(ii) USA (25 PJ)		

(continued)

Table 8.1 (continued)

Impact	Fraction of total global impact embodied in trade (absolute amount, year)	Largest exporter (i), largest importer (ii), largest bilateral trade flow (iii), gross flows, not net flows[a]	Method (name of database/model)	References
Raw materials	(a) **26 %** (15 Gt, 2005)	(a) n.p. for countries (i) OECD LD (5.5 Gt) (ii) OECD HD (9.9 Gt)	(a) IOT and bilateral trade analysis (GRAM/OECD)	(a) Bruckner et al. (2012)
	(b) **34 %** (22 Gt, 2007)	(b)(i) China (3.9 Gt) (ii) USA (3.5 Gt)	(b) MRIO analysis (GTAP, materialflows.net)	(b) Giljum et al. (2014)
	(c) **24 %** (16 Gt, 2008)	(c)(i) China (2.6 Gt) (ii) USA (2.8 Gt)	(c) MRIO analysis (WIOD)	(c) Arto et al. (2012)
	(d) **41 %** (29 Gt, 2008)	(d)(i) India (0.5 Gt biomass) China (5.2 Gt construction materials) Russia (1.2 Gt fossil fuels) Chile (0.7 Gt metal ores) (ii) USA (0.8 Gt biomass) USA (2.1 Gt construction materials) USA (1.3 Gt fossil fuels) USA (0.7 Gt metal ores)	(d) MRIO analysis (Eora)	(d) Wiedmann et al. (2015)
Metal ores	**62 % for iron ore** (1,380 Mt, 2008) **64 % for bauxite** (136 Mt, 2008)	(i) Brazil (315 Mt iron ore), Australia (44 Mt bauxite) (ii) China (350 Mt iron ore), USA (24 Mt bauxite)	MRIO analysis (Eora)	Wiedmann et al. (2014)

Ozone precursors emissions (NMVOC, CH$_4$, CO, NO$_x$)	28 % (109 Mt NMVOCe, 2008)	(i) China (17.4 Mt NMVOCe) (ii) USA (18.6 Mt NMVOCe)	MRIO analysis (WIOD)	Arto et al. (2012)
Acid emissions (NH$_3$, NO$_x$, SO$_x$)	26 % (2.1 Mt H$^+$e, 2008)	(i) China (0.65 Mt H$^+$e) (ii) USA (0.35 Mt H$^+$e)	MRIO analysis (WIOD)	Arto et al. (2012)
(a) Labour	(a) **18 %** (560 million persons-year equivalents, 2007)	(a) (i) China (130 mpeq) (ii) USA (115 mpeq) (iii) China to USA (27 mFTE, 2010, Alsamawi et al. (2014a))	(a) MRIO analysis (EXIOBASE)	(a) Simas et al. (2015)
(b) 'Bad' labour	(b) **16 %** for total labour, **15 %** for low-skilled labour, **17 %** for forced labour, **18 %** for occupational health damage, **19 %** for child labour, **19 %** for vulnerable employment, **20 %** for hazardous child labour and **38 %** for labour by women (all numbers for trade between seven world regions)	(b) (i) The APAC region is the largest exporter of all forms of (bad) labour, except for child labour and hazardous child labour for which Africa is the largest exporter (ii) n.p. (iii) APAC to Europe for all forms of (bad) labour, except for child labour and hazardous child labour for which Africa to Europe is the largest flow	(b) MRIO analysis (EXIOBASE)	(b) Simas et al. (2014)
Wages	**n.p.**	(iii) USA to Japan (112 US$bn, 2010)	MRIO analysis (Eora)	Alsamawi et al. (2014a)

^aSame year as fraction unless otherwise stated
^b*n.p.* not provided
^c*EEBT* emissions embodied in bilateral trade

Virtual water embodied in trade makes up between 22 and 30 % of total global water use, with the USA taking on a dual role of both largest exporter and importer of virtual water (though some studies suggest that China is the main exporter). When adjusting water use numbers with a factor for its scarcity in regions and countries of extraction, almost one third (32 %) of this 'scarce water' is associated with trade. India is the largest exporter of scarce water, the USA its largest importer.

Comparable numbers for the share of total impact embodied in trade are reported for other environmental indicators: 20–24 % for land use, 30 % for threatened species and 35 % for energy. Even higher is the share for raw materials: 41 % of all raw materials (biomass, fossil fuels, construction materials, minerals and metal ores) are extracted worldwide only in order to enable the export of goods and services from the country of extraction. And for metal ores the majority of extraction occurs due to export activities: 62 % of the global iron ore extraction and 64 % of the global bauxite mined are associated with trade. On average, only about one third of all raw materials actually leave the country of origin on a cargo ship, truck or plane. The rest are process wastes and auxiliary material flows that, whilst remaining near extraction sites, can still be attributed to the material footprint of other countries that import goods and services for their final consumption.

For most environmental impacts the direction of burden shifting (see Sect. 2.1) is from developed countries to developing countries, but not for all. An indirect threat to species through trade is experienced in countries such as Papua New Guinea, Madagascar or Indonesia, whereas air pollution and GHG emissions embodied in exports occur mostly in China. Russia exports embodied energy and emissions from traded fossil fuels as well as land. For the virtual use of land through trade, there are mixed results, depending on the type of land and on the characteristics of the model used for the analysis. In addition to Russia as the largest exporter of embodied land, China has been identified as exporting the most biologically productive land area and the USA as exporting the most cropland. Resource-rich countries that physically export large quantities of raw materials are also amongst the top exporters of embodied materials, e.g. India for biomass, Russia for fossil fuels, Chile for metal ores in general and more specifically Brazil for iron ore and Australia for bauxite. China virtually exports 39 % of all construction materials extracted worldwide (5.2 Gt of 13.3 Gt). Again, most of this material is not physically shipped abroad but used domestically in China to build up infrastructure for a highly export-oriented economy.

A strong driver of globalisation has been the move of production to places where wages and therefore total production costs are relatively low (Timmer et al. 2014). A large workforce in developing low-wage countries is employed to manufacture goods for exports, mostly to the developed world. Often working conditions are poor, and workers have low skills or are exposed to health and safety hazards. Sometimes children and other vulnerable persons are forced to work. Women often experience more detrimental conditions than men.

Industrial ecology research entered a new field when several studies were published in 2014 that investigated the 'labour footprint' of nations and the role of trade in employment conditions of exporting countries. On average, about 16–18 % of all labour in the world is embodied in trade (between seven world regions – the numbers would be higher when considering trade between all countries). Some forms of damaging labour conditions seem to be supported by trade, e.g. 20 % of all hazardous child labour is for exports. And 38 % of all work done by women became embodied in international trade. Asia is the largest exporting region of all forms of (bad) labour, except for child labour and hazardous child labour for which Africa is the largest exporter (Simas et al. 2014).

Wages on the other hand are highest in the developed world, and therefore trade flows of embodied wages take different paths to those for labour. The highest flows of wages embodied in exports are between developed countries, mostly from the USA to Japan (and backwards), Canada and Europe, but also to China.

The flow of money in trade has been studied extensively for a long time, but recently researchers have used newly available multi-region input-output (MRIO) models to study specific economic aspects of trade, such as fragmentation or value added (VA) in trade. Trade statistics are normally based on gross export values, thus double counting the VA along global supply/value chains (Kelly and La Cava 2013). Interest has therefore grown in VA as a 'trade commodity' that can become embodied in international trade flows, and methodological frameworks have been developed accordingly (e.g. Koopman et al. 2014). One study found that the foreign VA content of exports from Luxembourg was 61 % in 2011 (Foster-McGregor and Stehrer 2013). Interestingly, there seems to be a trend towards value being added by capital and high-skilled labour and away from less-skilled labour (Timmer et al. 2014). The capital share in the VA of emerging economies is rising, whilst the share of low-skilled labour in their VA is declining.

Meng et al. (2015) synchronously evaluate VA and CO_2 emissions in global trade. Their detailed analysis confirms the increasing fragmentation of international trade. They find that more than half (ca. 60 %) of China's emissions attributable to foreign final demand are embodied in the trade of intermediate goods (ca. 40 % of export emissions are embodied in the trade of final goods). Whether a country's emissions become embodied in the trade of final or intermediate goods depends on its position in the global value chain. Meng et al. (2015) demonstrate how CO_2 emissions from Poland's metal industry are associated with final demand in the USA: 90 % of these emissions are embodied in intermediate good trade (roughly half of which are traded directly between Poland and the USA, and the other half is traded by way of third countries).

2.2.2 Trends of Impacts Embodied in Trade

The results in Table 8.1 show clearly that trade is associated with a significant dislocation of environmental, social and economic factors, thus further separating impacts of production (both negative and positive) in one place from consumption elsewhere. Forty per cent of the national carbon footprint of the UK is exerted abroad (Hertwich and Peters 2009) and 75 % of its national water footprint (Hoekstra and Mekonnen 2012). The numbers presented in Table 8.1 are the latest available, but there has been a strongly increasing trend for the last few decades. For example:

- Land for the export production of crops grew rapidly by +2.1 % per year between 1986 and 2009 (Kastner et al. 2014a). At the same time, land supplying crops for direct domestic use remained almost unchanged.
- Flows of materials embodied in international trade are reported to have increased by 62 % between 1997 and 2007 (Giljum et al. 2014) and by 123 % between 1990 and 2008 (Wiedmann et al. 2015).
- Global trade in embodied iron ore has grown faster than its extraction, by a factor of 2.7 between 1990 and 2008 (Wiedmann et al. 2014). Trade of embodied bauxite has grown by a factor of 2.4.
- From 1995 to 2007 total global CO_2 emissions from production have increased by 32 %, whereas global emissions embodied in trade have increased by 80 % in the same period (from 4.6 Gt or 24 % of global production emissions to 8.3 Gt or 33 %) (Xu and Dietzenbacher 2014).
- In the most comprehensive study, Arto et al. 2012 present the trend of impacts embodied in trade from 1995 to 2008 for the following indicators: land +3.0 Mkm^2 (+22 %); raw materials +7.3 Gt (+80 %); blue, green and grey water +1.2 PL (+88 %); acid emissions +734 kt H^+e (+54 %); GHG emissions +4.7 Gt CO_2e (+83 %); and ozone precursors emissions +55.3 Mt NMVOCe (+103 %).

These examples show impressively how rapidly impacts associated with trade have grown in little more than 20 years, given that total global impacts have grown much slower (land +2 %, raw materials +43 %, water +37 %, acid emissions +12 %, GHG emissions +29 %, ozone precursors emissions +11 %; Arto et al. 2012).

3 Notes on Methodological Developments

This section briefly addresses some of the current issues surrounding the methods used to quantify impacts associated with trade. The list of topics discussed is not exhaustive but merely presents some of the highlights discussed in the literature and the industrial ecology community.

3.1 Merging of Disciplines

The analysis of social and economic indicators in the same way as for environmental issues – namely, from the viewpoint of trade embodiments and by using MRIO analysis – is a new and encouraging trend. It goes hand in hand with a similar development in life cycle sustainability assessment (LCSA) where social LCA increasingly complements the more traditional environmental impact and life cycle costing assessments (Kloepffer 2008; Parent et al. 2013).

Embracing and merging of data, metrics and methods from different disciplines is needed to address the fundamental questions of how a transition to sustainability can be achieved. Industrial ecology research greatly benefits from such an extension of its portfolio. After all, humans are part of the 'ecology' of industrial systems. Issues such as income inequality are of concern to both social and ecological sustainability (Alsamawi et al. 2014b). It is therefore important that socio-economic issues such as employment, wages, income inequality, occupational health, bad labour conditions, slavery, war casualties, etc. are monitored alongside environmental indicators.

It is to be hoped that the joint analysis of data from different fields supports a similar cooperation across different disciplines. The complexity of the sustainability challenge requires inter- and transdisciplinary solutions.

3.2 Assessing Actual Impacts and Their Unsustainability

As mentioned previously, most of the indicators described in this chapter represent pressures rather than impacts in the strict sense defined by LCA. Most footprint indicators (and consumption-based accounting studies) are designed to portray indirect pressures (Hoekstra and Wiedmann 2014), but there are recent attempts to introduce (environmental) impact assessment in footprint analysis.

This is perhaps most prominently the case for water footprinting where it has been argued that the scarcity of water needs to be incorporated into the metric (Chenoweth et al. 2014; ISO 2014; Kounina et al. 2013; Ridoutt and Pfister 2010). Some recent studies related to trade weight water use with data on water scarcity (e.g. Lenzen et al. 2013; Orlowsky et al. 2014).

A similar case can be made for the material footprint which sums up the mass of different raw materials into one number, thus reflecting an unweighted physical measure of pressure and *potential* impact (Wiedmann et al. 2015). Weighting according to *actual* environmental impacts has not been tried yet and is difficult, because different materials have different impacts, one material may have several impacts and characterisation data and models for localised impacts are not yet well developed. However, preliminary attempts of weighting resource footprints based on resource depletion have been presented (Fang and Heijungs 2014a).

Yet, there remains value in reporting footprints based on pressures alone. The pure mass or volume of resource use is practical information that relates to physical reality, i.e. how *much* actually is flowing. It allows, for example, to address questions of allocation of limited supplies or security of supply and sustainability of overall production and consumption. The ultimate goal of footprint accounting in general (and the assessment of impacts embodied in trade specifically) should be an evaluation of whether or not particular activities are sustainable (Fang and Heijungs 2014b). Environmental footprints measure human appropriation of natural resources and need to be interpreted in the context of maximum sustainable levels at the local and the global scale (Hoekstra and Wiedmann 2014). Exactly how high the sustainable thresholds of earth systems are is the subject of intense research (Steffen et al. 2015).

3.3 Addressing Uncertainty in MRIO Modelling

Currently the only tool to unravel the intricacies of international supply chains is MRIO modelling. The remarkable development in MRIO databases (Tukker and Dietzenbacher 2013; Wiedmann et al. 2011) has been accompanied by an equally impressive number of publications studying the impacts of globalisation and trade. Some studies have begun comparing the results obtained from different models, finding reasonable agreement as well as significant discrepancies for certain indictors (e.g. Peters et al. 2012 for CO_2 emissions and Wiedmann et al. 2015 and Schoer et al. 2013 for raw materials).

An important observation was made by Peters et al. (2012) in a pioneering comparative study: differences in consumption-based, embodied CO_2 emissions from different models were mostly due to the use of different territorial emission data and different definitions for allocating emissions to international trade. When adjusting for these issues, results were robust and in reasonable agreement. Larger discrepancies occur when different approaches are used for the calculations. Kastner et al. (2014b) find contradictory results for China's trade in embodied cropland when using physical instead of monetary input-output data. And Schoer et al. (2013) explore the differences of employing life cycle inventory data versus MRIO modelling for raw material equivalents embodied in EU27 imports.

A special issue of Economic Systems Research 2014 (26/3) was devoted to the question of uncertainty in MRIO analysis (Inomata and Owen 2014). Insights gained included the finding that the trade matrix structure (Leontief Inverse) is one major determinant of differences (Owen et al. 2014), likely due to assumptions made during its construction.

Further work remains to be done to improve the accuracy of MRIO models and to increase confidence in their results. This should include an increase in resolution, the use of specific process data in mixed units and hybrid LCA models and regular inter-comparison studies.

4 Is Trade Good or Bad? Some Final Thoughts

Is trade good or bad for sustainability? To answer this question conclusively would require comparing the status quo with the counterfactual of a world without trade. Alas, no one knows what this world would look like. It is easy enough to 'switch off' trade in models and to assume that the final demand is met by domestic production alone. But would final demand be the same? Would countries without trade consume the same amount of the same products? Most likely not. Many countries would certainly not be able to produce the products they import. What is clear, from the facts presented in the introduction, is that trade has been a strong driver of economic growth around the world. Had trade not happened, the GDP of all countries would very likely be much lower than it is today. Trade has also undoubtedly enabled and reinforced an increased exploitation of resources. Was it not for trade, many well-endowed countries would have extracted less materials for their own consumption.

Some studies have tried to quantify the effects of trade on GHG emissions. Using the domestic production assumption, López et al. (2013a) found that 1.1 Gt CO_2 were avoided trough trade between seven world regions in 2009, representing a reduction of 18 % of embodied emissions embodied in trade or 4.4 % of global production emissions. But according to Arto and Dietzenbacher (2014), the increase in trade at the global level and associated embodied emissions between 1995 and 2008 had little effect on total global GHG emissions. This was because 'Although domestically produced goods have been substituted by imports, the production abroad was on average as emission intensive as the production at home' (Arto and Dietzenbacher 2014, p. 5393).

For individual countries, the balance can be positive or negative, depending on the relative carbon intensity of their domestic production compared to the main trading partners. Trade between Spain and China, for example, is said to have increased global emissions by 30 Mt CO_2 in 2005 (López et al. 2013b). Arto et al. (2014), on the other hand, assert that overall Spain has been avoiding emissions through trade between 1995 and 2007.[4]

Undoubtedly, trade has had many economic and social benefits. Yet the economic growth spurred by trade has led to a corresponding rapid growth in physical activity with more raw material extractions, more throughput and more consumption in material terms. Any gains in efficiency achieved through technological advances were offset by this strong growth in demand. And even if domestic activities are strictly regulated – as, e.g. is the case for air pollution in western countries – global impacts are likely to rise further if policies do not address the issue of impacts embodied in trade. Kanemoto et al. (2014) have shown that emissions of air pollutants (SO_2 and NO_x) have been rising rapidly in developing countries, where regulation is missing or patchy. Parts of these emissions are embodied in exports to developing countries. In general, exports from developing and low-income nations are more ecologically intensive for GHG emissions, water, scarcity-weighted water, air pollution, threatened species, biomass, total material flow and ecological footprint than those from developed nations (Moran et al. 2013). This has been confirmed for SO_2 by Grether and Mathys (2013) who argue that trade imbalances tend to aggravate, rather than alleviate existing asymmetries in pollution intensities.

Policies aimed at increasing the sustainability of production and consumption need to go beyond domestic regulation and also target production technologies employed abroad. International cooperation on reducing trade-embodied and total impacts worldwide is the only way to tackle unsustainability at the national scale.

---

[4] Arto et al. (2014) estimated the net emissions avoided (NEA) by Spain through trade between 1995 and 2007 and found that a domestic technology assumption (DTA) based on physical values results in a three times higher estimate of NEA than a DTA based on monetary values. See also Tukker et al. (2013) for a discussion on how the DTA effects the estimation of CO_2 emissions embodied in imports to Europe.

Sato (2014) found that the lion's share of global carbon emissions embodied in trade is concentrated in a relatively small number of product categories of traded goods (amongst the top ten in 2006 were motor spirit (gasoline/petrol), steel, aluminium, motor vehicles, ships/boats and Portland cement). This suggests that focusing trade and mitigation policies on these products may be an effective strategy to tackle at least the pressing issue of global warming.

Acknowledgements I thank Angela Druckman and Roland Clift from the University of Surrey, UK, for the invitation to contribute this chapter and for their helpful comments.

References

Alsamawi, A., Murray, J., & Lenzen, M. (2014a). The employment footprints of nations: Uncovering master-servant relationships. *Journal of Industrial Ecology, 18*(1), 59–70. doi:10.1111/jiec.12104.

Alsamawi, A., Murray, J., Lenzen, M., Moran, D., & Kanemoto, K. (2014b). The inequality footprints of nations: A novel approach to quantitative accounting of income inequality. *PloS One, 9*(10), e110881. doi:10.1371/journal.pone.0110881.

Andrew, R. M., Davis, S. J., & Peters, G. P. (2013). Climate policy and dependence on traded carbon. *Environmental Research Letters, 8*(3). doi:10.1088/1748-9326/8/3/034011.

Arto, I., & Dietzenbacher, E. (2014). Drivers of the growth in global greenhouse gas emissions. *Environmental Science & Technology, 48*(10), 5388–5394. doi:10.1021/es5005347.

Arto, I., Genty, A., Rueda-Cantuche, J. M., Villanueva, A., & Andreoni, V. (2012). *Global resources use and pollution: Volume 1 – Production, consumption and trade (1995–2008)*. Joint Research Centre Technical Reports. Publications Office of the European Union, Luxembourg. Retrieved February 23, 2015, from http://ipts.jrc.ec.europa.eu/publications/pub.cfm?id=5860

Arto, I., Roca, J., & Serrano, M. (2014). Measuring emissions avoided by international trade: Accounting for price differences. *Ecological Economics, 97*, 93–100. doi:10.1016/j.ecolecon.2013.11.005.

Baldwin, R. (2013, July 10). *Misthinking globalisation*. Keynote Lecture by Richard E. Baldwin, Graduate Institute of International and Development Studies, Geneva. 21st international input-output conference of the International Input-Output Association (IIOA). Kitakyushu, Japan. Retrieved February 23, 2015, from http://www.iioa.org/Conference/21st/conference.html

Barrett, J., Peters, G., Wiedmann, T., Scott, K., Lenzen, M., Roelich, K., & Le Quéré, C. (2013). Consumption-based GHG emission accounting: A UK case study. *Climate Policy, 13*(4), 451–470. doi:10.1080/14693062.2013.788858.

Bergmann, L. (2013). Bound by chains of carbon: Ecological–economic geographies of globalization. *Annals of the Association of American Geographers, 103*(6), 1348–1370. doi:10.1080/00045608.2013.779547.

Bruckner, M., Giljum, S., Lutz, C., & Wiebe, K. S. (2012). Materials embodied in international trade – Global material extraction and consumption between 1995 and 2005. *Global Environmental Change, 22*(3), 568–576. doi:10.1016/j.gloenvcha.2012.03.011.

Chen, Z.-M., & Chen, G. Q. (2013). Virtual water accounting for the globalized world economy: National water footprint and international virtual water trade. *Ecological Indicators, 28*, 142–149. doi:10.1016/j.ecolind.2012.07.024.

Chenoweth, J., Hadjikakou, M., & Zoumides, C. (2014). Quantifying the human impact on water resources: A critical review of the water footprint concept. *Hydrology and Earth System Sciences, 18*(6), 2325–2342. doi:10.5194/hessd-10-9389-2013.

Dalin, C., Konar, M., Hanasaki, N., Rinaldo, A., & Rodriguez-Iturbe, I. (2012). Evolution of the global virtual water trade network. *Proceedings of the National Academy of Sciences, 109*(16), 5989–5994. Retrieved February 23, 2015, from http://www.pnas.org/content/109/16/5989.abstract

Davis, S. J., & Caldeira, K. (2010). Consumption-based accounting of CO_2 emissions. *Proceedings of the National Academy of Sciences, 107*(12), 5687–5692. doi:10.1073/pnas.0906974107.

Davis, S. J., Peters, G. P., & Caldeira, K. (2011). The supply chain of CO_2 emissions. *Proceedings of the National Academy of Sciences, 108*(45), 18554–18559. doi:10.1073/pnas.1107409108.

Dittrich, M., & Bringezu, S. (2010). The physical dimension of international trade, part 1: Direct global flows between 1962 and 2005. *Ecological Economics, 69*(9), 1838–1847. doi:10.1016/j.ecolecon.2010.04.023.

Fang, K., & Heijungs, R. (2014a). *Moving from the material footprint to a resource depletion footprint*. Concept Paper, Leiden University, Netherlands.

Fang, K., & Heijungs, R. (2014b). Rethinking the relationship between footprints and LCA. *Environmental Science & Technology, 49*(1), 10–11. doi:10.1021/es5057775.

Foster-McGregor, N., & Stehrer, R. (2013). Value added content of trade: A comprehensive approach. *Economics Letters, 120*(2), 354–357. doi:10.1016/j.econlet.2013.05.003.

Giljum, S., & Eisenmenger, N. (2008). North-south trade and the distribution of environmental goods and burdens: A biophysical perspective (report). In J. Martinez-Alier & I. Roepke (Eds.), *Recent developments in ecological economics* (Vol. I, pp. 383–410). Cheltenham: Edward Elgar.

Giljum, S., Bruckner, M., & Martinez, A. (2014). Material footprint assessment in a global input-output framework. *Journal of Industrial Ecology, 1*, XX–XX. doi:10.1111/jiec.12214

Grether, J.-M., & Mathys, N. A. (2013). The pollution terms of trade and its five components. *Journal of Development Economics, 100*(1), 19–31. doi:10.1016/j.jdeveco.2012.06.007.

Hellweg, S., & Milà i Canals, L. (2014). Emerging approaches, challenges and opportunities in life cycle assessment. *Science, 344*(6188), 1109–1113. Retrieved February 23, 2015, from http://www.sciencemag.org/content/344/6188/1109.abstract

Hertwich, E. G., & Peters, G. P. (2009). Carbon footprint of nations: A global, trade-linked analysis. *Environmental Science & Technology, 43*(16), 6414–6420. doi:10.1021/es803496a.

Hoekstra, A. Y., & Mekonnen, M. M. (2012). The water footprint of humanity. *Proceedings of the National Academy of Sciences, 109*(9), 3232–3237. doi:10.1073/pnas.1109936109.

Hoekstra, A. Y., & Wiedmann, T. O. (2014). Humanity's unsustainable environmental footprint. *Science, 344*(6188), 1114–1117. doi:10.1126/science.1248365.

Hoekstra, A. Y., Chapagain, A. K., Aldaya, M. M., & Mekonnen, M. M. (2011). *The water footprint assessment manual – Setting the global standard*. London: Routledge.

Inomata, S., & Owen, A. (2014). Comparative evaluation of MRIO databases. *Economic Systems Research, 26*(3), 239–244. doi:10.1080/09535314.2014.940856.

ISO. (2006). *International Standard ISO 14040: Environmental management – Life cycle assessment – Principles and framework. ISO 14040:2006(E)*, Second edition 2006-07-01. Geneva: International Organization for Standardisation. Retrieved February 23, 2015, from http://www.iso.org

ISO. (2014). *International Standard ISO 14046:2014 environmental management – Water footprint – Principles, requirements and guidelines*. Geneva: International Organization for Standardisation. Retrieved February 23, 2015, from http://www.iso.org/iso/catalogue_detail?csnumber=43263

Jakob, M., & Marschinski, R. (2013). Interpreting trade-related CO_2 emission transfers. *Nature Climate Change, 3*(1), 19–23. doi:10.1038/nclimate1630.

Jakob, M., Steckel, J. C., & Edenhofer, O. (2014). Consumption- versus production-based emission policies. *Annual Review of Resource Economics, 6*(1), 297–318. Retrieved February 23, 2015, from http://www.annualreviews.org/doi/abs/10.1146/annurev-resource-100913-012342

Kander, A., Jiborn, M., Moran, D. D., & Wiedmann, T. O. (2015). National greenhouse-gas accounting for effective climate policy on international trade. *Nature Climate Change, 5*(5), 431–435. http://dx.doi.org/10.1038/nclimate2555

Kanemoto, K., & Murray, J. (2013). Chapter 1: What is MRIO? MRIO benefits & limitations. In: J. Murray, M. Lenzen (Eds.), *The sustainability practitioner's guide to multi-regional input-output analysis, 2–11*. Common Ground Publishing, On Sustainability, Champaign, Illinois, USA. Retrieved February 23, 2015, from http://onsustainability.cgpublisher.com/product/pub.197/prod.10

Kanemoto, K., Moran, D., Lenzen, M., & Geschke, A. (2014). International trade undermines national emission reduction targets: New evidence from air pollution. *Global Environmental Change, 24*, 52–59. doi:10.1016/j.gloenvcha.2013.09.008.

Kastner, T., Erb, K. -H., & Haberl, H. (2014a). Rapid growth in agricultural trade: Effects on global area efficiency and the role of management. *Environmental Research Letters, 9*(3), 034015. Retrieved February 23, 2015, from http://stacks.iop.org/1748-9326/9/i=3/a=034015

Kastner, T., Schaffartzik, A., Eisenmenger, N., Erb, K.-H., Haberl, H., & Krausmann, F. (2014b). Cropland area embodied in international trade: Contradictory results from different approaches. *Ecological Economics, 104*, 140–144. doi:10.1016/j.ecolecon.2013.12.003.

Kelly, G., & La Cava, G. (2013). Value-added trade and the Australian economy. *Reserve Bank of Australia Bulletin*, March Quarter 2013, 29–38. Retrieved February 23, 2015, from http://www.rba.gov.au/publications/bulletin/2013/mar/4.html

Kloepffer, W. (2008). Life cycle sustainability assessment of products. *The International Journal of Life Cycle Assessment, 13*(2), 89–95. doi:10.1065/lca2008.02.376.

Koopman, R., Wang, Z., & Wei, S.-J. (2014). Tracing value-added and double counting in gross exports. *American Economic Review, 104*(2), 459–494. doi:10.1257/aer.104.2.459.

Kounina, A., Margni, M., Bayart, J.-B., Boulay, A.-M., Berger, M., Bulle, C., Frischknecht, R., Koehler, A., Milài Canals, L., Motoshita, M., Núñez, M., Peters, G., Pfister, S., Ridoutt, B., Zelm, R., Verones, F., & Humbert, S. (2013). Review of methods addressing freshwater use in life cycle inventory and impact assessment. *The International Journal of Life Cycle Assessment, 18*(3), 707–721. doi:10.1007/s11367-012-0519-3.

Lenzen, M., Moran, D., Kanemoto, K., Foran, B., Lobefaro, L., & Geschke, A. (2012). International trade drives biodiversity threats in developing nations. *Nature, 486*(7401), 109–112. doi:10.1038/nature11145.

Lenzen, M., Moran, D., Bhaduri, A., Kanemoto, K., Bekchanov, M., Geschke, A., & Foran, B. (2013). International trade of scarce water. *Ecological Economics, 94*, 78–85. doi:10.1016/j.ecolecon.2013.06.018.

Liu, X. (2010). *The silk road in world history*. Oxford: Oxford University Press. Retrieved February 23, 2015, from http://unsw.eblib.com/patron/FullRecord.aspx?p=547953

López, L. -A., Arce, G., & Kronenberg, T. (2013a). *Pollution haven hypothesis in emissions embodied in world trade: The relevance of global value chains*. Workshop: The wealth of nations in a globalizing world, 18–19 July 2013. Groningen, The Netherlands. Retrieved February 23, 2015, from http://www.rug.nl/research/ggdc/workshops/eframe/e-frame-workshop

López, L. A., Arce, G., & Zafrilla, J. E. (2013b). Parcelling virtual carbon in the pollution haven hypothesis. *Energy Economics, 39*, 177–186. doi:10.1016/j.eneco.2013.05.006.

Los, B., Timmer, M. P., & de Vries, G. J. (2015). How global are global value chains? A new approach to measure international fragmentation. *Journal of Regional Science, 55*(1), 66–92. doi:10.1111/jors.12121.

Marques, A., Rodrigues, J., Lenzen, M., & Domingos, T. (2012). Income-based environmental responsibility. *Ecological Economics, 84*, 57–65. doi:10.1016/j.ecolecon.2012.09.010.

Marques, A., Rodrigues, J., & Domingos, T. (2013). International trade and the geographical separation between income and enabled carbon emissions. *Ecological Economics, 89*, 162–169. doi:10.1016/j.ecolecon.2013.02.020.

Meng, B., Peters, G., & Wang, Z. (2015). *Tracing CO$_2$ emissions in global value chains*. IDE discussion paper no. 486. Institute of Developing Economies (IDE), JETRO, Chiba, Japan. Retrieved February 23, 2015, from http://www.ide.go.jp/English/Publish/Download/Dp/486.html

Minx, J. C., Baiocchi, G., Peters, G. P., Weber, C. L., Guan, D., & Hubacek, K. (2011). A "carbonizing dragon": China's fast growing CO_2 emissions revisited. *Environmental Science & Technology, 45*(21), 9144–9153. doi:10.1021/es201497m.

Moran, D. D., Lenzen, M., Kanemoto, K., & Geschke, A. (2013). Does ecologically unequal exchange occur? *Ecological Economics, 89*, 177–186. doi:10.1016/j.ecolecon.2013.02.013.

Orlowsky, B., Hoekstra, A. Y., Gudmundsson, L., & Seneviratne, S. I. (2014). Today's virtual water consumption and trade under future water scarcity. *Environmental Research Letters, 9*(7), 074007. Retrieved February 23, 2015, from http://stacks.iop.org/1748-9326/9/i=7/a=074007

Owen, A., Steen-Olsen, K., Barrett, J., Wiedmann, T., & Lenzen, M. (2014). A structural decomposition approach to comparing MRIO databases. *Economic Systems Research, 26*(3), 262–283. doi:10.1080/09535314.2014.935299.

Parent, J., Cucuzzella, C., & Revéret, J.-P. (2013). Revisiting the role of LCA and SLCA in the transition towards sustainable production and consumption. *The International Journal of Life Cycle Assessment, 18*(9), 1642–1652. doi:10.1007/s11367-012-0485-9.

Peters, G. (2008). From production-based to consumption-based national emission inventories. *Ecological Economics, 65*(1), 13–23. doi:10.1016/j.ecolecon.2007.10.014.

Peters, G. P., Minx, J. C., Weber, C. L., & Edenhofer, O. (2011). Growth in emission transfers via international trade from 1990 to 2008. *Proceedings of the National Academy of Sciences, 108*(21), 8903–8908. doi:10.1073/pnas.1006388108.

Peters, G. P., Davis, S. J., & Andrew, R. (2012). A synthesis of carbon in international trade. *Biogeosciences, 9*(8), 3247–3276. doi:10.5194/bg-9-3247-2012.

Ridoutt, B. G., & Pfister, S. (2010). A revised approach to water footprinting to make transparent the impacts of consumption and production on global freshwater scarcity. *Global Environmental Change, 20*(1), 113–120. doi:10.1016/j.gloenvcha.2009.08.003.

Sato, M. (2014). Product level embodied carbon flows in bilateral trade. *Ecological Economics, 105*, 106–117. doi:10.1016/j.ecolecon.2014.05.006.

Schoer, K., Wood, R., Arto, I., & Weinzettel, J. (2013). Estimating raw material equivalents on a macro-level: Comparison of multi-regional input-output analysis and hybrid LCI-IO. *Environmental Science and Technology, 47*(24), 14282–14289. doi:10.1021/es404166f.

Schütz, H., Bringezu, S., & Moll, S. (2004). *Globalisation and the shifting environmental burden. Material trade flows of the European Union*. Wuppertal: Wuppertal Institute.

Simas, M., Golsteijn, L., Huijbregts, M., Wood, R., & Hertwich, E. (2014). The "Bad Labor" Footprint: Quantifying the Social Impacts of Globalization. *Sustainability, 6*(11), 7514–7540. Retrieved February 23, 2015, from http://www.mdpi.com/2071-1050/6/11/7514

Simas, M., Wood, R., & Hertwich, E. (2015). Labor embodied in trade. *Journal of Industrial Ecology, 19*(3), 343–356. doi:10.1111/jiec.12187.

Steen-Olsen, K., Weinzettel, J., Cranston, G., Ercin, A. E., & Hertwich, E. G. (2012). Carbon, land, and water footprint accounts for the European Union: Consumption, production, and displacements through international trade. *Environmental Science & Technology, 46*(20), 10883–10891. doi:10.1021/es301949t.

Steffen, W., Richardson, J., Rockström, J., Cornell, S. E., Fetzer, I., Bennett, E. M., Biggs, R., Carpenter, S. R., de Vries, W., de Wit, C. A., Folke, C., Gerten, D., Heinke, J., Mace, G. M., Persson, L. M., Ramanathan, V., Reyers, B., & Sörlin, S. (2015). Planetary boundaries: Guiding human development on a changing planet. *Science, 347*(6223). http://www.sciencemag.org/content/early/2015/01/14/science.1259855.abstract

Timmer, M. P., Erumban, A. A., Los, B., Stehrer, R., & de Vries, G. J. (2014). Slicing up global value chains. *Journal of Economic Perspectives, 28*(2), 99–118. http://www.aeaweb.org/articles.php?doi=10.1257/jep.28.2.99

Tukker, A., & Dietzenbacher, E. (2013). Global multiregional input-output frameworks: An introduction and outlook. *Economic Systems Research, 25*(1), 1–19. doi:10.1080/09535314.2012.761179.

Tukker, A., de Koning, A., Wood, R., Moll, S., & Bouwmeester, M. C. (2013). Price corrected domestic technology assumption—A method to assess pollution embodied in trade using primary official statistics only. With a case on CO2 emissions embodied in imports to Europe. *Environmental Science & Technology, 47*(4), 1775–1783. doi:10.1021/es303217f.

Tukker, A., Bulavskaya, T., Giljum, S., Koning, A. D., Lutter, S., Simas, M., Stadler, K., & Wood, R. (2014). *The global resource footprint of nations – Carbon, water, land and materials embodied in trade and final consumption calculated with EXIOBASE 2.1.* Report from the EU FP7 Project CREEA. Leiden/Delft/Vienna/Trondheim. Retrieved February 23, 2015, from http://www.creea.eu/index.php/7-project/8-creea-booklet

Weinzettel, J., Hertwich, E. G., Peters, G. P., Steen-Olsen, K., & Galli, A. (2013). Affluence drives the global displacement of land use. *Global Environmental Change, 23*(2), 433–438. doi:10.1016/j.gloenvcha.2012.12.010.

Wiedmann, T. (2009). A review of recent multi-region input-output models used for consumption-based emission and resource accounting. *Ecological Economics, 69*(2), 211–222. doi:10.1016/j.ecolecon.2009.08.026.

Wiedmann, T., Lenzen, M., Turner, K., & Barrett, J. (2007). Examining the global environmental impact of regional consumption activities – Part 2: Review of input-output models for the assessment of environmental impacts embodied in trade. *Ecological Economics, 61*(1), 15–26. doi:10.1016/j.ecolecon.2006.12.003.

Wiedmann, T., Wilting, H. C., Lenzen, M., Lutter, S., & Palm, V. (2011). Quo Vadis MRIO? Methodological, data and institutional requirements for multi-region input-output analysis. *Ecological Economics, 70*(11), 1937–1945. doi:10.1016/j.ecolecon.2011.06.014.

Wiedmann, T. O., Schandl, H., Lenzen, M., Moran, D., Suh, S., West, J., & Kanemoto, K. (2015). The material footprint of nations. *Proceedings of the National Academy of Sciences, 112*(20), 6271–6276. http://dx.doi.org/10.1073/pnas.1220362110

Wiedmann, T., Schandl, H., & Moran, D. (2014, June 26). The footprint of using metals: new metrics of consumption and productivity. *Environmental Economics and Policy Studies*, published online, 1–20. doi:10.1007/s10018-014-0085-y

World Bank. (2015). World development indicators – Table 4.8: Structure of demand. The World Bank, Washington, DC, USA. Retrieved February 23, 2015, from http://wdi.worldbank.org/table/4.8

WRI. (2014). *Scaling our impact in urgent times – WRI's strategic plan 2014–2017.* World Resources Institute, Washington, DC, USA. Retrieved February 23, 2015, from http://www.wri.org/about/strategic-plan

WTO. (2013). *International trade statistics 2013 – Table A1a: World merchandise exports, production and gross domestic product, 1950–2012.* 15 November 2013. Geneva: World Trade Organization. Retrieved February 23, 2015, from http://www.wto.org/english/res_e/statis_e/its2013_e/its13_appendix_e.htm

Xu, Y., & Dietzenbacher, E. (2014). A structural decomposition analysis of the emissions embodied in trade. *Ecological Economics, 101*, 10–20. doi:10.1016/j.ecolecon.2014.02.015.

Yu, Y., Feng, K., & Hubacek, K. (2013). Tele-connecting local consumption to global land use. *Global Environmental Change, 23*(5), 1178–1186. doi:10.1016/j.gloenvcha.2013.04.006.

Zhang, C., Beck, M. B., & Chen, J. (2013). Gauging the impact of global trade on China's local environmental burden. *Journal of Cleaner Production, 54*, 270–281. doi:10.1016/j.jclepro.2013.04.022.

Chapter 9
Understanding Households as Drivers of Carbon Emissions

Angela Druckman and Tim Jackson

Abstract Households are accountable for nearly three quarters of global carbon emissions and thus understanding the drivers of these emissions is important if we are to make progress towards a low carbon future. This chapter starts by explaining the importance of using an appropriate consumption perspective accounting framework for assessing the carbon footprint of households. This contrasts from the more commonly used production perspective, as, for many Western countries in particular, once responsibility for emissions embedded in imported goods and services are taken into account, consumption emissions are often higher than production emissions.

The chapter then reviews findings concerning the determinants and composition of the carbon footprint of households, focusing on Western countries. One of the main determinants is income, with carbon footprints increasing with increasing incomes. However, other drivers, such as household size and composition, rural/urban location, diet and type of energy supply, also play a part. Studies show that the majority of an average carbon footprint arises from three domains: transportation, housing and food. Further analyses aimed at gaining a deeper understanding of the motivations behind the activities driving emissions, in particular those due to transportation and housing, show that recreation and leisure pursuits are responsible for a substantial portion of average carbon footprints. Studies indicate, for example, that activities such as spending time with friends and family in and around the home, which are generally low carbon and also enhance well-being, should be encouraged alongside the more mainstream strategies of improving systems of provision of energy, food, housing and transportation.

The finding that income is one of the principal drivers of carbon emissions is a challenging and important issue to address, as, for instance, incomes are arguably the driver of the rebound effect – a phenomenon that confounds attempts to reduce carbon footprints, making reducing emissions more of an uphill task than often acknowledged. This challenge leads us to a wider, whole-systems approach in which we view households as an integral part of the system of production and consumption.

A. Druckman (✉) • T. Jackson
Centre for Environmental Strategy, University of Surrey, Guildford, Surrey GU2 7XH, UK
e-mail: a.druckman@surrey.ac.uk

© The Author(s) 2016
R. Clift, A. Druckman (eds.), *Taking Stock of Industrial Ecology*,
DOI 10.1007/978-3-319-20571-7_9

In summary, industrial ecology, with its wide ranging systems approach as shown in this chapter, has a great deal to contribute to the quest to devise strategies to move towards lower carbon, fulfilling lifestyles.

Keywords Carbon footprint • Consumption-based accounting • Environmental input-output analysis • Household carbon-footprint • Personal carbon-footprint • Rebound effect • Time use • Work-time reductions

1 Introduction

Adam Smith stated that "consumption is the sole end and purpose of all production" (Smith 1904) thus putting consumption firmly in the field of industrial ecology. In this chapter we specifically focus on the carbon emissions caused by household consumption, as these have been estimated to be accountable for around 72 % of carbon emissions on a global basis (Hertwich and Peters 2009; Wilson et al. 2013). Thus the study of households[1] and how the environmental impacts for which they are responsible may be reduced is key to achieving a low carbon future.

Accordingly, in this chapter, we first examine the accounting perspective required to scrutinise the carbon emissions for which household consumption is responsible (Sect. 2). Section 3 reviews the evidence concerning the determinants of household carbon missions. In Sect. 4, we introduce the 'rebound effect' – a phenomenon that confounds attempts to reduce carbon footprints, making reducing emissions more of an uphill task than often acknowledged. In the final section (Sect. 5), we broaden the focus to look at households in the wider context of systems of production and consumption, and possibilities of win-win solutions that offer potential to reduce carbon while at the same time enhancing well-being.

Household consumption is a wide ranging topic and inevitably there are many limitations to this chapter. One of these is that we focus here on consumption by Western households. A second is that we do not review the prolific literature on the driving forces behind household consumption or the ways that household consumption may be reduced.[2] And a third limitation is our focus on 'carbon' emissions. However carbon is defined (see Sect. 2), use of carbon emissions as a single indicator can lead to policies that, while beneficial in terms of reducing global warming, may lead to unexpected and unintended detrimental consequences in terms of other environmental impacts. For example, Benders et al. (2012) analysed five environmental impact categories: global warming potential, acidification, eutrophication,

---

[1] We focus here on household carbon footprints as much of the consumption that gives rise to carbon emissions, such as energy use for space heating, arises at a household level. Estimation of per capita (personal) carbon footprints requires division by the number of people in a household with appropriate apportioning to children. Apportioning to children is generally done using equivalence scales (OECD 2015).

[2] For an overview of these literatures see Jackson (2005, 2006, 2009).

summer smog and land use. Combined analysis of the five impact categories found that food has the largest environmental impact,[3] whereas analysis of greenhouse gas emissions alone indicated that housing has the largest impact. Nevertheless, climate change caused by anthropogenic carbon emissions is currently accepted as the most urgent environmental threat (IPCC 2014) and is thus considered a useful indicator for the focus of this chapter.

2   Consumption Accounting and Carbon Footprinting

In this Section, we first set out the importance of the type of accounting framework used for exploring household carbon footprints, and explain how this is different from the default framework normally applied by governments in assessing their emissions and in international treaties. The framework is best introduced by posing the question: to what extent should Western consumers take responsibility for the things they buy? If, say, a UK consumer purchases a TV manufactured in China, which nation should take responsibility for the emissions incurred during its manufacture? This dilemma illustrates two different accounting approaches that must be untangled as we strive to devise strategies for a more sustainable future. According to accounting by the production perspective, China should take responsibility as the emissions arose on Chinese territory. This is the approach used in the Kyoto Protocol and is the most commonly used accounting approach (Bows and Barrett 2010; Wiedmann 2009). An alternative is the consumption perspective. According to this perspective, the UK should take responsibility, as export to the UK was the driving force motivating production, and a UK consumer is the primary beneficiary of the final product (Druckman et al. 2008; Peters and Hertwich 2008a; Peters et al. 2011; Lenzen 2008; Lenzen et al. 2007; Jackson et al. 2006).

The accounting perspective used is particularly important because accounting according to the production perspective shows that many Western economies are successfully reducing their carbon emissions. However, when the consumption perspective is used for accounting, not only are carbon emissions often found to be higher than compared to the production accounts, but they also tend to exhibit a rising trend (CCC 2013; Baiocchi and Minx 2010; Ahmad and Wyckoff 2003; Peters and Hertwich 2006a; Baiocchi et al. 2010). The reason for the differences shown between the two accounting perspectives is the quantity of carbon emissions embedded in trade, which is the subject of Wiedman's chapter (Chap. 8) in this book. An example of the importance of the carbon embedded in trade is given by Li and Hewitt (2008) who found that, through trade with China, the UK reduced its production based carbon dioxide emissions by approximately 11 % in 2004, compared with a non-trade scenario in which the same type and volume of goods are produced in the UK.

---

[3] They analysed 12 COICOP domains: Food, alcohol & tobacco, clothing, housing, furniture, health, transport, communication, recreation, education, restaurants, others.

Table 9.1 Recent definitions of a carbon footprint

"The carbon footprint is a measure of the exclusive total amount of carbon dioxide emissions that is directly and indirectly caused by an activity or is accumulated over the life stages of a product" (Wiedmann and Minx 2007: 4)
"A carbon footprint is equal to the greenhouse gas emissions generated by a person, organization or product" (Johnson 2008: 1569)
"A measure of the total amount of CO_2 and CH_4 emissions of a defined population, system or activity considering all relevant sources, sinks and storage within the spatial and temporary boundary of the population, system or activity of interest. Calculated as CO2e using the relevant 100-year global warming (GWP100)" (Wright et al. 2011: 69)
"Climate footprint: A measure of the total amount of CO_2, CH_4, nitrous oxide, hydrofluorocarbons, perfluorocarbons and sulfur hexafluoride emissions of a defined population, system or activity considering all relevant sources, sinks and storage within the spatial and temporal boundary of the population, system or activity of interest. Calculated as CO_2 equivalents using the relevant 100-year global warming potential" (Williams et al. 2012: 56)
"A measure of the amount of carbon dioxide released into the atmosphere by a single endeavour or by a company, household, or individual through day-to-day activities over a given period" (Collins English Dictionary 2012)

Source: Birnik (2013: 281)

While many studies explore the carbon emissions embedded in trade,[4] some studies focus specifically on the role of imported goods and services and their associated emissions in the carbon footprints of households[5] (Hertwich and Peters 2009; Lenzen et al. 2006; Munksgaard et al. 2005; Nijdam et al. 2005; Peters and Hertwich 2006b). Peters and Hertwich (2006a) put forward a general rule that countries with a high proportion of imports and relatively clean electricity generation are likely to have a significant proportion of their household carbon emissions attributed to imports. This means that, due to the supply chain emissions embedded in imported goods, households drive emissions in other countries as well as in their own country. For example, Weber and Mathews (2008) found that nearly 30 % of the carbon dioxide emitted to meet household demand in the US occurred outside the borders of the US.

Accounting according to the consumption perspective is commonly known as 'footprinting': this is the approach adopted in this chapter, and in particular the chapter is concerned with carbon footprinting. However, the definition of what is included in a carbon footprint is contentious, as shown in Table 9.1. In this chapter

---

[4] See for example: Davis and Caldeira (2010), Ahmad and Wyckoff (2003), Andrew et al. (2013), Atkinson et al. (2011), Cave and Blomquist (2008), Hertwich and Peters (2009), Lin and Sun (2010), Maenpaa and Siikavirta (2007), Munksgaard et al. (2005), Nakano et al. (2009), Peters and Hertwich (2008b), Peters et al. (2011), Shui and Hariss (2006), Weber and Peters (2009) and Knight and Schor (2014).

[5] Consumption accounting attributes carbon emissions to the 'final demand' of a country and is based on the UN System of National Accounts. According to this system, final demand is composed of government expenditure, capital investment, exports and household expenditure. Although there is an argument that government expenditure should be re-allocated to households, as government exists to serve households, it is generally kept as a separate category. A similar argument relates to investment (Hertwich 2011). In consumption accounting exports are excluded but imports are included.

we take a relaxed approach to what we mean by 'carbon' and include reviews of studies that range from assessing carbon dioxide emissions only to those that take a more comprehensive greenhouse gas approach. The main difference in results is that emissions due to food make up a larger portion of the carbon footprint of a household when the analysis is extended to a basket of greenhouse gases. What we are more stringent about in this chapter is that we take a whole supply chain, life cycle approach to assess the carbon emissions caused by households.

There are two basic categories of a household carbon footprint. First are direct emissions that arise due to direct energy use in the home (such as gas for space and water heating, and electricity for lighting and powering appliances and gadgets) and due to burning personal transportation fuels (petrol and diesel). Second is 'embedded' emissions, such as those that arise during our example of the manufacture of a TV made in China. Embedded emissions along supply chains (arising domestically and abroad) account for the majority (around 60–70 %) of the carbon footprints of Western households (Druckman and Jackson 2010; Dey et al. 2003; Bin and Dowlatabadi 2005; Baiocchi et al. 2010).[6]

Estimates of household carbon footprints are generally derived from expenditure data. Carbon emissions arising from expenditure on transportation fuels and energy use in the home are relatively easily estimated from information on prices, the carbon content of fuels and information from each country's Environmental Accounts. Estimation of carbon emissions embedded in other expenditures is harder and requires information on the technologies used to manufacture all products and services purchased, wherever in the world this may occur. This is generally done using Environmentally Extended Input-Output Analysis (EE-IOA) (Hertwich 2011; Munksgaard et al. 2005; Baiocchi et al. 2010; Weber and Matthews 2008; Weber and Perrels 2000; Lenzen et al. 2004; Wiedmann 2009). EE-IOA is a top-down methodology that combines information on the structure of the economy with environmental data (see Miller and Blair (2009)). There are some notable exceptions to this methodology. The first is hybrid analysis which combines process-based, bottom-up Life Cycle Assessment (LCA) with top down EE-IOA (Benders et al. 2012). Another exception is the work by Girod and de Haan (2009, 2010) who use a bottom-up LCA methodology only, based on physical functional units such as kg of food, person kilometres and living square meters.

3 What Makes a Household Carbon Footprint?

In this section we first examine the major socio-economic drivers of household footprints, we then explore the composition of average carbon footprints, and the final sub-section examines carbon footprints from the perspective of time-use.

---

[6] Estimates include 60 % for USA (Bin and Dowlatabadi 2005), 66 % and 70 % for the UK (Druckman and Jackson 2010 and Baiocchi et al. 2010, respectively), and 70 % for Australia (Dey et al. 2003).

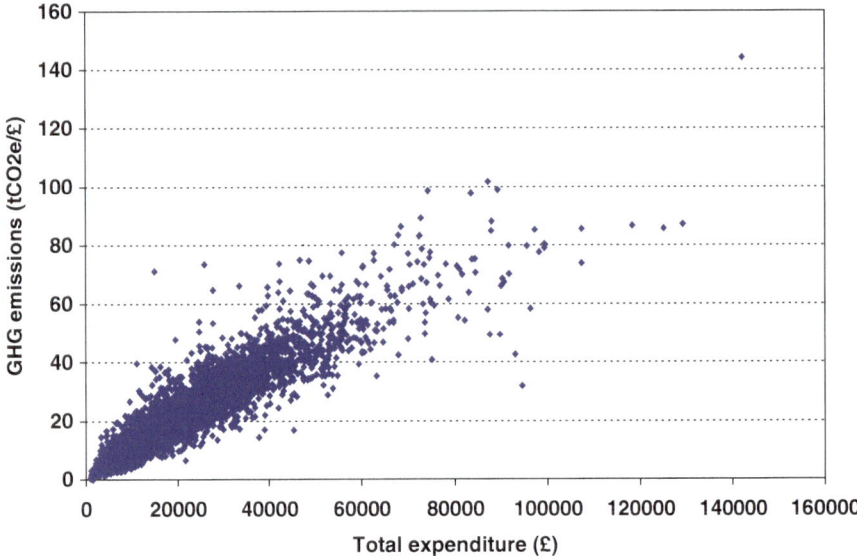

Fig. 9.1 The relationship between household carbon footprints and total household expenditure (2009) (Source: Chitnis et al. 2014, p. 21)

3.1 The Determinants of Household Carbon Footprints

One of the most important factors determining the carbon footprint of Western households is household income,[7] with (as illustrated in Fig. 9.1) household footprints generally increasing with income (Wier et al. 2001; Dey et al. 2003; Weber and Matthews 2008; Buchs and Schnepf 2013; Baiocchi et al. 2010; Gough et al. 2011; Kerkhof et al. 2009; Chitnis et al. 2014). As discussed by Baiocchi et al. (2010) and Dey et al. (2003), this finding dispels the 'Kuznets curve' theory according to which, as nations become more developed, incomes rise and emissions are hypothesized to fall.

Whereas the relationship between income and carbon footprint is strong, studies have shown that as incomes increase consumers tend to shift their expenditures away from carbon intensive 'necessities' towards discretionary expenditures that are generally less carbon intensive (Buchs and Schnepf 2013; Weber and Matthews 2008; Chitnis et al. 2014; Jones and Kammen 2011). For example, Fig. 9.2b shows that lower income US households tend to incur a greater proportion of their carbon emissions from 'necessities' such as food and home energy, and in particular the home energy emissions tend to arise from direct energy use (gas and electricity). Additionally, Chitnis et al. (2014), in a study of UK households, showed that high income households incur a higher proportion of their carbon due to 'Recreation and

---

[7] Total household expenditure is often used as a proxy for income, as total household expenditure is generally more accurately captured in surveys than income.

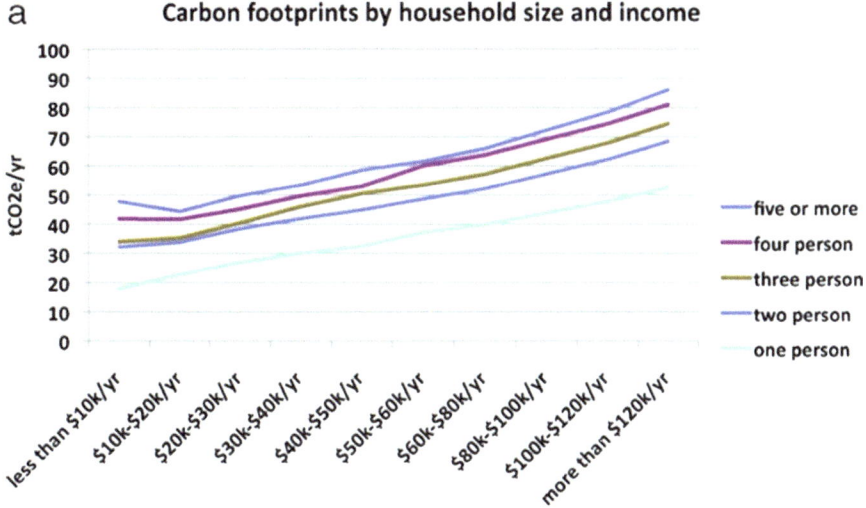

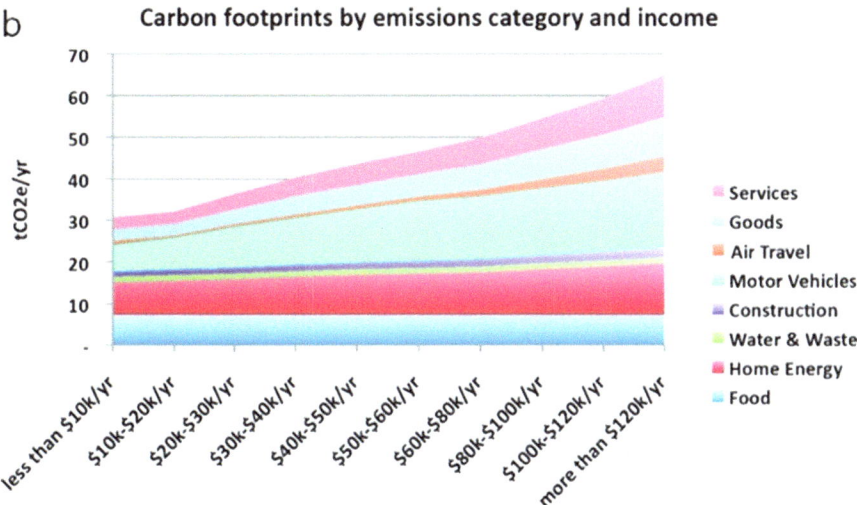

Fig. 9.2 (**a**) Carbon footprints by income bracket and household size; (**b**) Carbon footprints by category of emissions and income bracket for average household size of 2.5 persons (Source: Jones and Kammen (2011), Fig. 2, p. 4090)

Culture' than lower income households. Weber and Mathews (2008) also point out that there tends to be a higher diversity in the carbon footprint of households with higher incomes/total household expenditure.

Exceptions to the general pattern of low income households incurring a greater proportion of their carbon emissions from direct energy use than upper income households was found by Kerkhof et al. (2009) who compared four countries. While they found this pattern for UK and The Netherlands, they found reverse trends in

Sweden and Norway and attributed this mainly to the use of district heating in Sweden and the use of low carbon intensity electricity for heating in Norway.

Household size is generally found to be an important determinant of household carbon emissions (see Fig. 9.2a), as households with more people tend to benefit from economies of scale (Dey et al. 2003; Baiocchi et al. 2010; Jones and Kammen 2011; Weber and Matthews 2008; Tukker et al. 2010; Gough et al. 2011). As Tukker et al. (2010) explain, this is because people sharing a dwelling also share energy using appliances and cohabitants tend to require less living space than single occupants: this reduces the energy required for heating and cooling. Buchs and Schnepf (2013) note that economies of scale are less important for transport and indirect emissions, and Gough et al. (2011), in their analysis of different UK household types, found that younger single person households tend to emit relatively high amounts due to transport and personal services.

Gough et al. (2011) found statistically significant differences between the emissions of UK households according to employment status, with the working households exhibiting higher emissions when income and composition are controlled for, and the unemployed and unoccupied having lower emissions. The explanation Gough puts forward is that work-rich households tend to have higher emissions due to commuting and tend to substitute purchased goods and services for 'household production'. Buchs and Schnepf (2013) added to this by noting that workless households tend to have higher emissions due to home energy use.

Urban locations are generally more efficient in terms of direct emissions than rural locations (Wier et al. 2001; Jones and Kammen 2011; Buchs and Schnepf 2013; Baiocchi et al. 2010; Glaeser and Kahn 2010; Tukker et al. 2010). One reason for this is that urban transportation distances tend to be shorter with greater availability of public transport options. Another reason is that urban dwellings tend to be smaller and therefore more efficient to heat (Tukker et al. 2010; Wier et al. 2001; Baiocchi et al. 2010). Also the 'heat island effect' lowers energy required for space heating in urban locations[8] (EPA2014). However, as Baiocchi et al. (2010) point out, the general rule of urban households requiring less direct energy and hence having lower carbon footprints is, in some instances, counterbalanced by the fact that poorer rural households living in rural locations may not be able to afford a car or long recreational trips by aeroplane.

Households dwelling in extreme climates generally incur higher carbon emissions due to energy use for space heating and/or air conditioning (Tukker et al. 2010); however this effect is moderated by other factors, such as the type of energy supply and housing construction. For example, Kerkhof et al. (2009) attributed the higher household carbon emissions for space heating in the UK and the Netherlands than in Sweden and Norway to use of natural gas in the first two countries, district heating in Sweden and low carbon-intensity electricity in Norway. The carbon intensity of the electricity supply also effects household carbon footprints even if it is only used for powering lights, appliances and gadgets and not for heating, as intensities vary widely: for example, electricity from geothermal sources in Iceland

---

[8] This can reverse in hot climates, with urban locations needing more cooling.

has an intensity of just 0.00018 kg CO_2/kWh, and in Norway the intensity is 0.013 kg CO_2/kWh, compared to the EU average of 0.35 kgCO_2/kWh (DEFRA et al. 2014).

The construction of housing also, of course, affects the carbon footprint. For example, in the UK, much of the housing stock is hard to insulate adequately at reasonable costs (Hong et al. 2006) and hence occupiers of these dwellings tend to have relatively high carbon footprints. Also the control systems installed, such as thermostats, affect carbon footprints (Tukker et al. 2010).

The carbon emissions embedded in food products generally forms a substantial portion of a carbon footprint (see Sect. 3.2), in particular when analysed in terms of greenhouse gas emissions instead of carbon dioxide only (Dey et al. 2003; Nijdam et al. 2005; Tukker and Jansen 2006; Druckman and Jackson 2009, 2010). The type of diet has a high impact on this. In general, vegetarians and consumers who eat locally harvested seasonal food tend to have lower per capita environmental impacts from food consumption than individuals who rely on more traditional diets (Garnett 2013; Tukker et al. 2010).

Education has also been found to play a role in determining household carbon emissions, with high education being significant and positively related to emissions once income is controlled for (Baiocchi et al. 2010; Buchs and Schnepf 2013). Baiocchi et al. (2010) interpret this as support for justification for environmental education campaigns.

Other factors that influence household carbon footprints include social and cultural differences. This includes how people use their household control systems (Wood and Newborough 2007), whether, for example, it is the social norm to wear a jersey indoors during cold weather (Druckman et al. 2011b; Shove 2012), and the prevalence of a 'throwaway', consumerist culture, as opposed to a more thrifty way of living (Cooper 2010).

3.2 Composition of Household Carbon Footprints

In this sub-section, we look in more detail at the composition of average household carbon footprints of Western households. Generally the categories of transportation, housing and food make the largest contributions (Jones and Kammen 2011; Caeiro et al. 2012; Tukker 2006; Tukker and Jansen 2006). For example, Benders et al. (2012) found that these three domains account for nearly three quarters of carbon emissions and inclusion of the next largest category, recreation, accounted for around 85 % of average carbon footprints in The Netherlands. Jones and Kammen (2011) (see Fig. 9.3) assessed carbon emissions of an average US household in five main categories, with further sub-divisions and also making a distinction between direct and indirect emissions (blue and green in Fig. 9.3, respectively). While supporting the general findings that the broad categories of transportation, housing and food make up the majority of emissions, their analysis found that direct emissions from motor fuels was the largest sub-category, at around 20 % of the total, with electricity consumption coming next (15 %), followed by emissions due to meat consumption (5 %).

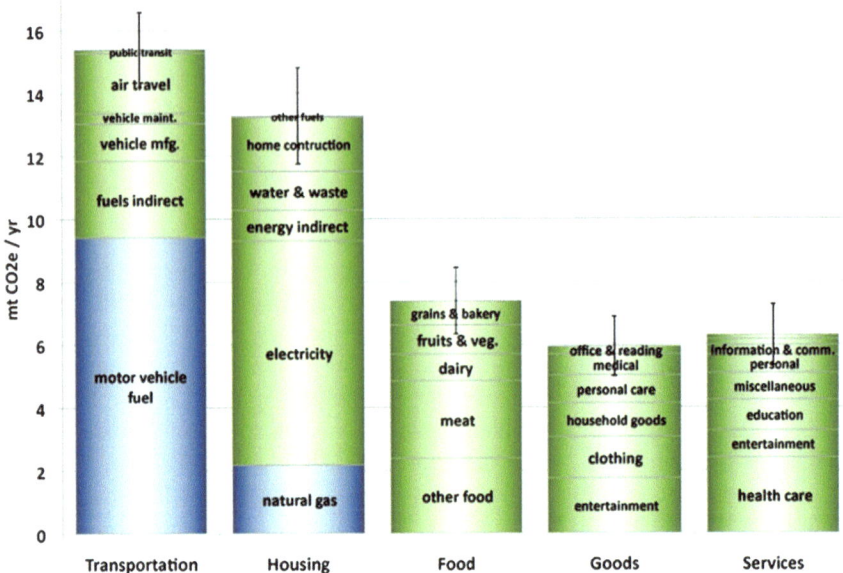

Fig. 9.3 Total carbon footprint of a typical US household (48 tCO₂e/yr) (Source: Jones and Kammen (2011), Fig. 1, p. 4090)

Studies vary in the number of categories used for analysing household carbon footprints, as shown in Table 9.2. Some studies use the top 12 categories of the Classification of Individual Consumption According to Purpose (COICOP) system which is part of the UN System of National Accounts (UN 2011).[9] COICOP categories are, however, primarily intended for economic rather than environmental analysis and so other researchers modify the categories to reveal the carbon implications of expenditures better. For example, Weber and Mathews (2008) add an extra category of 'Utilities/home energy'.

Travel is rarely undertaken as an end in itself, as it is generally undertaken to serve a purpose such as visiting friends, attending a football match or going to work. Similarly, water heated by gas may be used for food related activities such as washing up, or, for example, for health and hygiene purposes. Acknowledging this, and to further elucidate the activities that give rise to carbon emissions, Druckman and Jackson (2009, 2010) allocate carbon emissions to 'functional uses'. In this approach all carbon emissions that arise due to activities related to food (for example), such as emissions due to driving to supermarkets, energy used in preparing food, cooking and washing-up, emissions embedded in the production of food, and even those

---

[9] The 12 top COICOP categories are: Clothing and footwear; Housing, water, electricity, gas and other fuels; Furnishings, household equipment and routine household maintenance; Health; Transport; Communication; Recreation and culture; Education; Restaurants and hotels; Miscellaneous goods and services.

Table 9.2 A summary of selected studies on the carbon footprints of households

Source	Country	Number of categories	Carbon dioxide (CO_2) or greenhouse gases (GHG)?
Druckman and Jackson (2010)	UK	44	GHG
Jones and Kammen (2011)	USA	27	GHG
Dey et al. (2003)	Australia	17	GHG
Benders et al. (2012)	The Netherlands	12	GHG
Gough et al. (2011)	UK	5	GHG
Kerkhof et al. (2008)	The Netherlands	5	GHG
Jackson et al. (2006)	UK	27	CO_2
Bin and Bowlatabadi (2005)	USA	18	CO_2
Weber and Mathews (2008)	USA	13	CO_2
Baiocchi et al. (2010)	UK	12	CO_2
Kerkhof et al. (2009)	The Netherlands, UK, Sweden, Norway	12	CO_2
Druckman and Jackson (2009)	UK	9	CO_2

embedded in running the supermarkets, are attributed to the category 'Food and Catering'. The exception to this is emissions due to space heating which are included as a separate category as they account for such a high proportion of carbon emissions (13 %). Druckman and Jackson's (2010) analysis shows that there is an element of travel emissions in all categories apart from space heating. They find that while there are a great deal of carbon emissions tied up in the mundane activities of everyday life, such as keeping families warm ('Space Heating' 13 %), fed ('Food & Catering' 24 %), safe and secure ('Household'[10] 11 %) and clothed ('Clothing & Footwear' 8 %), 'Recreation & Leisure' is, however, the largest category at around 27 % (Druckman and Jackson 2010).

Understanding emissions due to recreation and leisure is important for a number of reasons: they arise due to 'discretionary' activities, and so this category may offer rich opportunities for reductions; this category accounts for a substantial proportion of the carbon footprint as described above (Druckman and Jackson 2009, 2010; Benders et al. 2012); emissions in this category are generally increasing, with energy intensive forms of leisure (such as flying on holidays) generally increasing whereas less energy intensive leisure activities, such as reading, are stable or decreasing (Aall et al. 2011). Also there has been an increasing 'materialisation' of leisure practices, whereby, for example there are increasing tendencies to buy specialist equipment and clothing for walking and other such pursuits (Aall et al. 2011).

---

[10] The Household category comprises the carbon emissions that are associated with constructing, occupying and running a dwelling.

In order to further understand emissions due to recreation and leisure, Druckman and Jackson (2010) divided recreation and leisure into 12 sub-categories with particular focus on holiday/non-holiday activities. They found that carbon emissions due to holidays account for around 10 % of an average UK household's entire carbon footprint. Of this, over half (52 %) of holiday emissions were found to be due to aviation, and when this was added to other holiday-related transport emissions, transportation accounted for nearly three quarters (74 %) of 'Holiday' emissions. Emissions due accommodation services in hotels were found to make up around just 16 % of 'Holiday' emissions. These figures give us an indication of how carbon emissions might be reduced, primarily in this case through reducing holiday travel emissions. Holidays are, however, a particularly difficult area to tackle: as Barr et al. (2010) said '*A holiday is a holiday*' during which people take a vacation from their environmental behaviour. Aviation emissions are, in particular, growing rapidly, and, due to political difficulties in introducing policies to restrict aviation demand, it is considered unlikely that this trend will be reversed (Macintosh and Wallace 2009).

3.3 Looking Through the Lens of Time-Use

How we use our time is a key determinant in the emissions for which we are responsible, in particular in the case of discretionary time-use, such as during recreation and leisure. Additionally, looking through the lens of time-use allows allocation of emissions due to space heating to functional uses.

Although researchers such as Minx and Baiocchi (2009) and Becker (1965) have laid out theoretical foundations for the relating how people use their time to sustainability, Godbey (1996) and Godbey et al. (1998) have explored the relationship between generation of municipal solid waste and time-use in USA, and Jalas (2002) have related direct and indirect energy use to use of time in Finland, to our knowledge the only study relating carbon emissions to time-use is Druckman et al. (2012). In their analysis of the 'carbon emissions per hour' of different activities for an average British household, Druckman et al. (2012) found the most carbon intensive uses of time are 'Personal Care' (which includes personal washing, clothes and care of clothing, and health care), 'Eating & Drinking' (which includes alcohol and eating out) and 'Commuting'. Apart from 'Sleep & Rest', the broad category of 'Leisure and Recreation' has the lowest intensity. However, 'Leisure and Recreation' is the second largest time-use category at 5.7 h per day on average, only exceeded by 'Sleeping and Resting' at 8.9 h per day (ONS 2006). Further analysis of leisure and recreation (see Fig. 9.4) showed clearly that activities in and around the home are the lowest in carbon emissions per hour, and that moving away from the home, thus incurring emissions due to transportation, increases emissions. Indeed, they found that emissions for 'Sports and Outdoor Activities' were nearly three times as carbon intensive as 'Spending time with family/friends at home'.

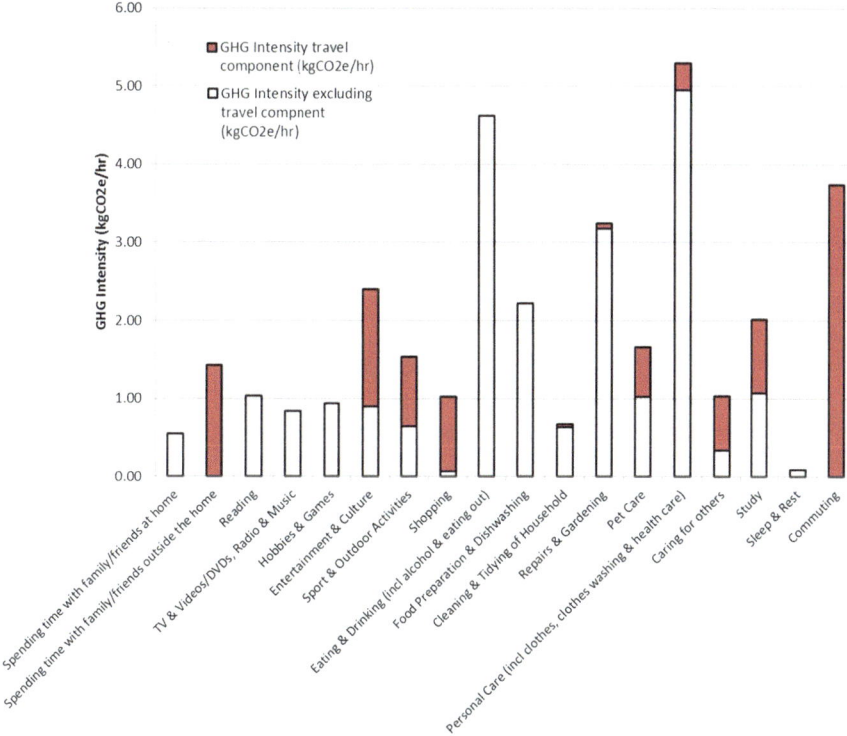

Fig. 9.4 Carbon intensity of time use for an average British household (2004) (Source: Druckman et al. (2012), Fig. 3a, p. 156)

4 The Rebound Effect

The discussions above lead to many suggestions concerning ways that the carbon footprints of households may be reduced, but this is beyond the scope of this chapter, as explained in Sect. 1. However, a systemic issue that works against many measures suggested for reducing emissions (such as installing loft insulation and travelling less) is the 'rebound effect'. The rebound effect, in relation to households, can be explained as follows (Sorrell 2007; Maxwell et al. 2011): When an action is carried out that is intended to save energy, it will often result in saving money also. However, a household always uses its income in some way or other. For example, when purchasing a car, suppose the purchaser decides to buy one that is more fuel efficient than the average car on the market. Knowing his normal mileage, he can calculate the fuel saved, and hence by how much he will expect to reduce his carbon footprint. However, as less fuel is now used for his normal journeys, less money is spent on this fuel. This freed up money might be spent on driving further, which will result in more carbon emissions. This is called the direct rebound effect. Alternatively, the money saved might be spent on something entirely different from motor vehicle fuel, such as taking a vacation. This will also give rise to more emissions, and this is known as the indirect rebound effect. Alternatively, he might decide, rather than to spend the money, to save it and therefore he puts it on deposit in a bank. The bank, however, then invests the money, and this investment, in turn, gives rise to carbon emissions. This is another example of the indirect rebound effect.

Another type of rebound effect that commonly arises is the 'embodied' rebound effect, and this is better illustrated through an example of loft insulation. In this example a person who installs loft insulation can calculate how much energy (and hence carbon emissions) will be saved through reduced fuel use. However, energy is used in the manufacture of the loft insulation, and, following the consumption accounting principle discussed in Sect. 1, carbon emissions from this energy use are the purchaser of the insulation material's responsibility. Hence these emissions off-set the expected savings, and this is known as the embodied rebound effect.

If a measure is expected to achieve a reduction of 100 $kgCO_2e$ then a rebound effect of 30 % implies that only 70 $kgCO_2e$ was saved, and a rebound effect of 100 % implies that no carbon was saved. A rebound effect greater than 100 % means that the measure resulted in more, not less, emissions, and, from this view, it would have been better not to have done the action at all. This is known as 'backfire'.

Until relatively recently, although the rebound effect was a well-known phenom-enon, there were few studies that had estimated to what extent it is a problem with respect to households. In the last few years, however, studies have been carried out to explore it focusing on various different countries. These include Lenzen and Dey (2002) and Murray (2013) for Australia; Alfredsson (2004) and Brännlund et al. (2007) for Sweden; Mizobuchi (2008) for Japan; Kratena and Wuger (2010) for Austria; and Thomas and Azevedo (2013) for US and Druckman et al. (2011a) and Chitnis et al. (2013, 2014) for the UK. These studies generally consider a variety of measures such as abatement actions (for example, reducing the amount of food wasted, reducing household room temperature thermostat settings and replacing short car journeys by walking or cycling) and energy efficiency measures (for exam-ple, installation of cavity wall insulation, loft insulation, condensing boiler, water tank insulation, energy efficient lighting and purchase of an efficient car). Chitnis et al. (2014), who estimated the rebound effect in terms of GHG emissions, found rebound to be around 0–32 % for measures affecting domestic energy use and around 25–65 % for measures affecting vehicle fuel. The possibility of backfire was found for measures that reduce food waste, with estimates being around 66–106 % (Chitnis et al. 2014). In general, rebound was found to be larger for lower income groups (with some exceptions) as they have a higher proportion of expenditure on direct energy (as discussed in Sect. 3.1) and this expenditure has relatively high income elasticities (Chitnis et al. 2014).

The conclusion from this rebound effect work is *not* that encouragement to carry out the abatement and energy efficiency actions should be abandoned: indeed, for all except food waste under certain conditions, considerable carbon emissions can be saved through these means and therefore it is imperative that such actions should be supported. However, it is vital that governments take into account the rebound effect when estimating reductions in carbon emissions that can be achieved, else they stand in danger of systemically missing their carbon reduction targets.

Nevertheless, efforts should be made to minimize the rebound effect wherever possible. The best way to do this is to encourage a wholesale shift in expenditure patterns towards low carbon goods and services. The rebound effect studies also highlight the importance of investment decisions, and Druckman et al. (2011a) show that in order to achieve zero rebound, the money saved through the abatement or efficiency actions should be invested in carbon neutral or reducing investments.

5   Concluding Comments

This chapter has explored the drivers and components of household carbon footprints. Evidence shows that 'hair-shirt' policies, particularly within the realm of recreation and leisure, are unlikely to gain enough traction to achieve the widespread changes needed (Soper 2008). The 'holy grail' is thus to devise low carbon lifestyles that achieve maximum happiness. However, economic growth (the policy goal of most governments[11]) aims to increase incomes. But it is generally found that as incomes increase, carbon footprints are likely to increase while well-being levels off (Lenzen and Cummins 2013; Jackson 2009). This raises the question: which policies enhance well-being, or at least do not reduce well-being, while being environmentally beneficial? Such activities represent win-win opportunities for encouraging activities which give rise to relatively low quantities of carbon emissions while at the same time enhancing well-being and happiness.

Reviews of the literature reveal that social activities such as conversing with friends and family, making love, reading and carrying out hobbies are low carbon activities that generally make people happy (Csikszentmihalyi 2006; Holmberg et al. 2012; Kahneman et al. 2004; Caprariello and Reis 2012; Nassen and Larsson 2015). For many of the activities that generally enhance happiness, the carbon emissions depend on how they are carried out. For example, being close to nature and physical activities such as walking, exercising and sport can be relatively low carbon if carried out without the use of personal transportation. Csikszentmihalyi (2006) talks about how goal-orientated activities can induce high levels of happiness. His theory is that when a person is carrying out an activity that is all-encompassing, in that the activity requires total concentration and focus (in other words, the person is "in the flow") then a high state of happiness can be achieved. Examples of this include playing a musical instrument or singing in a choir, both of which can be done in relatively low carbon ways, but one of Csikszentmihalyi's examples is the state of flow achieved during downhill skiing, and, depending on where one lives, this can be a very high carbon activity. Gatersleben et al. (2008) investigated how volunteering can yield high levels of happiness and, again, this may be carried out in high or low carbon ways. Shopping is an example of an activity that generally brings happiness, but is, arguably, rarely a particularly low carbon activity.

---

[11] The notable exception to this was Bhutan which has had for some years, the goal of increasing gross national happiness (Zurick 2006).

This discussion has highlighted some win-win approaches to reducing carbon emissions while increasing well-being, and these should be key components of strategies for moving towards a more sustainable future. But before closing this chapter it is worth taking stock and standing back to take a whole-systems approach.

A whole-systems approach requires looking at systems of production and consumption in which households play a central role. The economy is circular in nature: in simplistic terms, households earn wages from firms, and firms produce goods and services to sell to the households. Thus producers are consumers, and consumers are producers. Linking this understanding with the earlier discussion in which it was shown that one of the main determinants of a household's carbon footprint is income, and also that households spend or invest all their income, raises another possible win-win situation: that of working-hours reduction.

Reducing the average number of hours worked per week can have both a scale effect and a compositional effect (Gough 2013). Hypothetically, due to the scale effect of fewer hours at work, workers' incomes would be reduced, and thus expenditures and consumption would also be expected to be reduced. With each person working less, there is the possibility of increasing the number of people employed and thus reducing inequalities. High levels of inequality are associated with low levels of well-being (Wilkinson and Pickett 2009), and, furthermore, meaningful work is a generally found to be a positive factor in increasing well-being (Diener and Seligman 2004). Hence sharing the work may yield multiple benefits (Hayden 1999).

The compositional effect can be explained as follows: with lower incomes but less time at work, people's use of time outside work would be expected to change, as would the composition of their expenditure baskets. For example, rather than buying ready-meals, people may be more inclined to cook from raw ingredients. Now such changes in time and expenditure budgets might result in higher or lower carbon emissions. For example, with less time pressure, people might walk and cycle for short journeys rather than drive. On the other hand, some people may drive further and more often to visit friends. But if we look back to the graph in Fig. 9.1, we see that there is good evidence that lower incomes will, in general, result in lower carbon footprints.

Reducing the working week has been shown to enhance the work-life balance (Nassen and Larsson 2015; Kasser and Sheldon 2009; Eurofund 2013). For example, Hayden (1999) records how French employees reported overall improved quality of life when their working week was reduced to 35 h. In another investigation 400 Swedish employees who had their worktime reduced to 6 h per day for 18 months reported improved life satisfaction, health and a more equal gender-balance on time spent on housework (Bildt (2007) cited in Nassen and Larsson (2015)).

The suggestion of reducing working hours must be taken with an important warning concerning low income groups. Currently many low paid workers are struggling to meet their weekly household expenses (MacInnes et al. 2014; The Living Wage Commission 2014), and therefore any initiative to reduce the working week must be accompanied by special measures to protect them. If these are put in place, then work-time reduction offers a promising way to reduce unemployment by sharing the work, leading to reduced inequalities, while at the same time offering high prospects of increasing well-being and reducing environmental burdens (Hayden and Shandra 2009; Victor 2008; Jackson 2009; Coote et al. 2010; Knight et al. 2013; Pullinger 2014; Rosnick and Weisbrot 2007).

In conclusion, this chapter has reviewed the main determinants of Western household carbon footprints. What is clear from this body of work is that, seen from a consumption perspective, the majority of carbon impacts arise from transportation, food and housing. The need to improve systems of provision of food, energy and transportation and renovate or rebuild inefficient housing stock is therefore indisputable. However, where possible these measures should be supplemented by other approaches. For instance, through further analysis it is evident that recreation and leisure leads to the single highest proportion of household carbon emissions. Opportunities should therefore be sought for low carbon leisure activities which also enhance wellbeing. Such activities might include for instance spending time with friends and family in and around the home, or engaging in physical recreation in the local community.

One inescapable finding from this body of work is that income is one of the principal drivers of carbon emissions, with carbon footprints increasing with increasing incomes. Incomes also appear to drive the rebound effect. These understandings led us to a wider, whole-systems approach in which we view households as an integral part of the system of production and consumption. Policies on work-time reduction, with appropriate measures to safeguard low income households, can offer additional win-win opportunities that, to some extent, overcome this stumbling block. Ultimately, however, income growth is driven by economic structure. Approaches which tackle the structural implications of economic growth are also essential to a meaningful understanding of the potential to reduce carbon footprints. In summary, industrial ecology, with its wide ranging systems approach as shown in this chapter, has a great deal to contribute to the quest to devise strategies to move towards lower carbon, fulfilling lifestyles.

Acknowledgements We are grateful for support from the UK's Economic and Social Research Council (ESRC) for funding the Research Group on Lifestyles, Values and Environment (RESOLVE) (Grant Number RES-152-25-1004) and also for support from the UK Department of Environment, Food and Rural Affairs (DEFRA), the ESRC and the Scottish Government for funding the Sustainable Lifestyles Research Group (SLRG).

References

Aall, C., Klepp, I. G., Engeset, A. B., Skuland, S. E., & Stoa, E. (2011). Leisure and sustainable development in Norway: Part of the solution and the problem. *Leisure Studies, 30*, 453–476.

Ahmad, N., & Wyckoff, A. (2003). *Carbon dioxide emissions embodied in international trade of goods*. Paris: OECD.

Alfredsson, E. C. (2004). "Green" consumption–no solution for climate change. *Energy, 29*, 513–524.

Andrew, R., Davis, S. J., & Peters, G. (2013). Climate policy and dependence on traded carbon. *Environmental Research Letters, 8*(3). doi:10.1088/1748-9326/8/3/034011.

Atkinson, G., Hamilton, K., Ruta, G., & Van Der Mensbrugghe, D. (2011). Trade in 'virtual carbon': Empirical results and implications for policy. *Global Environmental Change, 21*, 563–574.

Baiocchi, G., & Minx, J. (2010). Understanding changes in the UK's CO2 emissions: A global perspective. *Environmental Science and Technology, 44*, 1177–1184.

Baiocchi, G., Minx, J., & Hubacek, K. (2010). The impact of social factors and consumer behavior on carbon dioxide emissions in the United Kingdom. *Journal of Industrial Ecology, 14*, 50–72.

Barr, S., Shaw, G., Coles, T., & Prillwitz, J. (2010). 'A holiday is a holiday': Practicing sustainability, home and away. *Journal of Transport Geography, 18*, 474–481.

Becker, G. (1965). A theory of the allocation of time. *Economic Journal, 75*, 493–517.

Benders, R. M. J., Moll, H. C., & Nijdam, D. S. (2012). From energy to environmental analysis. *Journal of Industrial Ecology, 16*, 163–175.

Bin, S., & Dowlatabadi, H. (2005). Consumer lifestyle approach to US energy use and the related CO2 emissions. *Energy Policy, 33*, 197–208.

Birnik, A. (2013). An evidence-based assessment of online carbon calculators. *International Journal of Greenhouse Gas Control, 17*, 280–293.

Bows, A., & Barrett, J. (2010). Cumulative emission scenarios using a consumption-based approach: A glimmer of hope? *Future Science, 1*, 161–175.

Brännlund, R., Ghalwash, T., & Nordström, J. (2007). Increased energy efficiency and the rebound effect: Effects on consumption and emissions. *Energy Economics, 29*, 1–17.

Buchs, M., & Schnepf, S. V. (2013). Who emits most? Associations between socio-economic factors and UK households' home energy, transport, indirect and total CO2 emissions. *Ecological Economics, 90*, 114–123.

Caeiro, S., Ramos, T. B., & Huisingh, D. (2012). Procedures and criteria to develop and evaluate household sustainable consumption indicators. *Journal of Cleaner Production, 27*, 72–91.

Caprariello, P. A., & Reis, H. T. (2012). To do, to have, or to share? Valuing experiences over material possessions depends on the involvement of others. *Journal of Personality and Social Psychology, 104*, 199.

Cave, L. A., & Blomquist, G. C. (2008). Environmental policy in the European Union: Fostering the development of pollution havens? *Ecological Economics, 65*, 253.

CCC. (2013). *Reducing the UK's carbon footprint and managing competitiveness risks*. London: Committee on Climate Change.

Chitnis, M., Sorrell, S., Druckman, A., Firth, S. K., & Jackson, T. (2013). Turning lights into flights: Estimating direct and indirect rebound effects for UK households. *Energy Policy, 55*, 234–250.

Chitnis, M., Sorrell, S., Druckman, A., Firth, S. K., & Jackson, T. (2014). Who rebounds most? Estimating direct and indirect rebound effects for different UK socioeconomic groups. *Ecological Economics, 106*, 12–32.

Collins English Dictionary. (2012). *Carbon footprint*. Retrieved March 29, 2015, from http://www.collinsdictionary.com/dictionary/english/carbon-footprint

Cooper, T. (2010). *Longer lasting products: Alternatives to the throwaway society*. Farnham: Surrey, UK.

Coote, A., Franklin, J., & Simms, A. (2010). *21 hours: Why a shorter working week can help us all to flourish in the 21st century*. London: New Economics Foundation.

Csikszentmihalyi, M. (2006). The costs and benefits of consuming. In T. Jackson (Ed.), *The Earthscan reader in sustainable consumption*. London: Earthscan.

Davis, S. J., & Caldeira, K. (2010). Consumption-based accounting of CO2 emissions. *Proceedings of the National Academy of Sciences of the United States of America, 107*(12), 5687–5692.

DEFRA, Ricardo-Aea & Carbon Smart. (2014). *Greenhouse gas conversion factor repository*. [Online]. Department for Environment and Rural Affairs. Retrieved March 29, 2015, from http://www.ukconversionfactorscarbonsmart.co.uk/Filter.aspx?year=38

Dey, C., Berger, C., Foran, B., Foran, M., Joske, R., Lenzen, M., & Wood, R. (2003). *Household environmental pressure from consumption: An Australian environmental atlas. Water, wind, art and debate: How environmental concerns impact on disciplinary research*. Australia: Sydney University Press.

Diener, E., & Seligman, M. E. P. (2004). Beyond money: Toward an economy of well-being. *Psychological Science in the Public Interest, 5*, 1–31.

Druckman, A., & Jackson, T. (2009). The carbon footprint of UK households 1990–2004: A socioeconomically disaggregated, quasi-multiregional input-output model. *Ecological Economics, 68*, 2066–2077.

Druckman, A., & Jackson, T. (2010, November). *An exploration into the carbon footprint of UK households* (RESOLVE Working Paper Series 02-10). Guildford: University of Surrey. Retrieved March 29, 2015, from http://resolve.sustainablelifestyles.ac.uk/sites/default/files/RESOLVE_WP_02-10.pdf

Druckman, A., Bradley, P., Papathanasopoulou, E., & Jackson, T. (2008). Measuring progress towards carbon reduction in the UK. *Ecological Economics, 66*, 594–604.

Druckman, A., Chitnis, M., Sorrell, S., & Jackson, T. (2011a). Missing carbon reductions? Exploring rebound and backfire effects in UK households. *Energy Policy, 39*, 3572–3581.

Druckman, A., Hartfree, Y., Hirsch, D., & Perren, K. (2011b). *Sustainable income standards: Towards a greener minimum?* York: Joseph Rowntree Foundation.

Druckman, A., Buck, I., Hayward, B., & Jackson, T. (2012). Time, gender and carbon: A study of the carbon implications of British adults' use of time. *Ecological Economics, 84*, 153–163.

EPA. (2014). *What is an urban heat island?* [Online]. Retrieved March 29, 2015, from http://www.epa.gov/heatisland/about/index.htm

Eurofund. (2013). *Third European quality of life survey – Quality of life in Europe: Subjective well-being*. Luxembourg: Publications Office of the European Union.

Garnett, T. (2013). Food sustainability: Problems, perspectives and solutions. *Proceedings of the Nutrition Society, 72*, 29–39.

Gatersleben, B., Meadows, J., Abrahamse, W., & Jackson, T. (2008). *Materialistic and environmental values of young volunteers in nature conservation projects* (RESOLVE Working Paper Series 07/08). Guildford: University of Surrey.

Girod, B., & De Haan, P. (2009). GHG reduction potential of changes in consumption patterns and higher quality levels: Evidence from Swiss household consumption survey. *Energy Policy, 37*, 5650–5661.

Girod, B., & De Haan, P. (2010). More or better? A model for changes in household greenhouse gas emissions due to higher income. *Journal of Industrial Ecology, 14*, 31–49.

Glaeser, E. L., & Kahn, M. E. (2010). The greenness of cities: Carbon dioxide emissions and urban development. *Journal of Urban Economics, 67*, 404–418.

Godbey, G. (1996). *No time to waste: Time use and the generation of residential solid waste* (PSWP Working Paper #4). New Haven: Yale School of Forestry and Environmental Studies.

Godbey, G., Lifset, R., & Robinson, J. (1998). No time to waste: An exploration of time use, attitudes toward time, and the generation of municipal solid waste. *Social Research, 65*, 101–140.

Gough, I. (2013). Carbon mitigation policies, distributional dilemmas and social policies. *Journal of Social Policy, 42*, 191–213.

Gough, I., Adbdallah, S., Johnson, V., Ryan-Collins, J., & Smith, C. (2011). *The distribution of total greenhouse gas emissions by households in the UK, and some implications for social policy* (LSE STICERD Research Paper No. CASE152). London: Centre for Analysis of Social Exclusion, London School of Economics, and New Economic's Foundation.

Hayden, A. (1999). *Sharing the work, sparing the planet*. London/New York: Zed Books Ltd.

Hayden, A., & Shandra, J. (2009). Hours of work and the ecological footprint of nations: An exploratory analysis. *Local Environment, 14*, 575–600.

Hertwich, E. G. (2011). The life cycle environmental impacts of consumption. *Economic Systems Research, 23*, 27–47.

Hertwich, E. G., & Peters, G. P. (2009). Carbon footprint of nations: A global, trade-linked analysis. *Environmental Science & Technology, 43*, 6414–6420.

Holmberg, J., Larsson, J., Nässén, J., Svenberg, S., & Andersson, D. (2012). *Low-carbon transitions and the good life* (Report 6495). Stockholm: Swedish Environmental Protection Agency.

Hong, S., Oreszczyn, T., & Ridley, I. (2006). The impact of energy efficient refurbishment on the space heating fuel consumption in English dwellings. *Energy and Buildings, 38*, 1171–1181.

IPCC. (2014). *IPCC fifth assessment synthesis report: Climate change 2014*. Geneva: International Panel on Climate Change.

Jackson, T. (2005). *Motivating sustainable consumption: A review of evidence on consumer behaviour and behavioural change*. London: Policy Studies Institute.

Jackson, T. (2006). *Earthscan reader in sustainable consumption*. London: Earthscan.

Jackson, T. (2009). *Prosperity without growth – Economics for a finite planet*. London: Earthscan.

Jackson, T., Papathanasopoulou, E., Bradley, P., & Druckman, A. (2006, June 1–2). *Attributing carbon emissions to functional household needs: A pilot framework for the UK*. In International conference on regional and urban modelling, 2006 Brussels, Belgium.

Jalas, M. (2002). A time use perspective on the materials intensity of consumption. *Ecological Economics, 41*, 109–123.

Johnson, E. (2008). Disagreement over carbon footprints: A comparison of electric and LPG forklifts. *Energy Policy, 36*, 1569–1573.

Jones, C. M., & Kammen, D. M. (2011). Quantifying carbon footprint reduction opportunities for U.S. households and communities. *Environmental Science & Technology, 45*, 4088–4095.

Kahneman, D., Krueger, A. B., Schkade, D. A., Schwarz, N., & Stone, A. A. (2004). A survey method for characterizing daily life experience: The day reconstruction method. *Science, 306*, 1776–1780.

Kasser, T., & Sheldon, K. (2009). Time affluence as a path toward personal happiness and ethical business practice: Empirical evidence from four studies. *Journal of Business Ethics, 84*, 243–255.

Kerkhof, A., Nonhebel, S., & Moll, H. C. (2008). Relating the environmental impact of consumption to household expenditures: An input–output analysis. *Ecological Economics, 68*, 1160–1170.

Kerkhof, A., Benders, R. M. J., & Moll, H. C. (2009). Determinants of variation in household CO2 emissions between and within countries. *Energy Policy, 37*, 1509–1517.

Knight, K., & Schor, J. (2014). Economic growth and climate change: A cross-national analysis of territorial and consumption-based carbon emissions in high-income countries. *Sustainability, 6*, 3722–3731.

Knight, K. W., Rosa, E. A., & Schor, J. B. (2013). Could working less reduce pressures on the environment? A cross-national panel analysis of OECD countries, 1970–2007. *Global Environmental Change, 23*, 691–700.

Kratena, K., & Wüger, M. (2010). *The full impact of energy efficiency on households' energy demand* [Online]. Austrian Institute of Economic Research (WIFO). Retrieved March 29, 2015, from http://www.wifo.ac.at/wwa/servlet/wwa.upload.DownloadServlet/bdoc/PRIVATE49458/WP_2010_356$.PDF

Lenzen, M. (2008). Consumer and producer environmental responsibility: A reply. *Ecological Economics, 66*, 547–550.

Lenzen, M., & Cummins, R. A. (2013). Happiness versus the environment – A case study of Australian lifestyles. *Challenges, 4*, 56–74.

Lenzen, M., & Dey, C. (2002). Economic, energy and greenhouse emissions impacts of some consumer choice, technology and government outlay options. *Energy Economics, 24*, 377–403.

Lenzen, M., Dey, C., & Foran, B. (2004). Energy requirements of Sydney households. *Ecological Economics, 49*, 375.

Lenzen, M., Wier, M., Cohen, C., Hayami, H., Pachauri, S., & Schaeffer, R. (2006). A comparative multivariate analysis of household energy requirements in Australia, Brazil, Denmark, India and Japan. *Energy, 31*, 181–207.

Lenzen, M., Murray, J., Sack, F., & Wiedmann, T. (2007). Shared producer and consumer responsibility – Theory and practice. *Ecological Economics, 61*, 27–42.

Li, Y., & Hewitt, C. N. (2008). The effect of trade between China and the UK on national and global carbon dioxide emissions. *Energy Policy, 36*, 1907–1914.

Lin, B., & Sun, C. (2010). Evaluating carbon dioxide emissions in international trade of China. *Energy Policy, 38*, 613–621.

Macinnes, T., Aldridge, H., Bushe, S., Tinson, A., & Born, T. B. (2014). *Monitoring poverty and social exclusion 2014*. York: Joseph Rowntree Foundation.

Macintosh, A., & Wallace, L. (2009). International aviation emissions to 2025: Can emissions be stabilised without restricting demand? *Energy Policy, 37*, 264–273.

Maenpaa, I., & Siikavirta, H. (2007). Greenhouse gases embodied in the international trade and final consumption of Finland: An input-output analysis. *Energy Policy, 35*, 128.

Maxwell, D., Owen, P., & Mcandrew, L. (2011). *Addressing the rebound effect – Final report*. European Commission DG ENV.

Miller, R. E., & Blair, P. D. (2009). *Input-output analysis: Foundations and extensions* (2nd Rev. ed.). Cambridge: Cambridge University Press.

Minx, J., & Baiocchi, G. (2009). Time-use and sustainability. In S. Suh (Ed.), *Handbook of input-output economics in industrial ecology*. Dordrecht: Springer.

Mizobuchi, K. (2008). An empirical study on the rebound effect considering capital costs. *Energy Economics, 30*, 2486–2516.

Munksgaard, J., Wier, M., Lenzen, M., & Dey, C. (2005). Using input-output analysis to measure the environmental pressure of consumption at different spatial levels. *Journal of Industrial Ecology, 9*, 169–186.

Murray, C. K. (2013). What if consumers decided to all "go green"? Environmental rebound effects from consumption decisions. *Energy Policy, 54*, 240–256.

Nakano, S., Okamura, A., Sakurai, N., Suzuki, M., Tojo, Y., & Yamano, N. (2009). *The measurement of CO2 embodiments in international trade: Evidence from the harmonised input-output and bilateral trade database* (OECD Science, Technology and Industry Working Papers, 2009/3). OECD Publishing. doi: 10.1787/227026518048OECD.

Nassen, J., & Larsson, J. (2015). Would shorter work time reduce greenhouse gas emissions? An analysis of time use and consumption in Swedish households. *Environment & Planning C: Government & Policy, 33*, 1–20. doi: 10.1068/c12239

Nijdam, D. S., Wilting, H. C., Goedkoop, M. J., & Madsen, J. (2005). Environmental load from Dutch private consumption: How much damage takes place abroad? *Journal of Industrial Ecology, 9*, 147.

OECD. (2015). *What are equivalence scales?* [Online]. Retrieved March 29, 2015, from http://www.oecd.org/eco/growth/OECD-Note-EquivalenceScales.pdf

ONS. (2006). *The time use survey 2005*. London: Office for National Statistics.

Peters, G., & Hertwich, E. (2006a). The importance of imports for household environmental impacts. *Journal of Industrial Ecology, 10*, 89–109.

Peters, G., & Hertwich, E. (2006b). Pollution embodied in trade: The Norwegian case. *Journal of Industrial Ecology, 16*, 379–387.

Peters, G., & Hertwich, E. (2008a). Post-Kyoto greenhouse gas inventories: Production versus consumption. *Climatic Change, 86*, 51–66.

Peters, G., & Hertwich, E. (2008b). CO2 embodied in international trade with implications for global climate policy. *Environmental Science & Technology, 42*, 1401–1407.

Peters, G. P., Minx, J. C., Weber, C. L., & Edenhofer, O. (2011). Growth in emission transfers via international trade from 1990 to 2008. *Proceedings of the National Academy of Sciences of the United States of America, 108*(21), 8903–8908. doi:10.1073/pnas.1006388108.

Pullinger, M. (2014). Working time reduction policy in a sustainable economy: Criteria and options for its design. *Ecological Economics, 103*, 11–19.

Rosnick, D., & Weisbrot, M. (2007). Are shorter work hours good for the environment? A comparison of US and European energy consumption. *International Journal of Health Services, 37*, 405–417.

Shove, E. (2012). Putting practice into policy: Reconfiguring questions of consumption and climate change. *Contemporary Social Science, 9*, 415–429.

Shui, B., & Harriss, R. C. (2006). The role of CO2 embodiment in US-China trade. *Energy Policy, 34*, 4063.

Smith, A. (1904). *An inquiry into the nature and causes of the wealth of nations, library of economics and liberty*. Retrieved March 16, 2015, from http://www.econlib.org/library/Smith/smWN18.html

Soper, K. (2008). Alternative hedonism, cultural theory and the role of aesthetic revisioning. *Cultural Studies, 22*, 567–587.

Sorrell, S. (2007). *The rebound effect: An assessment of the evidence for economy-wide energy savings from improved energy efficiency*. London: UK Energy Research Centre.

The Living Wage Commission. (2014). *Work that pays: The final report of the Living Wage Commission*. [Online]. Retrieved March 29, 2015, from http://livingwagecommission.org.uk/wp-content/uploads/2014/07/Work-that-pays_The-Final-Report-of-The-Living-Wage-Commission_w-4.pdf

Thomas, B. A., & Azevedo, I. L. (2013). Estimating direct and indirect rebound effects for U.S. households with input-output analysis part 1: Theoretical framework. *Ecological Economics, 86*, 199–210.

Tukker, A. (2006). Identifying priorities for environmental product policy. *Journal of Industrial Ecology, 10*, 1.

Tukker, A., & Jansen, B. (2006). Environmental impacts of products: A detailed review of studies. *Journal of Industrial Ecology, 10*, 159.

Tukker, A., Cohen, M. J., Hubacek, K., & Mont, O. (2010). The impacts of household consumption and options for change. *Journal of Industrial Ecology, 14*, 13–30.

UN. (2011). *Classification of Individual Consumption According to Purpose (COICOP)* [Online]. Retrieved March 29, 2015, from http://unstats.un.org/unsd/iiss/Classification-of-Individual-Consumption-According-to-Purpose-COICOP.ashx

Victor, P. (2008). *Managing without growth: Slower by design, not disaster*. Cheltenham/Northampton: Edward Elgar.

Weber, C. L., & Matthews, H. S. (2008). Quantifying the global and distributional aspects of American household carbon footprint. *Ecological Economics, 66*, 379–391.

Weber, C., & Perrels, A. (2000). Modelling lifestyle effects on energy demand and related emissions. *Energy Policy, 28*, 549.

Weber, C. L., & Peters, G. P. (2009). Climate change policy and international trade: Policy considerations in the US. *Energy Policy, 37*, 432–440.

Wiedmann, T. (2009). A review of recent multi-region input–output models used for consumption-based emission and resource accounting. *Ecological Economics, 69*, 211–222.

Wiedmann, T., & Minx, J. (2007). A definition of 'carbon footprint'. In C. C. Pertsova (Ed.), *Ecological economics research trends* (pp. 1–11). Hauppauge: Nova Science Publishers.

Wier, M., Lenzen, M., Munksgaard, J., & Smed, S. (2001). Effects of household consumption patterns on CO2 requirements. *Economic Systems Research, 13*, 259–274.

Wilkinson, R., & Pickett, K. (2009). *The spirit level: Why more equal societies almost always do better*. London: Allen Lane/Penguin Group.

Williams, I., Kemp, S., Coello, J., Turner, D. A., & Wright, L. A. (2012). A beginner's guide to carbon footprinting. *Carbon Management, 3*, 55–67.

Wilson, J., Tyedmers, P., & Spinney, J. E. L. (2013). An exploration of the relationship between socioeconomic and well-being variables and household greenhouse gas emissions. *Journal of Industrial Ecology, 17*, 880–891.

Wood, G., & Newborough, M. (2007). Energy-use information transfer for intelligent homes: Enabling energy conservation with central and local displays. *Energy and Buildings, 39*, 495–503.

Wright, L. A., Kemp, S., & Williams, I. (2011). 'Carbon footprinting': Towards a universally accepted definition. *Carbon Management, 2*, 61–72.

Zurick, D. (2006). Gross national happiness and environmental status in Bhutan. *Geographical Review, 96*, 657–681.

Chapter 10
The Social and Solidarity Economy: Why Is It Relevant to Industrial Ecology?

Marlyne Sahakian

One can no longer even imagine that there could be a single standard of value by which to measure things. The neoliberals (…) are singing the praises of a global market that is, in fact, the single greatest and most monolithic system of measurement ever created, a totalizing system that would subordinate everything – every object, every piece of land, every human capacity or relationship – on the planet to a single standard of value. – David Graeber (2001).

Toward an Anthropological Theory of Value:
The false coin of our own dreams, p. xi.

Abstract The goal of this contribution is to illustrate the linkages between industrial ecology (IE) and the social and solidarity economy (SSE), an economic paradigm that is robust in terms of conceptual and historical developments, and active around the world as a social movement. The SSE includes a range of activities, such as fair trade, community currencies and some forms of peer-to-peer sharing, to name but a few. The links and tensions between SSE and IE are considered first conceptually, by uncovering the theoretical frameworks attached to each field. Three 'solidarity' practices are then discussed in relation to industrial ecology activities, namely: aspects of the sharing economy, community currencies and forms of crowd-funding. A main finding is that the two fields of research and practice are compatible, as neither focus on economic growth and specifically profit as an ultimate aim; yet IE prioritizes biophysical considerations, whereas the SSE places more emphasis on people and power systems, as expected. One insight gleaned through this process is that more attention could be placed on labour conditions, power relations and governance systems in industrial ecology, building on previous and ongoing work in this area.

Four main fields of inquiry emerge: understanding whether 'solidaristic' cooperatives and enterprises could be more receptive to industrial ecology approaches and more adept at embracing resource exchanges such as in industrial symbiosis; ascertaining to what extent companies already involved in symbiotic relations might

M. Sahakian (✉)
Industrial Ecology Group, Faculty of Geosciences and the Environment,
University of Lausanne, Batiment Geopolis, CH-1015 Lausanne, Switzerland
e-mail: marlyne.sahakian@unil.ch

© The Author(s) 2016
R. Clift, A. Druckman (eds.), *Taking Stock of Industrial Ecology*,
DOI 10.1007/978-3-319-20571-7_10

also embody social and solidarity values, including notions of participative governance, limited profit-making, a focus on employee benefits, among others; considering certain forms of crowdfunding as an opportunity for abating economy-wide rebound effects through more socially just and environmentally sound investments; and finally, the potential for complementary currencies to work towards industrial ecology aims. One of the weaknesses of the social and solidarity economy has been that of scale, as SSE activities tend to take place on a micro-scale, with some notable exceptions. That being said, the SSE is well underway and expanding, in research and practice, presenting interesting synergies with IE and opportunities for further research and action. Bringing together IE and SSE ultimately brings to the fore a discussion around paradigms and associated values, including societal and environmental priorities which are not always aligned – raising questions around what values we wish to put forward in our economy, workplaces and society.

Keywords Community currencies • Crowd-funding • Reciprocity • Solidarity economy • Sharing economy

1 Introduction

In the past decade, a slew of terms have emerged to describe new economic models: the people-first or human economy (Ransom and Baird 2010; Hart et al. 2010), the new economy (Schor and Thompson 2014), the green economy (UNEP 2011), the sharing economy (Botsman and Rogers 2010; Gansky 2010), diverse economies (Gibson-Graham 2006; Gibson-Graham 2008) and the varying definitions of the circular economy (Yuan et al. 2006; Ellen MacArthur Foundation 2013), to name but a few. In some cases, these terms represent alternatives to the dominant market economy, understood here as being based on competition and private ownership; in most cases, what plays out in practice are economic activities that are at best complementary to and on the fringes of the market economy. Their emergence suggests a growing interest in finding new ways to engage in production, consumption, exchanges and financing, to better manage resources, and strive for greater prosperity than what is currently being achieved with the dominant economic model. More specifically, all of these approaches share the ambition of tackling the question of how economic systems can better serve – rather than be severed from – environmental or social aims. The goal of this contribution is to assess whether these models can bring new fields of reflection and action to the industrial ecology community.

One reason for this renewed interest in economic models is the heightened concern around recurring financial crises, the destabilization of natural cycles and widening inequalities. Particularly in times of economic depression, people look for alternative ways to access products and services, including financial services. In Switzerland in 1934, in the wake of the Great Depression and facing a credit crunch, a group of entrepreneurs created a new currency, the WIR,[1] which continues to be exchanged today – notably during economic recessions, where this complementary currency may have a stabilizing effect on the national economy (Studer 2006). Since the 2008–2009 financial crisis in Greece, international press has covered the emergence of new forms of barter, where services and products are traded based on units of exchange ranging from time to community currencies (Poggioli 2011; Lowen 2012). New peer-to-peer models for sharing resources and services are also very much in the spotlight today, propelled by information technologies that have made sharing and bartering activities available on a wider scale and across different sectors. The sharing economy touts lofty goals, such community-building, economic empowerment, creative expression, but also better resource management. Whether these goals are achieved remains to be evaluated in practice, particularly in relation to environmental aims: the notion that 'sharing' could lead to reduced energy and material throughputs are still up for debate (Cohen 2014). Kalamar (2013) coined the termed 'share-washing' for attempts by 'business as usual' activities to claim such social goals.

In this chapter, the social and solidarity economy (SSE) will be considered, an economic paradigm that is robust in terms of conceptual and historical developments, but also active around the world as a social movement. The SSE includes a range of activities, such as fair trade, community currencies, peer-to-peer activities, cooperative and mutual organizations, some forms of sharing, to name but a few. This chapter introduces the social and solidarity economy (SSE) as a theoretical framework, social movement and growing practice around the world. The goal is to not to contest the competitive market economy or promote the solidarity economy, *per se*, but rather place the social and solidarity economy in relation to industrial ecology (IE). In comparing the SSE to IE in theory and in practice, the chapter discusses how a different economic paradigm might be relevant to the industrial ecology community.

In the section that follows, the social and solidarity economy is defined, with a discussion around the conceptual links between this economic paradigm and industrial ecology; in the second section, three 'solidarity' practices will be discussed in relation to IE, namely: activities in the 'communal' sharing economy, community currencies and 'solidaristic' crowd-funding. By reflecting on the SSE and IE and understanding the opportunities and challenges this presents, the aim of this chapter is to further the work of industrial ecology community, both in research and practice.

---

[1] See Sahakian (2014) for an overview of the WIR as well as community currencies in Argentina and Japan. The WIR bank emerged from the *Wirtschaftsring-Genossenschaft* cooperative ('Economic Circle' in German, with a play on the word *wir*, which translates to *we* or *us*). Today, more than 60,000 SMEs exchange WIR currency, a unit in parity with the Swiss franc and more commonly circulated in German-speaking Switzerland. Given the small scale of WIR exchanges, there are some doubts as to whether the stabilizing effect can be attributed to the WIR alone; this claim would merit further investigation.

2   Conceptual Links

2.1   What Is the Social and Solidarity Economy?

The social and solidarity economy finds its roots in the social economy, which emerged in the early period of Western industrialization, in the nineteenth century, when poverty in urban centres was a central concern. The Welsh social reformer Robert Owen founded the cooperative movement at this time, and cooperatives and associations became 'the first line of defence' in rallying to address social ails (Lewis 1997 in Laville 2011). The social economy was relegated to the status of a 'third sector' in the post-war period, when the market economy was seen as responsible for regulating property and currencies, and where the welfare state was considered the primary vector for social action through the redistribution of wealth (Laville 1994). The social economy cast in this role of a 'third sector' represented all other forms of organization, including non-profit and non-governmental activities.

Neoliberal policies, recurring financial crises, as well as the failure of the welfare state to address social issues, all contributed to a rebirth of the social economy in the 1990s. As inequalities widened and environmental issues expanded to a local and global scale, the 'sustainable development' paradigm was brought into question in what has been called a 'crisis of values', leading to a renewed interest in the social economy (Laville and Cattani 2006). Almost 10 years after the first Earth Summit, the participants at the first World Social Forum in 2001 (Porto Alegre, Brazil) rallied around the phrase 'Another world is possible'; the social economy was seen as playing a prominent role in this new world order. For some, the notion of the 'social economy' was insufficient, as this term was too closely associated with the third sector at that time. Arguments were put forward to challenge a definition of the social economy solely based on the type of entity or legal status.[2] From this effort, the term 'civil and solidarity economy' was coined to account for numerous other initiatives that had emerged, particularly in Europe, that were neither non-for-profit nor non-governmental (Laville 2011). Private enterprises, for example, can be built on social values and have their place in what came to be known as the 'social and solidarity economy' (SSE), primarily in French, Spanish, Italian and Portuguese-speaking countries, in Europe and in the Americas.

As Fraisse (2003) rightly notes, the SSE is being interpreted in different ways around the world. In the United States, members of the SSE aim towards the sys-

---

[2]Historically in France, the definition of the social and solidarity economy had been based on organizational type: all mutual companies and cooperatives fall under this definition. In Western Switzerland, however, another definition is upheld, that of guiding principles and key criteria. A for-profit company can be part of the SSE, as long as it is aligned with the principles of the SSE and profit-making is not the primary goal. The debate about structure versus content of such organizations (Kawano 2009) has been focused on the distinction between theory and practice: some cooperatives may be seeking the economic benefit for their members above all, while some for-profits may be working towards greater social and environmental aims. Simply put, adopting a legal form does not guarantee that an entity becomes part of the social and solidarity economy (Swaton 2011; Defourny et al. 2000).

temic transformation of the economy as a whole, across 'all of the diverse ways that human communities meet their needs and create livelihoods together' (Miller in Kawano et al. 2009: 30), including the private and public sector, and as part of a 'counter-hegemonic political economy' (Satgar 2014) or post-capitalist agenda (Kawano 2013). In other contexts, such as Western Switzerland, the stance is more nuanced: SSE is complementary to the market economy, rather than an alternative; in the Philippines, the strategy is to create supply chains in the social and solidarity economy, mostly in rural areas (Sahakian and Dunand 2015). In some cases, the public sector also takes on the role of promoting the solidarity economy, as is the case in Luxembourg, which has appointed a Ministry of Labour, Employment and the Social and Solidarity Economy. In such cases, there is a risk that the public sector might instrumentalize the social and solidarity economy by outsourcing basic social services to third parties, under the guise of solidarity, an issue that has been hotly debated in SSE circles. As the late Bernard Eme also noted, the instrumentalization can go both ways: civil society actors can also uphold solidarity values in order to access state benefits (Eme (2005) in Laville and Cattani 2006).

As a social movement, the social and solidarity economy (SSE) has grown from local activities to regional and international networks of members, including researchers and practitioners. In the United States, the simpler 'solidarity economy' is the preferred term, which emerged from the U.S. Social Forum in 2007 and resulted in the launch of U.S. Solidarity Economy Network (SEN) (Kawano et al. 2009). In Latin America, SSE has been tied to the new 'development' paradigm termed *buen vivir* (the good life) (Giovannini 2014) and has been discussed in the literature by several authors (Lemaître and Helmsing 2012; Hillenkamp 2011, 2013; Arruda 2004; Singer 2002; among others). Less is known about SSE initiatives in Asia and Africa.[3]

Conceptually, the SSE is inspired by the notion of reciprocity put forward by Karl Polanyi, who famously argued that the economy is 'embedded' in the social realm; it has a social purpose and is subordinate to and inseparable from social relations (2001, originally published in 1944). Four ideal-type models are put forward in his work: (1) the market economy; and non-market economies including (2) self-sufficiency (including house-holding or relations between family members), (3) redistribution (usually through government) and (4) reciprocity. Several authors have argued for the need for the social and solidarity economy to be multidimensional, to include different types of entities, working progressively towards solidaristic goals, in a more plural economy and across these four ideal types (Laville 2003; Kawano et al. 2009). Fair trade is an example of the SSE in practice, which engages with Polanyi's different ideal types: the products traded are done so in solidarity between consumers and producers, but such activities can benefit from state support (e.g., redistribution), and also engage with the market economy and house-holding activities, such as craft making in the home.

---

[3] For a recent publication from South Africa, see Satgar (2014); for The Philippines, see Sahakian and Dunand (2015).

The notion of solidarity tends towards a definition of reciprocity as going beyond the duality of giving, receiving and the obligation to give in return, to exchanges across and between different subgroups (Polanyi 1957). Reciprocity in the solidarity economy entails complementary relations based on voluntary interdependence (Servet 2007: 264), or being 'invested with the potential of solidarity, consciously interdependent on others' (Servet 2006: 18). This is an important distinction in theory, as it means that the SSE is not solely about reciprocity in dire straits, out of necessity, towards the relations of 'master and slave', but rather out of an interest in the commons and the community. The work of Elinor Ostrom (2001/1990) is relevant here: collective action can be based on voluntary participation and an identification of needs, going beyond the individual need or that of a self-interested group, to broader social and environmental needs. Sahakian and Servet (in press) propose the term 'communal sharing' to describe forms of sharing that are aligned with the notion of collective action (Sahakian and Servet in press), a theme that will be further explored in the second part of this chapter.

SSE is generally understood as placing human beings at the centre of economic and social life (ISGC 1997). SSE seeks to foster solidarity by placing more importance on people, rather than the accumulation of capital or profit. What inspired early cooperative and associative movements of the nineteenth century and continues to drive efforts in the SSE today are the limitations on profit making: the financial gains for investors are subject to limits (Laville 2011). Another aspect is the emphasis placed on governance systems. The SSE promotes democratic processes within organizations: SSE entities are usually self-managed, self-organized and generally independent from State support. According to Laville (2003), SSE is ultimately about promoting democracy on the local level through economic activity, or the 'democratization' of the economy based on the participatory engagement of all citizens (Defourny and Develtere 1999; Fraisse et al. 2007). The vision is to include all types of people in economic life, engaging them to participate as economic actors, most often at the level of the community. In contexts such as the Global South, the focus is often on rural contexts and on engaging the poor as primary stakeholders in economic activities (Sahakian and Dunand 2015).

To summarize some of the guiding principles that are put forward by SSE actors, the goal of this economy is to place 'service to its members or to the community ahead of profit; autonomous management; a democratic decision-making process; the primacy of people and work over capital in the distribution of revenues' (Defourny et al. 2000). Activities could include some forms of social entrepreneurship; community currencies; micro-credit programs; as well certain worker, consumer and producer cooperatives; community gardens or community supported agriculture; social reinsertion programs; community-run exchange platforms; do-it-yourself initiatives; shared services and goods; insurance and financial services, among others. The extent to which these activities fall under SSE umbrella would depend on whether they are progressively seeking to participate in this economy (Swaton and Baranzini 2013), which would need to be evaluated not only in theory and discourse, but also in practice.

In Geneva, Switzerland, companies who want to be part of the SSE must sign a charter, and in doing so, they comply with transparency in reporting, engaging in activities that are of collective/public interest, being financially autonomous from subsidies or other forms of support and aiming at non-profit or limited profit-making. Based on a survey of 195 members in 2008, those employed in the SSE in Geneva experience a low rate of difference between the highest and lowest salary (by a factor of 1.3–2.3), higher average salaries for entry-level work and high rates of people working on flexible schedules (at an average 50–65 % rate of activity) (APRES-GE 2010). In Geneva, members must then commit to progressively put in place, over 2 years, an environmental management policy, participatory forms of management and social management policies that focus on employee diversity and welfare (Swaton and Sahakian 2014). All sectors of the economy are represented among the current 265 members, including health and social services, but also banking and insurance companies, housing cooperatives, food services, arts and leisure, training and education, among others.

Many SSE activities exist as the grassroots level, either marginalized by or hidden within the dominant market economy (Miller in Kawano et al. 2009). SSE tends to include smaller-scale activities, whether local or regional. In some contexts, the ambition is to link together different SSE activities in a supply-chain approach, as is the case in the Philippines where organic feedstock companies and dairy farms are coming together with fair trade associations (Sahakian and Dunand 2015). Examples of larger SSE enterprises and regional efforts are less common, raising questions of scalability. The classic example of a larger scale SSE activity is Mondragón, a cooperative based in the Basque region of Spain that is made up of 258 enterprises organized into a federated governance system including shared financial services and technical support (Kaswan 2014). The cooperatives in Trento county, northern Italy, are also on par with the Mondragón example in terms of scale and effectiveness, according to recent empirical research comparing the two (Prades 2013). The concept of federating cooperative efforts, sharing services and promoting democratic decision-making processes is underway in different contexts, including the *Conseil québécois de la co-opération et de la mutualité* (CQCM, Canada) and the Democracy Collaborative centred around the Evergreen Cooperatives in Cleveland, Ohio (ibid.). A magic number is usually upheld as representing the percentage of jobs that are involved in the social and solidarity economy: for both France and the City of Geneva, the SSE is said to represent approximately 10 % of salaried employment (INSEE 2012; Dunand 2012).

2.2   What Are the Conceptual Links between SSE and IE, and the Limits?

In the industrial ecology field, economic profit is not a central preoccupation. The payback of certain efforts in terms of return on investment is important in practice, where industrial ecology projects are concerned, but profit generating is not explicitly part of industrial ecology conceptual developments. The founding theoretical principles in IE draw from bio-economics, notably the link between natural laws and principles, such as thermodynamics and economic systems (Georgescu-Roegen 1971, 2006). In this perspective, economic activities draw from and are dependent on ecosystem services, suggesting that there are limits to economic growth (Daly 1977; Jackson 2009). Understanding patterns and trends from a biophysical perspective means assessing values based on environmental resources (material and energy), rather than solely price valuation.

IE and SSE therefore share the principle that economic activity should be subordinated to other factors. Yet industrial ecology privileges the biophysical dimension whereas the social and solidarity economy privileges the social dimension. These different priority areas could be problematic: Is it more important to aim for solidarity in social relations and governance structures, or to minimize energy and material throughputs? Can one be done at the expense of the other? This raises ethical issues in industrial ecology: What if optimal symbiosis is achieved in companies that exploit labour, for example? This also raises environmental sustainability issues among SSE enterprises: Can social goals be achieved to the detriment of environmental considerations?

Beyond the conceptual underpinnings that relate industrial ecology to bio-economics and ecological economics, the novelty of industrial ecology is to draw inspiration from natural systems. According to Erkman, 'the entire industrial system relies on resources and services provided by the biosphere, from which it cannot be dissociated' (1997: 1), yet material and energy throughputs could be better managed through a more holistic approach to organizing economic activities and industrial systems. Biomimicry in industrial ecology implies tending towards reduced resource throughputs and negative impacts. As a descriptive and analytical method, industrial ecology helps to uncover the 'metabolism' of systems, drawing from a comparison with living organisms (Ayres and Simonis 1994) towards understanding 'anthropogenic complex and coupled systems' (Fischer-Kowalski et al. 2009). In addition to describing and analysing, IE suggests how such a system might be 'restructured to make it compatible with the way natural ecosystems function' (Erkman 1997: 1) and is therefore also an operational tool.

Ehrenfeld (2000) goes a step further in distinguishing these practical features in the field of IE from its founding conceptual basis, which tends towards a normative context and can in turn shape paradigmatic thinking. Understood in *analogy* to natural systems, industrial ecology is a practical tool; in using natural systems as a *metaphor*, industrial ecology has the potential to go beyond prescription and techno-focused solutions to become transformative (Ehrenfeld 2003; Hess 2009). The analogy with natural systems allows IE to disengage with questions related to people and power relations. There have been efforts to embed industrial ecology in social relations (Boons and Howard-Grenville 2009). Some work has been done relating IE to fair employment (Alsamawi et al. 2014), legal considerations (Slone in Cohen-Rosenthal and Musnikow 2003) and the role of consumer culture and ethics (Hertwich 2005b; Ehrenfeld 2008; Sahakian and Steinberger 2011), but these aspects have not been sufficiently theorized to date. The work of the late E. Cohen-Rosenthal on environmental, labour and social issues is a key contribution in this area. His focus on workplace issues (Cohen-Rosenthal 1979), specifically in the industrial ecology community (Cohen-Rosenthal 2004), was consistently critical of reductionist strategies that would rely solely on engineering know-how, proper technologies and market-based incentives. For Cohen-Rosenthal, the focus should be on social processes, or the actual points of connection between material and energy flows, and specifically connections between people. Yet little attention is given in IE to the relation between interests, institutions and resources (Opoku 2004), and this dimension remains to be further conceptualized – although interest in this dimension has been growing in the past decade, as we will now turn to.

IE engages with various tools, which are now considered a part industrial ecology approaches, including Input–Output Accounting and Material Flow Analysis, as well as Life Cycle Assessments. The value of LCA has been to extend beyond the 'end of pipe' perspective, which only addresses final outputs, in terms of pollution and environmental degradation. Rather, inputs and outputs are described and can be quantified and qualified across the production-consumption chain, from extraction of natural resources, through manufacturing, distribution, usage and final disposal. In more recent years, social LCA has developed to consider the social impacts in the life cycle of products and processes (UNEP 2009; Andrews et al. 2009; Hauschild et al. 2008), with more recent work in developing a social hotspots database that includes a consideration of labour intensity and worker rights (Benoit-Norris et al. 2012) and on integrating occupational health and safety standards (Scanlon et al. 2015). One aspect of the social and solidarity economy is that the primacy of people over profit is not an 'end of pipe' gesture, but rather embedded across the economic activity in question, from its guiding principles and mission, to how salaries and employee benefits are organized, to how profit is shared. We will come back to this notion in the next section of this chapter, when we consider forms of 'end of pipe sharing' to distinguish more robust solidarity economy activities from 'business as usual'.

The notion of reciprocity, which is an important concept underpinning the SSE, is interesting to explore in relation to IE. If we continue with the notion of IE as being based on a metaphor, can we identify any forms of reciprocity in nature? In other terms, do we see cooperation or competition in natural systems? Ehrenfeld (2000) suggests that a balance is needed between both. Does altruism exist in nature? Perhaps something closer would be mutualism, which would suggest cooperation between members that lead to mutually beneficial or even neutral outcomes. For Ehrenfeld, '…the power of the concept of industrial ecology lies in its normative context and in its potential to shape paradigmatic thinking. It is normative in the sense that (…) three features of the ecological metaphor – community, connectedness, and cooperation – are characteristics we should strive for in designing our worlds.' (Ehrenfeld 2000: 238). These three features, of community, connectedness and cooperation, are closely related to the social and solidarity economy and would merit being further conceptualized. We could draw on the work of Pierre-Joseph Proudhon (1809–1865) here, the father of mutualism in economic theory, who suggested that human activities could be mutualized towards the creation of a collective good.

In terms of governance systems, the notion of voluntary engagement in complementary relations, which is important in the SSE literature, is not explicit in the IE literature. Forms of auto-organization are usually upheld as the preferred way of achieving symbiosis, defined as system exchanges that convert 'negative environmental externalities in the form of waste that used to be discarded into positive environmental externalities such as the spillover benefits of decreased pollution and reduced need for raw material imports' (Chertow and Ehrenfeld 2012: 15). Industrial symbiosis research points to the relevance of self-organizing systems, working in mutually beneficial forms of cooperation, with mobilization capacity, and on a local and regional scale (Massard et al. 2014; Chertow and Ehrenfeld 2012; Boons and Spekkink 2012; Chertow 2007), which is very much the terrain of the social and solidarity economy. As different cooperative movements begin to federate and create new models in the SSE, the industrial ecology community could gain from understanding how services are shared and maximized between these entities. Cooperatives are not the usual stomping ground of industrial ecology activities, but could represent an interesting starting point for discussions that bring together social and biophysical dimensions.

Given the important role of small-to-medium enterprises (SMEs) in developing countries and following the fourth World Social Forum in 2004 (Mumbai, India), a report focusing on the social responsibility of SMEs also sought to make a link between industrial ecology and the solidarity economy. In that report, the late Ramesh Ramaswamy – co-author of a novel book on applied industrial ecology in developing countries and particularly India (Erkman and Ramaswamy 2003) – suggested that SMEs could gain competitive advantages in the global market by organizing production into 'complementary clusters' particularly in the agricultural sector (Asian Coalition 2004: 17). The report suggested industrial ecology could enhance the 'social responsibility' of SMEs in four ways, by addressing: 'producers (greater consciousness in using biodegradable and recyclable materials, lesser waste, more profits); consumers (better quality products); the environment (lesser toxic wastes); and society at large (more socially responsible enterprises, healthier environment)' (ibid.).

In theory, industrial ecology principles should be retained, no matter the organizational structure or system at hand, be it a for-profit SME or a cooperative. Here industrial ecology joins recent interpretations of the social and solidarity economy, in focusing more on how guiding principles can be applied in practice, rather than on institutional settings. That being said, there tends to be a bias towards democratic systems in the industrial ecology literature; the same is true for the social and solidarity economy, which upholds democracy as the key governance system. On this level, the two fields are compatible.

3 Linkages Between the SSE and IE in Practice

3.1 The Sharing Economy vs. End of Pipe Giving: Applicability to IE

The sharing economy is generally understood as involving 'sharing' in the creation, financing, production, distribution and consumption of a variety of goods and services. Certain activities that fall under this banner are part of the current dominant market economy, based on the maximization of utility and the meeting of needs towards pecuniary goals. On the other hand, the sharing economy could be seen as part of the solidarity economy, including activities that aim towards solidarity and mutual support. In the latter reading, 'sharing' would be less about maximizing unused resources and more about placing resources 'in the commons', drawing from theories in environmental governance and alluding to forms of voluntary, collective action around resource usage (Ostrom 2001/1990). A key distinction here is to understand sharing not as the distribution of pieces of a pie, albeit in collaborative ways, but rather 'sharing' based on an understanding of identified and recognized societal needs, through democratic processes, and with regards for current and future generations (Servet 2014) – or as part of the social and solidarity economy.

In either definition of sharing – as part of the competitive market economy or the solidarity economy – the notion of optimizing resources is highly compatible with industrial ecology principles. The de-materialization of the economy has been a central theme in industrial ecology these past years. Placing value on under- or unused resources is also promoted in the 'sharing' economy. Industrial ecology approaches could help consider the wider system, beyond a unit or transaction, to determine whether there are any rebound effects associated with such forms of sharing. By renting fashionable clothing, for example, does overall private acquisition of consumption increase or decrease? Has car sharing led to increases in private car ownership or increases in public transportation, and over what scale? In terms of describing activities within the sharing economy, industrial ecology could help provide a lens through which to understand where 'sharing' is actually taking place: is there only 'end of pipe sharing', once a resource has already been developed and privately owned; or is 'sharing' built into the entire process, in how a product or service is conceived, delivered and maintained? As an operational tool, industrial ecology could help suggest how certain practices within a sharing economy could be restructured, towards a normative goal of more efficient resource usage and lessened environmental impact. Industrial ecology has a role to play in helping to better define whether the sharing economy is leading to a more optimal system overall and to what extent, based on biophysical considerations.

The sharing economy, in all different forms, is well underway and provides an opportunity for industrial ecology practitioners to break with the notion of 'production and consumption' to consider new ways of maximizing resources, from peer-to-peer exchanges, as well as new business models. The 'sharing' taking place in the social and solidarity economy could therefore be a rich terrain for further study and practice. Industrial ecology methodologies could be applied for better evaluating and guiding activities towards more efficient 'resource' sharing; but perhaps more importantly, advances in sharing in the social and solidarity economy could further inform the industrial ecology community towards a more holistic reading of socio-industrial ecosystems – integrating questions of current and inter-generational solidarity and mutual support, as well as questions related to power, trust, wellbeing and governance, towards collaborative industrial eco-systems.

The emphasis on maximizing local resources towards symbiosis could benefit from some of the approaches being developed in the sharing economy, not least the use of information and communication technologies that distinguish from more traditional forms of sharing. Tools for 'sharing' are being developed through new smart phone apps and web sites, using participative methods in their design and deployment, and based on establishing relations of trust through peer review systems. Trust has been a central theme in symbiosis studies, within the industrial ecology community, and much could be learned from the 'sharing' economy in this respect. One aspect of the sharing economy that is currently gaining the attention of researchers and practitioners working on labour relations is the question of casualized labour or the precarity of labour in the sharing economy. In considering the future of environmental reporting, attention should also be placed on labour issues as well as human rights (Fatkin in Sarkis 2001). The work of Cohen-Rosenthal would be relevant here, to place a focus once more on labour conditions and the workplace, as a way to contribute to employee and societal wellbeing – as discussed in the Chap. 8 by Wiedmann in this volume.

Finally, the maximization of resources through 'sharing' could be more effective in some cases when taken out of the market economy and placed into the solidarity economy. As Guillaume Massard suggests, founder of SOFIES industrial ecology consulting group, 'the solidarity economy in Geneva allows the development of business models for material reuse with other conditions than the typical market conditions, making it attractive to recycle certain materials that would not be collected if subjected to the market economy prices' (Massard 2015). In Geneva, social reinsertion programs are part of the SSE, whereby unemployed people are given positions in enterprises as part of their training, some of which focus on recycling materials such as electronic products. In the case where the State does not offer subsidies to an enterprise directly but allows for this form of subsidized labour, the business model for recycling such products can become more attractive. 'Everything that goes beyond market profitability, this is where the solidarity economy can introduce different biases, such as complementary currencies or subsidized labour,' according to Massard.

3.2   Community Currencies: Idea of Démurage and Applicability to IE

National currencies are a fairly recent invention: centralizing money was first conceived by European royalty, in attempts to limit feudal power, then reinforced by empires seeking tighter control of the colonies, and finally by the modern nation-states. Money is far from being neutral: it affects the kind of transactions we make, and the kinds of relations we establish with those exchanges and within society (Lietaer and Kennedy 2008); the value of money is ultimately a social construct (Graeber 2001). Historically, diverse monetary systems always existed in parallel, around the world, from Europe to Indonesia (Sahakian 2014). The thirteenth century Republic of Venice had two types of currencies for external commerce, the *ducat* (silver) and *zecchino* (gold), and two other currencies in less precious metals for local exchanges, the *nasoni* and *cavalotti* (Lietaer and Kennedy 2008). In making a parallel to industrial ecology concepts, having different types of currencies working together in a complementary manner tends towards ecosystem diversity. Our national and supra-national currencies, by contrast, are similar to monoculture farming; a weak euro impacts the entire euro-zone along with international trade, with no other currencies in place to provide a stabilizing effect. The argument for complementary currencies rests on the need for system diversity in our monetary systems.

Beyond national currencies, other forms of monetary units exist and are widely exchanged as complementary currencies, such as airline mileage (The Economist 2005) and Bitcoins (McMillan 2014). One definition of complementary currencies is that they exist at the nexus of unsatisfied needs and under-utilized resources (Lietaer and Kennedy 2008). In 2010, an estimated one million people worldwide engaged in complementary currency systems in over 4,000 associations in over forty countries (Blanc 2010); no doubt this number has increased since then. Currencies can have aims other than facilitating exchange, creating wealth, promoting brand loyalty or allowing for the redistribution of wealth through taxation. This is where the notion of community currencies comes in, as a subset of complementary currencies, and tied to the guiding principles of the social and solidarity economy (SSE). Community currencies are often designed towards social or environmental aims, are generated in and spent in a given region and not tied to national currencies, thus sheltered from the whims of international financial markets. The advocates of community currencies point to the need to diversify the local economic system and harness the potential of regional wealth creation and related expenditures. For Blanc (2010), one of the main objectives of such currencies is to 'resist globalization' and encourage the use of local income for local production and consumption; a second objective is to benefit local populations through a fairer distribution of wealth, rather than wealth accumulation among an elite; third, such currencies should aim at transforming the nature of trade and solidifying social relations based on trust, proximity between producer and consumer and the notion of producer as consumer (or 'prosumer'). Handbooks designed to guide those interested in stimulating regional economies are available (Lietaer and Kennedy 2008).

Examples of community currencies have flourished in the past two decades and around the world through what are called *Systèmes d'échanges locaux (*SEL) or Local Exchange Trading Systems (LETS). The first LETS in the UK was created in Norwich in 1985, growing by 2001 to include 300 trading schemes, involving 22,000 people and an annual turnover equivalent of £1.4 million (Williams et al. 2001 in Seyfang 2007). The local exchange systems can involve the trading of different products and services, but also time as a resource. This relates to time banking, a form of exchange based on the egalitarian notion that each member's time is equivalent to another's. Services such as baby-sitting or painting can be exchanged for computer programming or legal advice, with no distinction between the type of service offered; it is an hour of time that is being exchanged. Time banks are increasingly the subject of academic research: in the recent edited volume, *Sustainable lifestyle and the quest for plenitude* (Schor and Thompson 2014), the up- and downsides of a Boston time bank are discussed, based on empirical research. The reason LETS have remained small and marginal, in relation to the dominant market economy, may have to do with a number of factors, which might include the question of quality of skills being traded, the availability of staple goods and services and government regulations that count LETS earnings as equivalent to cash income, among others (Seyfang 2007).

Japan was a forerunner in community currency experiments: in the 1970s, a volunteer labour bank allowed for inter-generational care, but was modestly successful (Hirota 2011). In the late 1990s, the Director of Service Industries Division of the Ministry of Economics, Trade and Industry (METI), Toshiharu Kato, founded the Eco-Money network to support regional experiments in new currencies across Japan.[4] By 2003, 25 projects were implemented by 55 different organizations across the country. The main roles of Eco Money projects are to enhance community integrity, foster public participation, create a sustainable economic environment and maintain a viable natural environment (Okuno 2004). In the Chiba prefecture, the "peanut" was introduced as a regional currency, where one "peanut" is equal in value to one yen; one hour of work is equal in value to 1,000 peanuts, with all transactions recorded as "plus and minuses" on individual record sheets. For example, peanuts can be gained by helping someone build a web site and driving a neighbour to the hospital, which can then be used towards paying for a language class (Okuno 2004). Visitors to the 2005 World Exposition in Aichi, Japan, could earn Expo Eco-Money points through pro-environmental actions (e.g., bringing your own bag to the store, for example), which were then exchanged for services or products or used to make donations to environmental projects. In another example in the Yasu-Cho community in Shiga prefecture, community members came together to help protect local forests and raise environmental awareness. An *eco-yama* (eco-mountain) card was issued to encourage people to earn credit by helping to maintain the forest or to help develop local renewable energies (solar and biomass). Local stores and businesses agreed to accept these credits, creating a local currency in solidarity between local commerce, residents and the natural environment.

Levine (2003) suggests that one of the limits of the IE-natural system analogy is that products play a central role in the economy: interactions or exchanges in industrial ecology are mutualistic when they create a positive feedback loop in terms of product value and related economic benefits, yet there is no 'analogy' with a socially constructed 'product value' in natural systems. We might imagine new ways of assigning value to under-utilized material and energy resources, by creating new ways of trading such products and services outside of the capitalist market economy – by incorporating social and environmental values, for example. In relation to

---

[4] According to Lietaer (2004), eco-monies were introduced as part of a regional development strategy following Kato's investigations in regional development elsewhere. His study of high-tech development models in the United States, including Silicon Valley, lead him to conclude that regional learning clusters should be promoted, involving entrepreneurs and small corporations, alongside ecological, economic, and community-driven initiatives (Okuno 2004). Complementary currencies were seen as part of this design.

climate change, one proposal is to reward low-carbon investments through the creation of a monetary unit based on the social cost of carbon (SCC): in this example, 'the SCC is neither a market price, nor the tax incorporated in the prices of goods. It is a notional price defined as the social value of avoided CO_2 emissions' (Aglietta et al. 2015: 4), which relies on a strong independent body to calculate exactly what that price might be. In this approach, the monetary unit is assigned a politically negotiated value, and not a value based on the whims of a capitalist marketplace.[5]

To further this example around a SCC currency, the notion of *démurage* could be of interest, defined as a negative incentive against the accumulation of notes.[6]*Démurage* is a familiar term in complementary currencies, as it stimulates the constant exchange of complementary currencies, against the accumulation of wealth and towards its distribution. When a currency has been assigned *démurage*, the more money you accumulate over time, the lesser the value of that money. In the example of a monetary unit based on the social cost of carbon, this could incentivise the holders of this currency to accelerate their investments in low-carbon projects – rather than hoard carbon credits for the purpose of financial speculation. To return to the founding metaphor of IE, the accumulation of benefits (i.e., endless profits) is not apparent in ecosystems, which tend towards equilibrium. *Démurage* could limit the effects of accumulation for a complementary currency that strives towards environmental goals. Whether this type of system would be state and independently regulated (as Aglietta et al. suggest in relation to this version of a monetary unit based on the SCC), driven by enterprises, or managed by everyday people remains to be debated, along with an assessment of the resulting environmental impacts.

3.3   Crowdfunding in the Solidarity Economy: towards IE Principles

Rifkin claims in his latest book (2014) that we are facing a transformation in modes of production and operation, largely due to new technologies, and in a transition from a capitalist to collaborative marketplace. The financial sector is no exception to this trend: promoted through the Internet and mobile payment services, new

---

[5] There is much debate around the social cost of carbon (SCC), as well as the notion of a more versatile shadow price of carbon (SPC) used internally by companies in their strategic planning (CDP 2013) or in the policy arena (Price et al. 2007). The point here is not to discuss these developments in depth but rather illustrate the example of how a monetary unit could operate outside of the marketplace, based on values assigned through policy negotiations – assuming democratic processes and a strong independent regulatory body.

[6] The notion of *démurage* emerged in 1933, when the Austrian village of Wörgl introduced a new currency designed with a negative incentive against the accumulation of the notes. The National Bank of Austria closed down this experiment within a few months, fearing it would be replicated to other regions and ultimately challenge the national currency (Sahakian 2014). The concept is still alive today, however, with various community currencies building in *démurage* to insure the circulation of notes.

opportunities for raising donations, credit and loans abound. Given the credit crunch in certain contexts, some of these online platforms offer more favourable interest rates than traditional lending mechanisms. Certain financial platforms are 'business as usual', facilitated by new technologies; however, some of these online services could fall under the umbrella of the social and solidarity economy. In Sahakian and Servet (in press), the distinction is made between crowdfunding that aims towards 'communal sharing' *versus* self-interest. Related to the former definition, crowdfunding that tends towards solidarity implicates people coming together to address a broader need, serving either social or environmental aims.

Certain crowdfunding platforms attempt to achieve these aims (e.g., SPEAR and Kisskissbankbank), while others focus specifically on promote investments in renewable energies (e.g., Solar Mosaic and Wiseed).[7] Certain platforms propose both 'business as usual' projects as well as projects that aim at a social good. Take for example Kickstarter: raising funds for the customizable smart watch 'Pebble'[8] does not aim towards achieving a greater environmental or social common good (although this may be a matter of perspective), whereas raising funds for favela painting does aim to transform an under-privileged area of Rio. In the case of Smart Angels or Unilend,[9] the projects financed tend to fall in the 'business as usual' category and these platforms are also based on capitalistic notions, in that they are privately owned and seek pecuniary gains. The type of institutional framework governing a platform does not determine its mission of purpose. There is currently very little information publicly available on how such platforms operate in practice, including their governance systems, and this merits further study, particularly in relation to notions of democratic governance systems.

Assessing the level of 'communal sharing' and 'solidarity' that takes place through crowd-funding entails considering a range of factors, including the motivations for setting up such a platform, for proposing projects and contributing funds. Such projects would need to be evaluated over time, to ensure that they deliver on their promises. Crowdfunding could benefit the industrial ecology community in one obvious way, in raising funds for interesting and novel projects. In other sectors, crowdfunding has been used to test the viability or public support for certain initiative. Crowdfunding for a new biogas facility for example could demonstrate the interest and acceptance among the general public for such a project.

There is yet another way in which crowdfunding could work in synergy with industrial ecology: as an investment tool. The economic gains made through reducing consumption and via efficiency measures can lead to decreases in consumption that are lower than expected through a direct rebound effect (Hertwich 2005a). More challenging to measure is 'indirect rebound', where for example money saved through reduced energy or material consumption in one consumption area (e.g.

---

[7] See associated websites: www.spear.fr, www.kisskissbankbank.com, joinmosaic.com, www.wiseed.com

[8] See Pebble project page on the Kickstarter website: www.kickstarter.com/projects/597507018/pebble-e-paper-watch-for-iphone-and-android (Retrieved March 3, 2015).

[9] See: www.smartangels.fr, www.unilend.fr

switching from car journey to bicycles) leads to funds spent on other energy- and material-intensive activities (e.g., a long distance flight). Economy-wide rebound considers the effects such rebounds might have across the economy (Druckman et al. 2011). Using the savings generated from increased efficiencies to invest in projects that aim towards industrial ecology goals could be one way to counter such rebounds (as also discussed by Druckman and Jackson, Chap. 9 in this volume), with certain crowdfunding platforms providing the opportunity for such investments (Sahakian and Servet in press). While quantifying and qualifying rebounds is a difficult exercise, allowing more opportunities for people to invest more directly in environmentally sound and socially just activities, seems like a logical way forward.

4   Conclusion

In this chapter we have considered the links and tensions between the fields of the social and solidarity economy (SSE) and industrial ecology (IE), highlighting new directions for reflection, research and practice in industrial ecology. This exercise raises more questions than answers. The first main conclusion is about value systems and paradigms: the SSE tends to place people over profit; the IE field tends to put planet over profit, both therefore go beyond pecuniary interests yet are based on differing normative goals. By placing these two fields side by side, it becomes apparent that the IE field could further theorize questions related to social context and power structures. The notion of solidarity in relation to the workplace is becoming increasingly topical in discussions around the sharing economy: ethical issues are emerging around questions of employment security, health and safety standards, employee personal development and employee sharing in value creation and related wealth (Conway 2014). The question of labour precarity will no doubt increase in the coming years, thus the relevance of Cohen-Rosenthal's contribution to the IE literature as well as more recent work on employee wages, health and safety. In addition to the increasing focus on consumption and consumers in the IE field over the past decade, more attention could also be given to production and specifically labour relations: would a 'world of work' based on solidarity contribute to greater prosperity?

Beyond considering how IE might integrate principles from the SSE, another approach would be to question whether solidarity, in a community or between different stakeholders in an economic activity, might be a promising foundation for industrial symbiosis (see Chaps. 5 by Chertow and Park, and 19 by Bailey in this volume). Rather than 'create' an eco-community or industrial eco-park through top-down policies, or hope that bottom-up forms of symbiosis will emerge from self-organizing entities, one interesting new avenue for the industrial ecology community would be to assess where the SSE is already active in certain fields – at the level of cooperatives, SMEs, larger enterprises, neighbourhoods or regions – then build on the existing relations of trust and solidarity, to maximize resource efficiency.

Whether 'solidaristic' enterprises would be more receptive to such efforts and more adept at embracing such resource exchanges remains to be seen and would be an interesting area of future research. Another field of inquiry would be to ascertain to what extent companies already involved in symbiotic relations also embody social and solidarity values. Chertow and Ehrenfeld found that 'one of the most distinctive elements of industrial symbiosis is that, while all industrial actors seek to reduce private costs and increase private benefits, those in the symbiotic networks that have been studied also participate in the creation of public environmental benefits' (2012: 18). Here, the SSE can contribute through an analysis of the profit structure, governance system, societal aims and, more generally, in better understanding the culture and values within such systems. Two other fields of inquiry involve the consideration of 'solidaristic' crowdfunding as an opportunity for abating economy-wide rebound effects through the promotion of more socially just and environmentally sound investments; and finally, the potential for complementary currencies to work towards industrial ecology aims, such as reducing global carbon emissions.

One of the weaknesses of the social and solidarity economy has been that of scale: although various institutions exist at the level of cities, countries and regions to federate activities across sectors, the actors within the SSE typically operate on a more micro scale. Over the past few years, there has been increasing attention to macroeconomic thinking in the industrial ecology and ecological economics community, questioning notions of wellbeing and prosperity, the inadequateness of current models and the need for transformative investments in the future (Røpke in Cohen et al. 2013; Jackson 2009; Victor 2008). As Tim Jackson put it 'The truth is that there is as yet no credible, socially just, ecologically sustainable scenario of continually growing incomes for a world of 9 billion people.' (2009: 86). The social and solidarity economy is not a magic wand solution and operates at the margins of the dominant capitalist economy, yet it is well underway and expanding, in research and practice. Reflecting on how the SSE and industrial ecology community might come together towards more macro-level perspectives would be a worthy exercise. One main outcome of this analysis is that transactions and exchanges are important flows, but more attention should be placed on the underlying values holding together our everyday practices and put forward in our economy, workplaces and society as a whole – including social and environmental values that are not always aligned.

Acknowledgements The author would like to thank the editors for shepherding this chapter into this volume and for their comments. I am indebted to the Industrial Ecology Group at the University of Lausanne for their generous contribution of ideas, including Suren Erkman, Loïc Leray, Frederic Meylan, Ignes Contreiras, Theodore Besson and Vincent Moreau. I thank particularly Jean-Michel Servet for his detailed comments in relation to the solidarity economy. I thank Guillaume Massard for his insights, as well as Maurie Cohen and Sophie Swaton for fruitful exchanges around sharing. The term 'end-of-pipe' sharing emerged from a lively discussion with Nedal Nassar at the 2014 GRC on Industrial Ecology.

References

Aglietta, M., Espagne, E., & Fabert, B. P. (2015). A proposal to finance low carbon investment in Europe. *Finance Stratégie, 24*, 1–7.

Alsamawi, A., Murray, J., & Lenzen, M. (2014). The employment footprints of nations. *Journal of Industrial Ecology, 18*(1), 59–70.

Andrews, E. S., Barthel, L.-P., Tabea, B., Benoît, C., Ciroth, A., Cucuzzella, C., Gensch, C.-O., Hébert, J., Lesage, P., Manhart, A., & Mazeau, P. (2009). *Guidelines for social life cycle assessment of products. C. Benoît*. Paris: UNEP/SETAC Life Cycle Initiative.

APRES-GE. (2010). *Etude statistique: Photographie de l'économie sociale et solidaire à Genève*. Geneva: Chambre de l'Economie Sociale et Solidaire.

Arruda, M. (2004). *Humanizar o Infra-Humano. A formação do ser humano integral: Homo evolutivo, práxis e economia solidária*. Petropolis: Vozes/Pacs.

Ayres, R., & Simonis, U. E. (1994). *Industrial metabolism: Restructuring for sustainable development*. Tokyo: United Nations University Press.

Benoit-Norris, C., Cavan, D. A., & Norris, G. (2012). Identifying social impacts in product supply chains: Overview and application of the social hotspot database. *Sustainability, 4*, 1946–1965.

Blanc, J. (2010). *Community and complementary currencies. The human economy: A citizen's guide* (pp. 303–312). Cambridge/Malden: Polity Press.

Boons, F., & Howard-Grenville, J. (Eds.). (2009). *The social embededness of industrial ecology*. Cheltenham/Northampton: Edward Elgar.

Boons, F., & Spekkink, W. (2012). Levels of institutional capacity and actor expectations about industrial symbiosis: Evidence from the Dutch stimulation program 1999–2004. *Journal of Industrial Ecology, 16*(1), 61–69.

Botsman, R., & Rogers, R. (2010). *What's mine is yours: The rise of collaborative consumption*. New York: HarperCollins Publishers.

CDP. (2013). *Use of internal carbon price by companies as incentive and strategic planning tool: A review of findings from CDP 2013 disclosure*. New York: CDP Worldwide.

Chertow, M. R. (2007). "Uncovering" industrial symbiosis. *Journal of Industrial Ecology, 11*(1), 11–30.

Chertow, M., & Ehrenfeld, J. (2012). Organizing self-organizing systems. *Journal of Industrial Ecology, 16*(1), 13–27.

Coalition, A. (2004). *Report on the seminar on social responsibility of SMEs for development of people's economy. Asian coalition for SME development*. Mumbai: World Social Forum.

Cohen, M. J. (2014). *Some reflections on the sharing economy: Where is the sharing in car "Sharing"?* Retrieved February 2, 2015, from http://ssppjournal.blogspot.ch/2014/02/some-reflections-on-sharing-economy.html

Cohen, M. J., Brown, H. S., & Vergragt, P. J. (Eds.). (2013). *Innovations in sustainable consumption: New economics, socio-technical transitions and social practices. Advances in ecological economics*. Cheltenham/Northampton: Edward Elgar.

Cohen-Rosenthal, E. (1979). Enriching workers' lives. *Change, 11*(5 - Education and Work: Two Worlds or One?), 64–66.

Cohen-Rosenthal, E. (2004). Making sense out of industrial ecology: A framework for analysis and action. *Journal of Cleaner Production, 12*, 1111–1123.

Cohen-Rosenthal, E., & Musnikow, J. (Eds.). (2003). *Eco-industrial strategies: Unleashing synergy between economic development and the environment*. Sheffield: Greenleaf publishing.

Conway, M. (2014). *The downside to lower labor costs in the sharing economy*. Huff Post Business. Retrieved March 3, 2015, from http://www.huffingtonpost.com/maureen-conway/the-downside-to-lower-labor-costs_b_5759122.html

Daly, H. E. (1977). *Steady-state economics: The economics of biophysical equilibrium and moral growth*. San Francisco: W.H. Freeman and Company.

Defourny, J., & Develtere, P. (1999). *Origines et contours de l'économie sociale au nord et au sud. L'économie sociale au Nord et au Sud* (pp. 25–50). Brussels: De Boeck Université.

Defourny, J., Develtere, P., & Fonteneau, B. (Eds.). (2000). *Social economy North and South.* Leuven and Liège, HIVA and Centred' Economie Sociale.

Druckman, A., Chitnis, M., Sorrell, S., & Jackson, T. (2011). Missing carbon reductions? Exploring rebound and backfire effects in UK households. *Energy Policy, 39,* 3572–3581.

Dunand, C. (2012). *L'économie sociale et solidaire: une troisième voie.* Retrieved February 2, 2015, from http://www.apres-ge.ch/node/32911

Ehrenfeld, J. R. (2000). Industrial ecology: Paradigm shift or normal science? *American Behavioral Scientist, 44*(2), 229–244.

Ehrenfeld, J. (2003). Editorial: Putting a spotlight on metaphors and analogies in industrial ecology. *Journal of Industrial Ecology, 7*(1), 1–4.

Ehrenfeld, J. R. (2008). *Sustainability by design: A subversive strategy for transforming Our consumer culture.* New Haven/London: Yale University Press.

Ellen MacArthur Foundation. (2013). *Towards the circular economy: Economic and business rationale for an accelerated transition.* Retrieved February 3, 2015, from http://www.ellenmacarthurfoundation.org/circular-economy/circular-economy/towards-the-circular-economy

Erkman, S. (1997). Industrial ecology: An historical view. *Journal of Cleaner Production, 5*(1–2), 1–10.

Erkman, S., & Ramaswamy, R. (2003). *Applied industrial ecology: A new platform for planning sustainable societies: Focus on developing countries with case studies from India.* Bangalore: Aicra Publishers.

Fischer-Kowalski, M., Hertwich, E., & Lifset, R. (2009). *Strategies and tools for refashioning the social metabolism: IE as a key to the transition to sustainability.* ISIE conference, Lisbon, Portugal.

Fraisse, L. (2003). Quels projets politiques pour l'économie solidaire? *Cultures en mouvement, 62,* 4.

Fraisse, L., Guérin, I., & Laville, J.-L. (2007). Economie solidaire: des initiatives locales à l'action publique. Introduction. *Revue Tiers Monde, 190,* 245–253.

Gansky, L. (2010). *The mesh: Why the future of business is sharing.* London: Portfolio Penguin.

Georgescu-Roegen, N. (1971). *The entropy law and the economic process.* Cambridge, MA: Harvard University Press.

Georgescu-Roegen, N. (2006). *La Décroissance: Entropie, Ecologie, Economie (1979).* Paris: Sang de la Terre.

Gibson-Graham, J. K. (2006). *A postcapitalist politics.* Minneapolis: University of Minnesota Press.

Gibson-Graham, J. K. (2008). Diverse economies: Performative practices for 'other worlds'. *Progress in Human Geography, 32*(5), 613–632.

Giovannini, M. (2014). Indigenous community enterprises in Chiapas: a vehicle for buen vivir? *Community Development Journal*, Advance access online March 3. doi:10.1093/cdj/bsu019.

Graeber, D. (2001). *Toward an anthropological theory of value: The false coin of our own dreams.* New York: Palgrave.

Hart, K., Laville, J.-L., & Cattani, A. D. (Eds.). (2010). *The human economy: A citizen's guide.* Cambridge/Malden: Polity Press.

Hauschild, M. Z., Dreyer, L. C., & Jørgensen, A. (2008). Assessing social impacts in a life cycle perspective—Lessons learned. *CIRP Annals – Manufacturing Technology, 57,* 21–24.

Hertwich, E. G. (2005a). Consumption and the rebound effect: An industrial ecology perspective. *Journal of Industrial Ecology, 9*(1–2), 85–98.

Hertwich, E. G. (2005b). Editorial: Consumption and industrial ecology. *Journal of Industrial Ecology, 9*(1–2), 1–6.

Hess, G. (2009). Forum: L'écosystème industriel. Difficulté épistémologique d'une telle analogie. *Natures Sciences Sociétés, 17,* 40–48.

Hillenkamp, I. (2011). Solidarités, marché et démocratie: éclairages boliviens. *Finance & The Common Good/Bien commun II–III*(37–38), 76–95.

Hillenkamp, I. (2013). *L'économie solidaire en Bolivie: entre marché et démocratie.* Geneva: Editions Karthala.

Hirota, Y. (2011). What have complementary currencies in Japan really achieved? Revealing the hidden intentions of different initiatives. *International Journal of Community Currency Research, 15*(D), 22–26.

INSEE. (2012). *L'économie sociale en 2012.* Retrieved February 2, 2015, from http://www.insee. fr/fr/themes/detail.asp?reg_id=99&ref_id=eco-sociale

ISGC (1997). *Lima Declaration.* International Solidarity Globalization Conference, Lima, Peru.

Jackson, T. (2009). *Prosperity without growth: Economics for a finite planet.* London: Routledge/ Earthscan.

Kalamar, A. (2013). *Sharewashing is the new greenwashing.* Retrieved February 2, 2015, from http://www.opednews.com/articles/Sharewashing-is-the-New-Gr-by-Anthony-Kalamar-130513-834.html

Kaswan, M. J. (2014). Developing democracy: Cooperatives and democratic theory. *International Journal of Urban Sustainable Development, 6*(1), 190–205. doi:10.1080/19463138.2014.9510 48.

Kawano, E. (2009). Into the light: The emerging solidarity economy movement in the United States, *Cayapa, 10*: 50–64. Mérida, Venezuela: Centro Internacional de Investigación e Información sobre la Economía Pública Social y Cooperativa.

Kawano, E. (2013). *Social solidarity economy: Toward convergence across continental divides.* UNRISD News & Views. Retrieved March 3, 2015, from http://www.unrisd.org/unrisd/website/newsview.nsf/%28httpNews%29/F1E9214CF8EA21A8C1257B1E003B4F65?OpenDocument

Kawano, E., Masterson, T. N., & Teller-Elsberg, J. (Eds.). (2009). *Solidarity economy I: Building alternatives for people and planet.* Amherst: Center for Popular Economics.

Laville, J.-L. (Ed.). (1994). *L'économie solidaire: une perspective internationale.* Paris: Desclée de Brouwer.

Laville, J.-L. (2003). A new European socioeconomic perspective. *Review of Social Economy, LX*(3), 389–405.

Laville, J. -L. (2011). *What is the third sector? From the non-profit sector to the social and solidarity economy: Theoretical debates and European reality* (EMES European Research Network Working Paper 11/01).

Laville, J.-L., & Cattani, A. D. (Eds.). (2006). *Dictionnaire de l'autre économie.* Paris: Folio actuel.

Lemaître, A., & Helmsing, A. H. J. B. (2012). Solidarity economy in Brazil: Movement, discourse and practice analysis through a Polanyian understanding of the economy. *Journal of International Development, 24,* 745–762.

Levine, S. H. (2003). Comparing products and production in ecological and industrial systems. *Journal of Industrial Ecology, 7*(2), 33–42.

Lietaer, B. (2004). Complementary currencies in Japan today: History, originality and relevance. *International Journal for Community Currency Research, 8,* 1–23.

Lietaer, B., & Kennedy, M. (2008). *Monnais Régionales: De nouvelles voies vers une prosperité durable.* Paris: Charles Léopold Mayer.

Lowen, M. (2012). Greece bartering system popular in Volos. *BBC Europe, News.* Retrieved February 3, 2015, from http://www.bbc.com/news/world-europe-17680904

Massard, G. (2015, January 20). Personal Interview by M. Sahakian. Geneva

Massard, G., Jacquat, O., & Zürcher, D. (2014). *International survey on ecoinnovation parks. Learning from experiences on the spatial dimension of eco-innovation. Environmental studies.* Bern: Federal Office for the Environment and the ERANET ECO-INNOVERA.

McMillan, R. (2014). The Fierce battle for the Soul of Bitcoin. *WIRED Business.* Retrieved February 23, 2015, from http://www.wired.com/2014/03/what-is-bitcoin

Okuno, S. (2004). Enhancing sustainable communities with local currencies: Eco-money experiment in Japan. In *World forum proceedings of the International Research Foundation for Development*.

Opoku, H. N. (2004). Policy implications of industrial ecology conceptions. *Business Strategy and the Environment, 13*, 320–333.

Ostrom, E. (2001/1990). *Governing the commons: The evolution of institutions for collective action*. Cambridge: Cambridge University Press.

Poggioli, S. (2011). *Modern Greeks return to ancient system of barter*. Retrieved Nov 8, 2013, from http://www.npr.org/2011/11/29/142908549/modern-greeks-return-to-ancient-system-of-barter

Polanyi, K. (1957). The economy as instituted process. In K. Polanyi, C. M. Arensberg, & H. W. Pearson (Eds.), *Trade and market in the early empires* (pp. 243–270). Glencoe: Free Press.

Polanyi, K. (2001). *The great transformation: The political and economic origins of our time (1944)*. Boston: Beacon.

Prades, J. (2013). *Comment résister au capitalisme? Tous en coopératives!* Aubiet: Le vent se lève.

Price, R., Thornton, S., & Nelson, S. (2007). *The social cost of carbon and the shadow price of carbon: What they are, how to use them in economic appraisal in the UK* (Defra evidence and analysis series). London: Department for Environment, Food and Rural Affairs.

Ransom, D., & Baird, V. (Eds.). (2010). *People first economics*. Oxford: New Internationalist Publications.

Rifkin, J. (2014). *The zero marginal cost society: The internet of things, the collaborative commons, and the eclipse of capitalism*. New York/London: Palgrave Macmillan.

Sahakian, M. (2014). Complementary currencies, what opportunities for sustainable consumption in times of crisis and beyond? Sustainability: Science. *Practice & Policy, 10*(1), 4–13.

Sahakian, M. D., & Dunand, C. (2015). The social and solidarity economy towards greater 'sustainability': Learning across contexts and cultures, from Geneva to Manila. *Community Development Journal, 50*, 403–417.

Sahakian, M., & Servet, J.-M. (in press). Separating the wheat from the chaff: Sharing versus self-interest in crowdfunding. In D. Assadi (Ed.), *Strategic approaches to crowdfunding*. Hershey: IGI Global.

Sahakian, M., & Steinberger, J. K. (2011). Energy reduction through a deeper understanding of household consumption: Staying cool in metro manila. *Journal of Industrial Ecology, 15*(1), 31–48.

Sarkis, J. (Ed.). (2001). *Greener manufacturing and operations: From design to delivery and back*. Sheffield: Greenleaf.

Satgar, V. (Ed.). (2014). *The solidarity economy alternative: Emerging theory and practice*. Pietermaritzburg: University of KwaZulu-Natal Press.

Scanlon, K., Lloyd, S., Gray, G., Francis, R., & LaPuma, P. (2015). An approach to integrating occupational safety and health into life cycle assessment: Development and application of work environment characterization factors. *Journal of Industrial Ecology, 19*(1), 27–37.

Schor, J. B., & Thompson, C. J. (Eds.). (2014). *Sustainable lifestyles and the quest for plenitude case studies of the new economy*. New Haven: Yale University Press.

Servet, J.-M. (2006). Towards an alternative economy: Reconsidering the market, money and value. In C. Hann & K. Hart (Eds.), *Market and society: The great transformation today*. Cambridge: Cambridge University Press.

Servet, J.-M. (2007). Le principe de la réciprocité chez Karl Polanyi, contribution à une définition de l'économie solidaire. *Revue Tiers Monde, 2*(190), 255–273.

Servet, J. -M. (2014). *De nouvelles formes de partage: la solidarité au delà de l'économie collaborative*. Veblen Institute. Retrieved March 3, 2015, from http://www.veblen-institute.org/spip.php?page=imprimer&lang=fr&id_article=190

Seyfang, G. (2007). *Personal carbon trading: Lessons from complementary currencies*. CSERGE working paper ECM 07–01. Retrieved March 3, 2015, from http://cserge.ac.uk/sites/default/files/ecm_2007_01.pdf

Singer, P. (2002). *Introduçaon à Economia Solidaria*. Sao Paolo: Fondacion Perseu Editora.

Studer, T. (2006). *WIR and the Swiss National Economy*. Rohnert Park: Lulu Online Publishing.

Swaton, S. (2011). *Une entreprise peut-elle être "sociale" dans une économie de marché?* Charmey: Editions de l'Hèbe.

Swaton, S., & Baranzini, L. (2013). Définir la nouvelle économie sociale par les critères plutôt que par les statuts? Une analyse théorique à partir des critères retenus en Suisse par Après-Ge. *Cahiers du CIRTES, Presses Universitaires de Louvain, 9*, 53–68.

Swaton, S., & Sahakian, M. (2014). *The social and solidarity economy: What is it?* SCORAI workshop on inter- trans-disciplinarity. Lausanne.

The Economist. (2005, December 20). *Frequent-flyer miles: Funny money*. Retrieved February 23, 2015, from http://www.economist.com/node/5323615

UNEP. (2009). *Guidelines for social life cycle assessments of products. B. M. Catherine Benoît.* Paris: United Nations Environment Programme.

UNEP. (2011). *Towards a GREEN economy. Pathways to sustainable development and poverty eradication.* A Synthesis for Policy Makers. Retrieved March 3, 2015, from http://www.unep.org/greeneconomy/Portals/88/documents/ger/GER_synthesis_en.pdf

Victor, P. A. (2008). *Managing without growth: Slower by design, not disaster*. Cheltenham/Northampton: Edward Elgar.

Yuan, Z., Bi, J., & Moriguichi, Y. (2006). The circular economy: A new development strategy in China. *Journal of Industrial Ecology, 10*(1–2), 4–8.

Chapter 11
Industrial Ecology in Developing Countries

Megha Shenoy

Abstract Sustainable development is not a simple, singular and well-tested path. It needs an interdisciplinary examination of resource use patterns, ecological heritage, demographics and cultural values. Industrial ecology, owing to its emphasis on using a holistic approach, can provide a valuable platform to draw out sustainable strategies and policies for developing countries to implement. It can offer a paradigm within which IE methods and tools can inform responses to local development challenges. Within this paradigm, sustainable industrial, rural and urban development strategies and policies in developing countries should follow from IE research and analyses.

A SWOT analysis of IE in developing countries highlights strengths of high economic growth and threats from outdated policies and inadequate industrial ecology awareness in the policy making and governing spheres. Examination of the IPAT equation in the context of developing countries highlights the role that new technological hubs such as China and India can play, the significance of increasing affluence among "new consumers" in the developing world and the role of population in managing resources sustainably.

Research in IE since its introduction to the global south around the mid-1990s has primarily focused on two concepts of IE – cleaner production and eco-industrial parks – largely due to the impetus of development organizations. Other studies using the IE lens and tools have shown the potential of the IE paradigm in developing countries. These studies have highlighted the importance of focussing on scarce resources such as water, examining the possibilities of using well-tested technologies and evaluating the long-term maintenance of new technologies and practices before recommending their implementation. New policies in the developing world can gain from the IE community in terms of assistance in simplifying and downsizing data requirements, application of solutions to contemporary sustainability challenges and framing effective policies based on IE concepts.

M. Shenoy (✉)
Independent Sustainability Research Consultant,
SJ 102, Shriram Spandhana, Chalghatta, Bangalore 560037, India
e-mail: shenoymegha@gmail.com

© The Author(s) 2016
R. Clift, A. Druckman (eds.), *Taking Stock of Industrial Ecology*,
DOI 10.1007/978-3-319-20571-7_11

Keywords Developing countries • Eco-industrial parks • Industrial ecology •
Sustainability • Sustainability policy

1 Introduction

1.1 Benefits of IE for Developing Countries

The topic 'industrial ecology (IE) in developing countries' is as vast as the ocean
and I would not wish to attempt to raft through its entirety with this chapter. Using
my field glasses, I attempt to provide an overview, highlight areas of industrial ecology that have been examined in developing countries and two-way streets for developing countries to benefit from IE and *vice versa*. Developing countries[1] can use the
concepts and tools of IE to ensure that the improvements they make to the quality of
life for their citizens is achieved in harmony with improving the health of ecological
systems, while investing effort, time and resources into a resilient economy. Regions
with higher population densities are also ones whose populations are most at risk
due to climate change and other environmental disasters caused by unsustainable
industrialization. Some of these countries (especially BRICS[2] countries) have high
GDP growth rates, owing to relatively recent industrialization, and are at a point
where they could redefine their "development" paradigm and vision towards
embracing sustainable progress, rather than focusing on narrowly defined economic
expansion (Fig. 11.1).

1.2 GDP Fixation

Despite the overwhelming focus on improving a developing country's GDP, its citizens have to realize, as renowned environmental and policy analyst Vaclav Smil
(1996) reveals, that there is little worth of China's impressive 10 % GDP growth if
the true cost of environmental damage caused by this GDP increase is about 15 %
of its GDP. The out-dated practice of "pollute now, clean up later" will only degrade
the country's environment and quality of life and increase economic expenditure on
future remediation efforts (Erkman and Ramaswamy 2003; Chiu and Yong 2004).
Moreover, once a country's development paradigm and infrastructure are built on a
foundation of modern consumerism reliant on profligate use of fossil fuels, corrective action to move towards sustainability will be expensive, complex and challenging to navigate. Some of the reasons for the resistance in developed countries to

---

[1] Defined as those with a lower Human Development Index (HDI) and lower standard of living
relative to developed countries.

[2] "BRICS countries" refers to Brazil, Russia, India, China and South Africa.

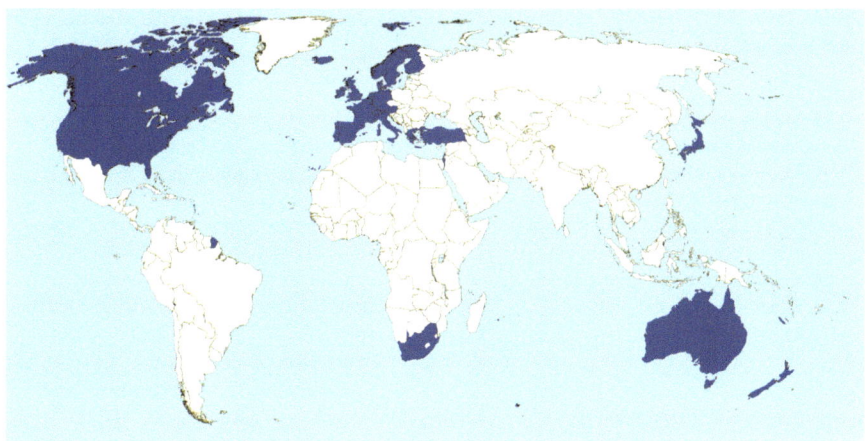

Fig. 11.1 Developed and developing countries (CIA 2013; Augusti 2008)

redefine their development paradigms lie in their high accumulated debt and the enormous investments they have made in infrastructure that is not designed for sustainability (see Chaps. 6 and 7). When examined further, these aspects may reveal a potentially more optimistic picture for developing countries.

1.3 Previous Studies on IE in Developing Countries

Sustainable development is certainly not a simple, singular and well-tested path. Many different interest groups are in conflict over which environmental and social challenges to tackle first, as well as their solutions. As previous reviews on industrial ecology in developing countries suggest, IE offers an umbrella paradigm, a sort of panoramic vision within which individual local crises could be approached with pragmatic solutions. Industrial ecology based solutions offer the advantage of simultaneously addressing several other interlinked problems to yield concurrent benefits for all stakeholders (Chiu and Yong 2004; Lowe 2006). A SWOT analysis of IE in developing countries, based on a previous study that looked at IE in Asian developing countries (Chiu and Yong 2004), is presented in Table 11.1. Some strengths, such as rapid economic growth, can also be viewed as a weakness, opportunity and threat.

In the developing world context, characterized by immediate development challenges brought on by rapid industrialization and urbanization, Chertow (2008) suggests applying a narrower focus to the word "industry" in IE, rather than using it in its broad sense to cover a range of anthropogenic activities. This suggestion may hold well in contexts such as in Eco-Industrial Parks (EIPs) where resources are cycled between industrial firms and collective benefits are realized to reduce environmental and social impacts. However, even in these contexts it is important to realize that, in

Table 11.1 SWOT analysis for the potential of IE in developing countries (Adapted from Chiu and Yong 2004)

Strengths	Weaknesses
1. *Economic growth*: Most developing countries are growing fast with large foreign direct investments and domestic industrialization, especially in the BRICS countries. This situation can provide the economic impetus for funding industrial ecology research and implementation in policy and industrial innovation.	1. *Developing country specific IE based models and data*: Models specific for understanding flows of resources and their interaction with socio-economic groups in developing countries need to be developed, especially in the unorganized/informal sector. Specific metric and indicators more suited for developing countries should be identified. Background data especially that of life cycle inventories (LCIs) for life cycle assessments (LCAs) lacking for most developing countries.
2. *Human resources*: Most developing countries have high population densities with demographics emphasizing youth populations, especially in countries like India.	2. *Scarcity in financial resources*: There is meager funding for research and development of IE in developing countries.
3. *Research*: In some developing countries, the research and academia have been exposed to industrial ecology, making this an ideal stage for the setting up IE research and education centres.	3. *Dearth of education programs*: There are very few specialized educational programs in IE in developing countries. Most of these programmes are in China
4. *Awareness*: People are aware about sustainability and are looking for methods and tools to implement solutions for it. This awareness is present amongst the corporates, citizens and the government.	4. *Inadequate clarity in the role of different governing bodies*: In several developing countries, there is a lack of clarity in the roles of governing bodies for resource management. In many cases, there are gaping gaps in governance and in some cases there is overlap in the responsibilities of public sector institutions. There is a lack of an integrated and collaborative approach to resource management.
5. *Role of government*: The government in some developing countries have shown interest in IE and implemented IE based policies, especially in China. Political cooperation can lead to growth of IE in groups of these countries such as South Asian Association for Regional Cooperation (SAARC), Commonwealth countries, etc.	5. *Insufficient data*: Data necessary to make content informed policy decisions are insufficient and at times unreliable.
	6. *Scarcity of green tech*: Lack of innovation and access to green and clean industrial technologies.
	7. *Insufficient enforcement*: Lack of enforcement of sustainable policies and for management of resources.
Opportunities	Threats
1. *Redefine sustainable development*: Have an opportunity to redefine their development paradigm and polices to maximize social welfare while limiting environmental impacts of development focused on consumerism.	1. *Strong focus on economic growth from rapid industrialization*: There is strong focus on industrialization and economic growth rather than increasing social welfare. This thrust has already damaged the ecological health of developing countries to a great extent.

(continued)

Table 11.1 (continued)

Opportunities	Threats
2. *International co-operation*: Several international institutions are collaborating with partners in developing countries to investigate systems using IE concepts and tools.	2. *Inadequate industrial ecology awareness*: Especially in the policy making and governing spheres and the public domain.
3. *Development of new models and tools*: Can contribute to the development of new models and tools in IE.	3. *Insufficient data*: Lack of sufficient macro and micro level data to inform policies on sustainable management of resources.
4. *Global political arena*: Several developing countries are allying with one another to further their negotiations in global agreements such as those regarding climate change. They can use these political collaborations to cross collaborate on IE based policies and strategies.	4. *Focus on remediation*: Stuck in the "industrialization – pollution – remediation" running wheel to further development. Lack insights into how to transition towards sustainable policy development and enforcement. IE may therefore be viewed merely as a technical "add on" or "fix" to remediate pollution, caused by inefficient management of resources and insufficient lack of enforcement of environmental policies.
	5. *Outdated policies*: Policies in some cases prevent effective IE implementation. For example "Zero discharge policy" in India disallows water cascading among industries. In some countries laws inhibits the formation of waste exchange networks and industrial symbiosis.
	6. *High externalities of industrialization*: In several countries, the externalities associated with industries are tremendous as proper working conditions, environmental protection and social benefits to affected communities are not included in the cost of production.

the absence of an overarching sustainable development paradigm, industrial residues that are expensive for industries to recycle will be disposed of in the cheapest and most often not the cleanest of ways. Moreover, polluting industries such as coal fired power plants in EIPs may be further locked-in in industrial networks, making it difficult to replace them with cleaner technologies such as plants based on renewable resources. On the policy front, it is important for developing countries to be aware of these interlinked complications and think of ways to avoid net long-term damage to their ecological, social and economic health, as explored in Chap. 6.

1.4   IE in the Policy Context

Around the world, national policies for environmental protection have evolved from a perception of industrial processes as linear chains, rather than viewing them as cyclic ones. These policies, therefore, are aimed at cleaning up pollution at the end of a chain, rather than avoiding its creation. Furthermore, policies have been artificially compartmentalized for protection of naturally interconnected systems of air, water, forests, agricultural lands and urban settlements via separate air, water and waste acts. More than 30 years after its initiation, industrial ecology has evolved to inform progressive sustainable policies in the developed world. These policies include embedding life-cycle thinking in the European Union's legislation and intelligent design of infrastructure (Chertow et al. 2015). Learning from these significant initiatives, developing countries need to progress towards this next generation of sustainable policymaking whose overarching vision should be to initiate, support and enforce sustainable conditions.

IE provides an overarching framework for individual polices and schemes that support sustainable urban, industrial and rural development while ensuring weeding out, rather than locking in, of polluting technologies. Periodic monitoring of these policies will ensure that progress is not made in a fragmented and unsystematic fashion. For instance policies that focus on improving fuel efficiency of automobiles may be blind to the requirements of increasing the amount of land utilized for road transportation, the huge investment required for this infrastructure and the way it locks development into favouring road transport and ownership of automobiles. Such short-sighted policies may yield some benefits to the quality of life in the short term but harm it in the long run.

Considering what has been achieved by industrial ecology in the developed world and learning from experience is important to avoid reinventing the wheel and losing out on one of our most precious resources – time. However, this is not the only direction for learning to take place; this chapter highlights the fact that developing countries not only offer valuable, relatively uncharted landscapes for enriching our understanding of how industrial ecology can inform sustainable development but also help uncover unexplored practices and technologies that can provide a similar level of social welfare by using far less resources.

2   What Has Been Achieved by IE in the Global South?

Taking the glass is half full" approach, let us begin examining what the glass contains with a brief history of IE in developing countries. Industrial ecology was first systematically introduced into developing countries in 1995, when Erkman and Ramaswamy initiated a collaboration to disseminate and experiment on applying IE in the Indian context (ROI 2010). This collaboration resulted in the publication of case studies (Erkman and Ramaswamy 2003) and the establishment of the Resource Optimization Initiative (ROI). In 1999, an Industrial Ecology conference was organized at the Indian Institute of Management, Ahmedabad (Erkman 2015, Origins of industrial ecology in developing countries. Personal Communication via Email). Around the same time, in 1997, the faculty of the Dalian University of Technology (DUT) began its IE work in China.

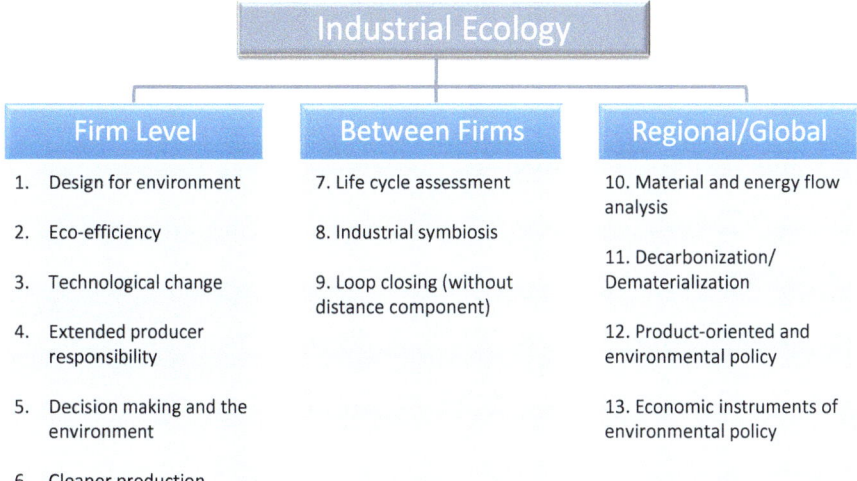

Fig. 11.2 Industrial ecology concepts and tools at the firm level, between firm and regional/global level (Adapted from Lifset and Graedel 2002)

2.1 Hotspots of IE in the Global South

Available research papers and reports highlight the main streams of IE that have been researched and implemented in developing countries. Analysing 131 documents (comprising 83 peer-reviewed journal articles, 27 books and book chapters and 21 reports from development organizations, institutions and companies) that examined various concepts and tools of IE (see Fig. 11.2) in the Indian context from 1997 to 2009 (Shenoy and Chertow 2009) revealed that Cleaner Production was a "hot spot" being explored in a majority of cases (see Fig. 11.3).

Examination of industrial ecology in other countries has revealed that almost two decades since its first introduction in developing countries, IE has developed in two relatively large branches: (1) Cleaner Production and (2) Eco-Industrial Development. Apart from these two areas, several cases that use Material Flow Analysis, Life Cycle Assessment, Extended Producer Responsibility and dematerialization have been examined in diverse developing world contexts.

2.2 Cleaner Production

Cleaner production (CP) is an IE strategy that has been implemented in several developing countries since the early 1990s. The Chinese National Cleaner Production Centre (CNCPC) was established in December 1994 (UNIDO/UNEP n.d.). In 1995, the world-wide UNIDO/UNEP National Cleaner Production Centre

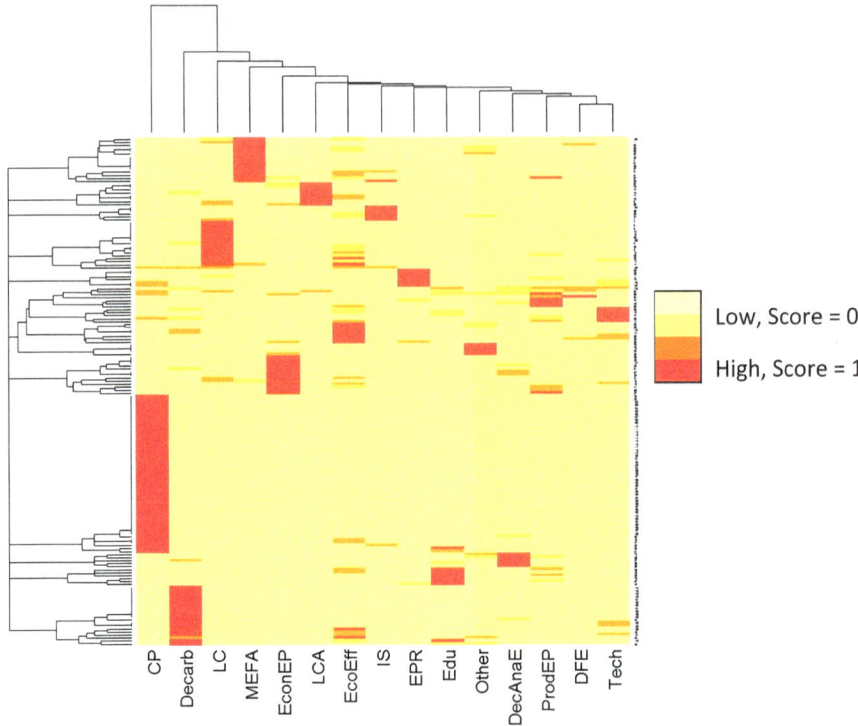

Fig. 11.3 "Heat map" of 131 documents that examined various concepts and tools of IE in India (*CP* cleaner production, *Decarb* decarbonisation, *LC* loop closing (without distance component), *MEFA* material and energy flow analysis, *EPR* extended producer responsibility, *IS* industrial symbiosis, *LCA* life cycle assessment, *EcoEff* eco-efficiency, *Other* includes IE related topics, such as social network analysis, social capital, ecological economics, sustainability modelling, scenario analysis, international environmental agreements and treaties, etc., *DecAnaE* decision making and the environment, *Tech* technology interventions for sustainability, *DFE* design for environment, *ProdEP* product oriented environmental policy)

(NCPC) programme included the CNCPC and established NCPCs in Brazil, India, Mexico, Tanzania, Tunisia and Zimbabwe (Nishikawa 2009). Since then the programme has expanded to include 47 developing and transition countries (UNIDO/ UNEP 2010). In each of these countries, training programmes on CP have been conducted and in-plant CP assessments have been completed (UNIDO/UNEP 2010). In the 1990s, the World Bank sponsored a project focused on "Environmental Management Capacity Building" that resulted in CP promotion efforts in India (Rathi 2003).

In 2002, CP was included in policy in the Cleaner Production Promotion Law that was passed by the Chinese government in 2002. This law defines and sets goals and targets for clean industry, clearly specifies implementation responsibilities and outlines ways to measure successful implementation of CP in industries (Mol and Liu 2005). Other developing countries are yet to frame policies to support and facilitate CP. Currently, the lack of specific policies to facilitate and sustain CP, funding and capacity for implementation, lax enforcement of regulations and insufficient external social pressure demanding change are some of the significant barriers to CP implementation and expansion (Muduli et al. 2013).

2.3   Eco-Industrial Development

Applying the concept of CP at the scale of an industrial area gives rise to eco-industrial networks that exchange materials and realize collective sustainable benefits (see Chap. 5). In the broadest sense, industrial areas that are either designed or remodelled for this purpose are called Eco-Industrial Parks – EIPs. The first EIP in the developing world was set up in 2000 by the Chinese Research Academy on Environmental Science, an affiliated institution of the China State Environmental Protection Administration (Chiu and Yong 2004). Currently as many as 60 parks have been approved under the national pilot EIP programme in China (Zhang et al. 2010). In 2009, China framed and enforced the Circular Economy Promotion Law that supports the development of EIPs, via specific regulations and schemes to raise resource recycling rates in production, circulation and consumption cycles (WB 2009). In India and in most other developing countries, national or regional policies that facilitate EIPs are lacking (Ashton and Shenoy 2015). In India, over the past few years, the German development agency, GIZ, has been involved in the establishment of EIPs in a few states in India (GIZ n.d.).

Examination of industrial symbiosis in India has revealed high potential for spontaneously evolved waste exchange networks (Bain et al. 2010). This observation reveals potential for a bottom-up approach to develop EIPs such as that carried out by the UK National Industrial Symbiosis Program (NISP) (Boons et al. 2011), further facilitated by policies and financial assistance that encourage industries to implement recycling strategies (especially for materials that do not have established markets). In addition, there is a significant need to (1) develop new methods to quantify material flows in the large informal sector, and (2) examine power relationships and negotiating authority between partners involved in residue exchanges, especially between large well-established companies and informal players such as farmers' co-operatives, individual farmers and waste recyclers. Such findings were revealed when uncovering the residue exchanges involving ash granted by large-scale companies to individual small land-holding farmers in south India (Bain et al. 2010).

In other developing countries, including Cambodia, Vietnam, Egypt, Namibia, South Africa, Colombia and Peru (UNIDO in prep.; Chertow et al. 2008), eco-industrial development is at varying stages of development. Most studies that report and examine EIPs in developing countries have focused on the physical flows of matter and energy. However, there is a need for a deeper understanding of crucial aspects of EIPs such as (1) systems for inter-organizational networking opportunities, (2) stakeholder participation and (3) measurement of life cycle environmental and social impacts of EIPs for their long term sustainability (Eckelman and Chertow 2013; Ashton and Shenoy 2015).

Other than CP and EIPs, case studies that examine resource flows in systems in India have highlighted that in the developing world context it is important to (1) focus on scarce resources, such as water, including mapping its distribution over many sources and users, a majority of which are unorganized or informal; (2) redefine the pollution problems to highlight scarcity and imbalanced use of these resources; (3) examine the possibilities of using well tested, off-the-shelf technologies that are used for other purposes before exploring new technologies and (4) evaluate possibilities and solutions for the long-term maintenance of new technologies and practices before recommending their implementation (Erkman and Ramaswamy 2000, 2003; ROI 2005; Shenoy et al. 2010).

3   Current Issues

Continuing with the approach of 'the glass is half full' and asking 'with what?', we now examine the current situation: how and with what to fill the rest of the glass. For IE to yield truly sustainable initiatives in the developing world, we need to view the current environmental crises as symptoms of a particular development paradigm (Prins et al. 2010), founded on profligate use of fossil fuels and consumerist attitudes with exorbitant embedded energy and resource demands. Developing countries need to realize the value of this perspective and not follow the same development pathway (Shenoy 2010). A new development paradigm which places sustainability above economic growth has been pioneered by Bhutan in its concept of Gross National Happiness (GNH) (Ura et al. 2012). In 2011, the UN adopted the Gross National Happiness (GNH) and is now examining ways to measure this index in countries around the world (Kelly 2012). However, most developing countries place economic gains above sustainability in their development, owing to which they have undergone tremendous environmental damage in the recent past (GFW 2012).

3.1   Impact of Technology

Industrial ecology offers insights on ways to measure and manage impacts (environmental and social) so as to track progress on a sustainable development pathway. In the IPAT equation,

$$\text{Total Environmental Impact} = \text{Population} \times \frac{\text{GDP}}{\text{Person}} \times \frac{\text{Environmental impact}}{\text{Unit of per capita GDP}}$$

Ehrlich and Holdren (1971) define the third term – the technology term – as "a measure of how much each unit of production or consumption pollutes". Graedel and

Allenby (1995) optimistically place the responsibility of sustainable development on this technology term to encourage sustainable technological innovation by individual companies and corporates. Given that developing countries such as China and India are now emerging as leading technology hubs, technological innovation can contribute significantly to our sustainable development. Several companies from developing countries, members of the World Business Council for Sustainable Development (WBCSD), Global Reporting Initiative, Greening of Industry Network (GIN) and Asia Pacific Roundtable for Cleaner Production (APRCP), are including sustainability and the triple bottom line in their growth strategies. Some of the IE-based technology solutions that these companies can adopt are explored in other chapters in this book. They include (1) greening of the supply chain, (2) extending producer responsibility, (3) environmental certification and (4) dematerializing the economy. Despite the acceptance of these approaches, it would be naïve to entrust corporates entirely with the responsibility for sustainable development. Although technology and corporates can play a significant role, there is a definite need for the presence of overarching policies and government funding to facilitate sustainable technology development.

3.2   Impact of Population and Affluence

From the perspective of a developing country with the world's highest population density, it is apparent that the two other terms in the IPAT equation – GDP/person (also called the Affluence term) and Population – need to take on equal and sometimes even larger responsibility in shaping a sustainable future. The environmental impact of the rich and affluent and rural to urban migration in developing countries can be very significant. For example, in 1990 in India, the collective CO_2 emissions of "new consumers" was found to be 15 times greater than that of the rest of the population (Myers and Kent 2004). Analysing the environmental impact of a person's lifestyle with respect to their personal (disposable) income would also be extremely important for developing countries, to measure and limit the impacts of increasing affluence and population.

3.3   Policy Development and Funding

Learning from studies that have examined ways for IE to inform policy development, developing countries need to defragment environmental policies across supply chains and across artificially compartmentalized environmental areas. For example, policies that focus on environmental protection of water more than land can lead to treatment of waste water only to end up with hazardous sludge that will continue to contaminate landfills and eventually leach into ground water. In addition, developing countries require an approach of not simply applying or adapting

IE concepts and tools but new ways of framing their problems and hence finding solutions (Erkman and Ramaswamy 2000).

If resources are cycled efficiently, as advocated by the IE paradigm, then we can expect to devote progressively less financial resources into future remediation. This argument provides a strong case for the financial value of IE, supporting the allocation of public funding for (1) industrial ecology research and education; (2) providing financial assistance to micro-, small-scale and cottage-scale industries to invest in efficient, clean technologies (Erkman and Ramaswamy 2000) and (3) monitoring environmental parameters and measures of human development, a task which is made more complex by the significance of the unorganized sector (Erkman and Ramaswamy 2003).

4 What Can IE Give to the Global South?

Although there is some financial impetus for developing countries to adopt IE, this need not be the only motivation for adopting IE. Recent research has found that high levels of human development can be achieved at moderate energy consumption levels and, more importantly, that increasing energy consumption does not necessarily contribute to higher living standards (Fig. 11.2) (Steinberger and Roberts 2010) (Fig. 11.4).

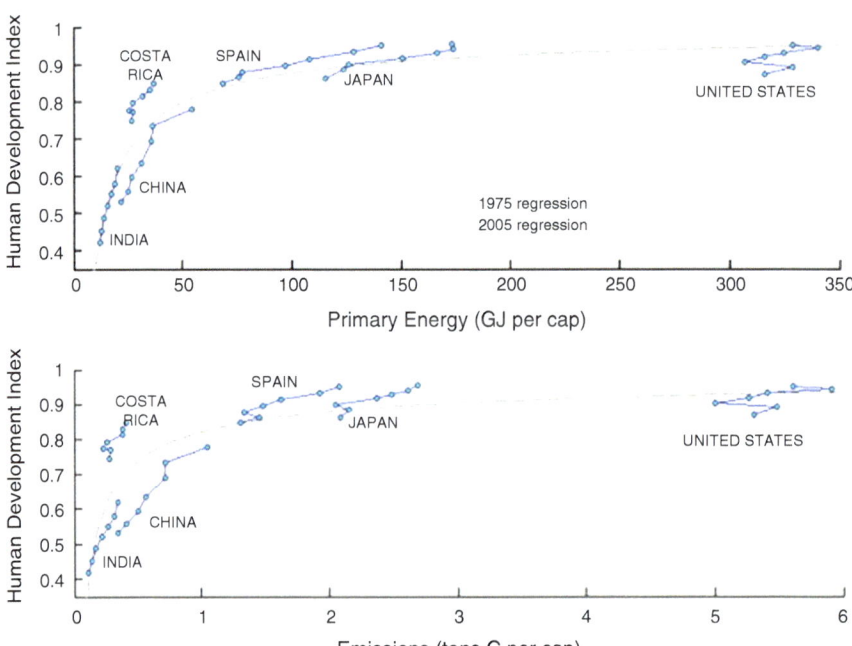

Fig. 11.4 Human development Index for specific countries vs. energy and carbon emissions from 1975 to 2005. Regression curves for 1975 and 2005 are shown for reference (From Steinberger and Roberts 2010)

4.1 Challenges, Metrics and Models

This motivation to improve human development standards comes with significant challenges specific to developing countries, due to limited data availability. The IE community can provide significant assistance to developing countries by simplifying and downsizing data requirements to a point which yields sufficiently accurate results to inform policy. It would indeed be counterproductive if developing countries were to wait for economic growth to fuel complex data gathering operations that can in turn inform their sustainable development policies. Some of the relatively simple metrics developed in IE that can be used to inform specific policies are ratios of different materials to measure resource efficiency. These ratios along with the caveats that need to be kept in mind while incorporating them into policies are in Table 11.2.

In addition, IE research has developed frameworks and models that capture the complexity of real systems by integrating several IE-based tools, for example the social-ecological-infrastructural systems (SEIS) framework and LCA analysis in EIPs (Ramaswami et al. 2012; Eckelman and Chertow 2013). An example of the socio-economic metabolism approach is advocated in Chap. 6. The SEIS framework is currently being used to assess environmental impacts of emerging cities in USA, China and India (Ramaswami et al. 2015). A similar framework is the Integrative Regional Action Planning (IRAP) framework that integrates planning across undeveloped land, rural and urban regions (Jaderi et al. 2014; Van Zeijl-Rozema and Martens 2011; Huynen et al. 2004; Lowe 2006). Such integrative frameworks call for cooperation between various stakeholders and institutions for a comprehensive understanding of regions, their impacts and solutions. These inte-

Table 11.2 Metrics, policy guidance and caveats for some ratios for understanding ways to optimize of resource flows

Metrics	Policy guidance and caveats
$\dfrac{\text{Virgin Materials}}{\text{Recycled Materials}}$	Incentivize companies to pursue closing of loops rather than disposal of industrial residues. Policies need to clearly define virgin, recycled materials and by-products to avoid misrepresentation
$\dfrac{\text{Actual Recycled Materials}}{\text{Potential Recycled Materials}}$	Industrial cluster level, city level and state level metrics to favour those that recycle more
$\dfrac{\text{Renewable Fuel Sources}}{\text{Fossil Fuel Sources}}$	Individual companies, city and state level metrics to favour renewable energy sources. Although, in most cases a higher ratio would indicate lower environmental impacts, this may not always be the case. For e.g. biodiesel from palm oil extracted from plantations grown on destroyed forest land can have much higher life cycle impacts compared to fossil fuels (Crutzen et al. 2008; Fargione et al. 2008)
$\dfrac{\text{Economic Output}}{\text{Material Input}}$	This measure can be used to improve resource efficiency

grative frameworks may need to be adapted to suit developing country contexts and, in some cases, new models will need to be developed. For example, while smart cities developed in the global North may be commendable in their ability to capture complex impacts and employ ecological design, they may not address the economic and social constraints and challenges in the developing world.

The IE community can also offer solutions to contemporary challenges, including how to address the problems of rebound that arises from improving energy efficiency (Gillingham et al. 2013) and avoiding "lock-in" of polluting technologies in eco-industrial networks (Boons et al. 2011; Shi et al. 2010). Analyses on how to include these effects into policy are needed, not only in countries that already have energy efficiency and EIP policies in place but also those that are developing new policies.

5   How Can the Global South Contribute to IE?

Industrial ecology concepts and tools have developed primarily in the developed world. However, owing to their low energy and material consumption patterns, developing countries can provide unexamined contexts for the developed world to learn from. Some of these contexts where the IE lens has provided valuable insights are (1) informal recycling by the unorganized sector in several developing countries (Medina 1997; Wilson et al. 2006); (2) provisioning of fresh vegetables and fruits on a regular basis to reduce food wastage due to inefficient management of household food inventories (Sahakian et.al. in press); and (3) water balance of cities in India (Eckelman et al. 2010).

In some cases, historic management of resources in developing countries has been postulated as more sustainable than the way resources are currently managed around the world; examples are flood management along the Brahmaputra in West Bengal (Rasid and Paul 1987) and community-based construction and maintenance of water tanks in the pre-British era in various regions in India (Mosse 1997). IE can offer qualitative and quantitative measures to understand and monitor such initiatives.

6   Conclusions

Some developing countries have realized the benefits of IE approaches and have used its concepts for sustainable growth of communities; others are yet to realize the value of this approach. The benefits of IE for developing countries include strategy elaboration and policy framing for sustainable development. Developing countries need to re-examine their development strategies and make important decisions to improve social welfare and build healthy economies, while protecting their ecological heritage. Redefining a country's development pathway is a complex process that needs an interdisciplinary examination of cultural values, ecological heritage, resource use patterns and demographics. Due to its emphasis on a holistic approach, industrial ecology can provide a valuable platform to draw out this redefinition in ways that can be implemented.

References

Ashton, W., & Shenoy, M. (2015). A culture of closed loops: Industrial ecology in India. In P. Deutz, D. Lyons, & J. Bi (Eds.), *International perspectives on industrial ecology*. Cheltenham: Edward Elgar.

Augusti, R. G. D. (2008). *Map of Developed Countries (DCs) as described by the CIA world factbook*. Retrieved January 27, 2015, from http://www.wikiwand.com/en/The_World_Factbook_list_of_developed_countries

Bain, A., Shenoy, M., Ashton, W., & Chertow, M. (2010). Industrial symbiosis and waste recovery in an Indian industrial area in south India. *Resources, Conservation and Recycling, 54*(12), 1278–1287.

Boons, F., Spekkink, W., & Mouzakitis, Y. (2011). The dynamics of industrial symbiosis: A proposal for a conceptual framework based upon a comprehensive literature review. *Journal of Cleaner Production, 19*(9–10), 905–911. doi:10.1016/j.jclepro.2011.01.003.

Chertow, M. (2008). Industrial ecology in a developing context. In C. Clini, I. Musu, & M. Gullino, (Eds.), *Sustainable development and environmental management* (pp. 335–349). Dordrecht: Springer. doi:10.1007/978-1-4020-6598-9_24.

Chertow, M., Ashton, W. S., & Espinosa, J. C. (2008). Industrial symbiosis in Puerto Rico: Environmentally related agglomeration economies. *Regional Studies, 42*(10), 1299–1312. doi:10.1080/00343400701874123.

Chertow, M., Fisher-Kowalski, M., Clift, R., & Graedel, T. E. (2015). *Industrial ecology*. A note from the Presidents: International Society for Industrial Ecology. Retrieved January 27, 2015, from http://www.is4ie.org/A-Note-from-the-Presidents

Chiu, A. S. F., & Yong, G. (2004). On the industrial ecology potential in Asian Developing Countries. *Journal of Cleaner Production, 12*(8–10), 1037–1045. doi:10.1016/j.jclepro.2004.02.013.

CIA. (2013). *Advanced developing countries: Appendix B. International Organizations and Groups*. Retrieved January 27, 2015, from https://www.cia.gov/library/publications/the-world-factbook/appendix/appendix-b.html

Crutzen, P. J., Mosier, A. R., Smith, K. A., & Winiwarter, W. (2008) N2O release from agro-biofuel production negates global warming reduction by replacing fossil fuels. *Atmospheric Chemistry and Physics, 8*, 389–395. doi:10.5194/acp-8-389-2008.

Eckelman, M., & Chertow, M. (2013). Life cycle energy and environmental benefits of a US industrial symbiosis. *International Journal of Life Cycle Assessment, 18*(8), 1524–1532. doi:10.1007/s11367-013-0601-5.

Eckelman, M., Shenoy, M., Ramaswamy, R., & Chertow, M. (2010). Applying industrial ecology tools to increase understanding of demand-side water management in Bangalore, India. *Asian Journal of Water, Environment and Pollution, 7*(4), 71–79.

Ehrlich, P. R., & Holdren, J. P. (1971). Impact of population growth. *Science, 171*(3977), 1212–1217. doi:10.1126/science.171.3977.1212.

Erkman, S., & Ramaswamy, R. (2000). *Cleaner production at the system level: Industrial ecology as a tool for development planning (Case studies in India)*. Paper presented at the UNEP's 6th International High-level Seminar on Cleaner Production, Montreal, Canada, October 16–17, 2000.

Erkman, S., & Ramaswamy, R. (2003). *Applied industrial ecology: A new platform for planning sustainable societies*. Bangalore: Aicra Publishers.

Fargione, J., Hill, J., Tilman, D., Polasky, S., & Hawthorne, P. (2008) Land clearing and the biofuel carbon debt. *Science, 319*, 1235–1238.

GFW. (2012). Countries with highest ratio of tree cover loss to gain (2001–2012). *Global Forest Watch*. Retrieved January 28, 2015, from http://www.globalforestwatch.org/countries/overview

Gillingham, K., Kotchen, M. J., Rapson, D. S., & Wagner, G. (2013). Energy policy: The rebound effect is overplayed. *Nature, 493*(7433), 475–476.

GIZ. (n.d.). *Eco industrial development in India*. Retrieved January 27, 2015, from http://www.ecoindustrialparks.net/

Graedel, T. E., & Allenby, B. R. (1995). *Industrial ecology*. Englewood Cliffs: Prentice Hall.

Huynen, M., Martens, P., & De Groot, R. S. (2004). Linkages between biodiversity loss and human health: A global indicator analysis. *International Journal of Environmental Health Research, 14*(1), 13–30. doi:10.1080/09603120310001633895.

Jaderi, F., Ibrahim, Z. Z., Jaafarzadeh, N., Abdullah, R., Shamsudin, M. N., Yavari, A. R., & Nabavi, S. M. B. (2014). Methodology for modeling of city sustainable development based on fuzzy logic: A practical case. *Journal of Integrative Environmental Sciences, 11*(1), 71–91. doi:10.1080/1943815X.2014.889719.

Kelly, A. (2012). Gross national happiness in Bhutan: The big idea from a tiny state that could change the world. *The Guardian*. Retrieved April 2, 2015, from http://www.theguardian.com/world/2012/dec/01/bhutan-wealth-happiness-counts

Lifset, R., & Graedel, T. E. (2002). Industrial ecology: Goals and definitions. In R. U. Ayres & L. W. Ayres (Eds.), *A handbook of industrial ecology* (pp. 3–15). Cheltenham: Edward Elgar.

Lowe, E. (2006). *Integrative Regional Action Planning (IRAP)*. Indigo development. Retrieved January 10, 2015, from http://www.indigodev.com/IRAPsum.html

Medina, M. (1997). *Informal recycling and collection of solid wastes in developing countries: Issues and opportunities*. In: UNU/IAS working paper (Vol. 24). UNU/IAS.

Mol, A. P. J., & Liu, Y. (2005). Institutionalising cleaner production in China: The cleaner production promotion Law. *International Journal of Environment and Sustainable Development, 4*(3), 227–245.

Mosse, D. (1997). The symbolic making of a common property resource: History, ecology and locality in a tank-irrigated landscape in south India. *Development and Change, 28*(3), 467–504.

Muduli, K., Govindan, K., Barve, A., & Geng, Y. (2013). Barriers to green supply chain management in Indian mining industries: a graph theoretic approach. *Journal of Cleaner Production, 47*, 335–344. doi:http://dx.doi.org/10.1016/j.jclepro.2012.10.030.

Myers, N., & Kent, J. (2004). *The new consumers: The influence of affluence on the environment*. Washington/Covela/London: Island Press.

Nishikawa, T. (2009). *National Cleaner Production Centre Programme*. Presentation at UNCRD Inaugural Regional 3R Forum in Asia, Session 5. Retrieved April 2, 2015, from http://www.uncrd.or.jp/content/documents/Session5_Nishikawa-rev.pdf

Prins, G., Galiana, I., Green, C., Grundmann, R., Korhola, A., Laird, F., Nordhaus, T. Jnr, R. P., Rayner, S., Sarewitz, D., Shellenberger, M., Stehr, N., & Tezuko, H. (2010). *The Hartwell paper: A new direction for climate policy after the crash of 2009*. Institute for Science, Innovation & Society, University of Oxford; LSE Mackinder Programme. London: London School of Economics and Political Science.

Ramaswami, A., Weible, C., Main, D., Heikkila, T., Siddiki, S., Duvall, A., Pattison, A., & Bernard, M. (2012). A social-ecological-infrastructural systems framework for interdisciplinary study of sustainable city systems. *Journal of Industrial Ecology, 16*(6), 801–813. doi:10.1111/j.1530-9290.2012.00566.x.

Ramaswami, A., Russell, A., Chertow, M., Weible, C., & Romero-Lankao, P. (2015). *PIRE: Developing low-carbon cities in the US, China, and India through integration across engineering, environmental sciences, social sciences, and public health*. Grantome.

Rasid, H., & Paul, B. K. (1987). Flood problems in Bangladesh: Is there an indigenous solution? *Environmental Management, 11*(2), 155–173.

Rathi, A. K. A. (2003). Promotion of cleaner production for industrial pollution abatement in Gujarat (India). *Journal of Cleaner Production, 11*(5), 583–590.

ROI. (2005). Projects of ROI. Retrieved January 27, 2015, from http://www.roionline.org/ongoing_projects.htm

ROI. (2010). *ROI 2010 activity report*. Bangalore: Resource Optimization Initiative.

Sahakian, M., Saloma, C., & Erkman, S. (in press). *(Un)sustainable food consumption dynamics: Changing practices and patterns in Asia's cities*. London: Routledge.

Shenoy, B. (2010). India can be a shining example to solve energy crisis by Gandhian approach. *Energy Manager, 3*(3), 59–62.

Shenoy, M., & Chertow, M. (2009). *Industrial ecology in India: Past, present and future*. Paper presented at the 5th international conference on industrial ecology: Transition towards sustainability. Lisbon, Portugal, 21–24 June.

Shenoy, M., Kumari, R., & Lokanath, S. (2010). *Challenges for improving energy efficiency, water recycling and reducing carbon di-oxide emissions for cottage-scale industries in southern India*. Paper presented at the Summer Symposium on Sustainable Systems (4S), Sannäs, Finland, 14–17 June.

Shi, H., Chertow, M., & Song, Y. (2010). Developing country experience with eco-industrial parks: a case study of the Tianjin Economic-Technological Development Area in China. *Journal of Cleaner Production, 18*(3), 191–199. doi:10.1016/j.jclepro.2009.10.002.

Smil, V. (1996). *Environmental problems in China: Estimates of economic costs* (East-West Center Special Reports, Vol. 5). Honolulu: East-West Center (EWC).

Steinberger, J. K., & Roberts, J. T. (2010). From constraint to sufficiency: The decoupling of energy and carbon from human needs, 1975–2005. *Ecological Economics, 70*(2), 425–433. doi:10.1016/j.ecolecon.2010.09.014.

UNIDO. (in preparation). *Eco-industrial parks in emerging and developing countries: Achievements, good practices and lessons learned in planning, development and management of eco-industrial parks*. A project under the global RECP Programme. UNIDO.

UNIDO/UNEP. (2010). Taking Stock and Moving Forward: The UNIDO–UNEP National Cleaner Production Centres. Austria: UNIDO Cleaner and Sustainable Production Unit and UNEP Business and Industry Unit, Sustainable Consumption and Production Branch.

UNIDO/UNEP. (n.d.). *China national cleaner production centre: Case study in good organization, management and governance practices*. Retrieved April 2, 2015, from https://www.unido.org/fileadmin/user_media/Services/Environmental_Management/Contacts/Contacts/CNCPC%20web.pdf

Ura, K., Alkire, S., Zangmo, T., & Wangdi, K. (2012). *A short guide to gross national happiness index*. Thimphu: The Centre for Bhutan Studies.

Van Zeijl-Rozema, A., & Martens, P. (2011). Integrated monitoring of sustainable development. *Sustainability the Journal of Record, 4*(4), 199–202.

WB. (2009). *Circular economy promotion law*. Public-Private Partnership in Infrastructure Resource Center for Contracts, Laws and Regulation (PPPIRC).

Wilson, D. C., Velis, C., & Cheeseman, C. (2006). Role of informal sector recycling in waste management in developing countries. *Habitat International, 30*(4), 797–808. doi:10.1016/j.habitatint.2005.09.005.

Zhang, L., Yuan, Z., Bi, J., Zhang, B., & Liu, B. (2010). Eco-industrial parks: National pilot practices in China. *Journal of Cleaner Production, 18*(5), 504–509. doi:10.1016/j.jclepro.2009.11.018.

Chapter 12
Material Flow Analysis and Waste Management

Yuichi Moriguchi and Seiji Hashimoto

Abstract Material flow analysis (MFA, also known as Material Flow Accounting) has become one of the basic tools in industrial ecology, since its pioneering development by Ayres. This chapter reviews progress in MFA with emphasis on the use of MFA to support waste management and recycling policy.

Waste statistics are compiled in most developed and some developing countries, but the basis is insufficiently standardized so that care is needed in making comparisons between countries. This also applies to recycling flows, which are difficult to define and quantify. Waste arising from demolition can be predicted by dynamic modeling which also predicts future resource demand, but the discrepancies between predicted and reported waste quantities can be large due to "missing" or "dissipated" stock. Metals represent an important and valuable component of waste; metals in end-of-life vehicles and e-waste in particular need to be quantified for their recycling and ecological and human health impact assessment. MFA has also been applied to international trade of secondhand products containing metals. MFA studies on phosphorus have revealed potential ways to increase recovery that go beyond recycling from obvious wastes. Analysis of stocks must be an important topic in coming decades.

Policies designed to move the economy towards "circularity" have been promoted in some countries, including China and Japan, as practical manifestations of the industrial ecology paradigm. In China, the link to MFA was only recognized some years after the introduction of the policy, whereas in Japan MFA was accepted from the outset.

Measures are being advocated, for example by the OECD, to improve the comparability of MFA across different data sources. Input-output analysis is increasingly applied to estimate and represent material flows. In general, MFA has matured to the point where it is now mandated as a tool for national and international policy. But further expansion and integration are expected.

Y. Moriguchi (✉)
Department of Urban Engineering, The University of Tokyo,
7-3-1 Hongo Bunkyo-ku, Tokyo 113-8656, Japan
e-mail: yuichi@env.t.u-tokyo.ac.jp

S. Hashimoto
Department of Environmental Systems Engineering, Ritsumeikan University,
1-1-1 Noji-higashi, Kusatsu, Shiga 525-8577, Japan
e-mail: shashimo@fc.ritsumei.ac.jp

© The Author(s) 2016
R. Clift, A. Druckman (eds.), *Taking Stock of Industrial Ecology*,
DOI 10.1007/978-3-319-20571-7_12

Keywords Material flow analysis • Material flow accounting • MFA • Waste management and recycling • Statistics • Input-output analysis • Socio-economic metabolism • Circular economy

1 Introduction – Historical and Institutional Perspectives

This chapter reviews the recent progress in methodologies and applications of material flow analysis (MFA) with emphasis on waste management and recycling perspective. An overview of the progress of economy-wide MFA can be found in a recent review by Fischer-Kowalski et al. (2011).

Robert Ayres and Allen Kneese presented the first version of what would become MFA of national economies as early as in 1969 (Ayres and Kneese 1969) (see Chap. 1). Ayres subsequently edited two books on resource accounting (Ayres and Ayres 1998, 1999) as well as a handbook of industrial ecology (Ayres and Ayres 2002). The section on MFA of this handbook (Bringezu and Moriguchi 2002) begins with the sentence, "understanding the structure and functioning of the industrial or societal metabolism is at the core of industrial ecology," followed by the definition that "Material flow analysis (MFA) refers to the analysis of the throughput of process chains comprising extraction or harvest, chemical transformation, manufacturing, consumption, recycling and disposal of materials." Approaches for quantifying the metabolism of physical economies were comprehensively reviewed by Daniels and his colleagues (Daniels and Moore 2001; Daniels 2002).

Looking back on the last two decades, ConAccount, a network of institutions and experts working on MFA, has played a key role in the progress of MFA (Moriguchi 2007). ConAccount started in May 1996 as a concerted action entitled "Coordination of Regional and National Material Flow Accounting for Environmental Sustainability." ConAccount convened several times, mainly in Europe. More recently, it was integrated into the International Society of Industrial Ecology (ISIE) as a section, now entitled Socio-Economic Metabolism (SEM) (see also Chap. 6).

In addition to national accounting of material flows, MFA has been increasingly used as a basis for analyzing and planning waste management and recycling systems. By search on the Web of Science ("material flow analysis" and "waste or recycling"), more than 300 articles that used an MFA approach for analyzing waste and recycling issues were found. Publications in these areas started in the late 1990s and the number has gradually increased, to reach more than 50 in 2014. The citations in each year have also increased up to 800 in 2014 with about 11 average citations per article. The number of articles and their citation is expected to expand further. In addition to Journal of Industrial Ecology, many articles are found in journals such as *Resources, Conservation and Recycling*, *Environmental Science and Technology*, *Waste Management*, and *Waste Management and Research*.

The following sections review the empirical studies of MFA for waste management and recycling from the viewpoint of target wastes and systems and MFA-based policies and concepts for sustainable resource and waste management. The chapter goes on to discuss the outlook for possible integration of methodologies and future developments in MFA.

2   Review of Empirical Studies from the Viewpoint of Target Wastes and Systems

2.1   Waste in General

In many developed countries, waste statistics are available (e.g. OECD 2005) as a basis for framing waste management and recycling policy (see Chap. 14). Institutionalization of the compilation of waste statistics is essential for appropriate waste management in developing countries (see Chap. 11). However, one of the issues in waste statistics is that definition and coverage of waste streams (e.g. municipal/industrial, hazardous/non-hazardous) vary considerably across countries. Therefore, comparison of waste indicators needs careful interpretation. The same applies to indicators such as recycling rate.

It is interesting and useful to capture recycling flows as parts of more comprehensive picture of material flows: such flows are accounted for in some national MFAs. However, defining recycling flows is not easy. The first version of the methodological guide for economy-wide MFA (Eurostat 2001) says "Recycling flows are not part of the material balance," because "First, data on materials recycled within statistical units are not normally available. Second, the definition and measurement of recycling flows is difficult." In this regard, Hashimoto and Moriguchi (2004) categorized the forms of recycling and proposed alternative indicators that can avoid double-counting and/or inflated recycling flows or rates. Concerning recycling ratio, Graedel et al. (2011) gave definitions of various metrics related to end-of-life recycling. Further, Bailey et al. (2008) proposed input–output cycling metrics that can measure cycling of both direct and indirect flows in a complex system while traditional metrics only account for direct flows. Further discussion and development are needed on the issue of defining and measuring recycle flows.

2.2   Construction and Demolition Waste

A number of MFA studies are available related to construction and demolition waste for countries and regions such as China (Shi et al. 2012; Huang et al. 2013), Japan (Hashimoto et al. 2007, 2009), The Netherlands (Müller 2006), Norway (Bergsdal et al. 2007), and Taiwan (Hsiao et al. 2002); for cities such as Beijing (Hu, D. et al. 2010; Hu, M. et al. 2010); and for specific infrastructures such as highway traffic system (Wen and Li 2010).

The dynamic model presented by Müller (2006) is now often used to estimate future resource demand and waste generation. The feature of this model is that service provided by stocks, determined by population and lifestyle, is the driver of future service demand and related resource demand (see Chaps. 6 and 7). This is a reasonable assumption when we want to foresee long-term trends of material flows.

Modeling future demolition waste generation is one objective of these MFA studies. However, Hashimoto et al. (2007) showed that there can be very large discrepancies between the amounts estimated in the studies and the statistical quantities reported. One possible reason is that considerable amounts of construction materials do not emerge as wastes. Hashimoto et al. (2007) referred to this as "missing stock" or "dissipated stock" and then proposed a framework for estimating potential wastes accumulated within an economy (Hashimoto et al. 2009). Materials input into an economy include dissipatively used materials, such as crushed stone used for leveling the ground and reclaiming ground, and permanent structures, such as tunnels and dams with a low probability of being demolished. This point should be considered when we model future generation of demolition waste and its recyclability.

Demolition waste is also important from the viewpoint of disaster waste management because it is a major portion of the waste to be managed following a disaster. Tanikawa et al. (2014) estimated such waste as "lost material stock," taking the great east Japan earthquake as a case study. Methodological development for quickly estimating the amount of disaster waste is important for the IE community, because planning waste management and recycling is one of the first steps for recovery from disaster.

2.3 End-of-Life Vehicles and e-Waste

End-of-life vehicles contain many valuable materials that should be recovered (see Chap. 18). An objective of MFA studies is, therefore, to capture flows of resources contained in end-of-life vehicles, such as aluminum (Cheah et al. 2009; Mathieux and Brissaud 2010; Modaresi and Müller 2012; Hatayama et al. 2012), steel, copper, lead, and zinc (Fuse et al. 2009; Yano et al. 2014). Further, Richa et al. (2014) analyzed lithium-ion battery waste flows from electric vehicles in the future.

Material flows of e-waste have been studied in many countries to support its management and recycling, e.g. Brazil (Araújo et al. 2012), China (Liu et al. 2006; Yang et al. 2008; Chung 2012; Zhang et al. 2012; Habuer et al. 2014; Li et al. 2015), Chile (Steubing et al. 2010), Czech Republic (Polak and Drapalova 2012), Germany (Walk 2009), Hong Kong (Chung et al. 2011; Lau et al. 2013), India (Dwivedy and Mittal 2010a, b), Indonesia (Andarani and Goto 2014), Iran (Rahmani et al. 2014; Alavi et al. 2015), Japan (Yamasue et al. 2007; Oguchi et al. 2008; Yoshida et al. 2009), Nigeria (Osibanjo and Nnorom 2008; Nnorom and Osibanjo 2008), South Korea (Lee et al. 2007; Kim et al. 2013), Spain (Gutierrez et al. 2010), USA (Kang and Schoenung 2006; Leigh et al. 2007; Kahhat and Williams 2012; Lam et al.

2013; Schumacher et al. 2014), and in the global economy (Yu et al. 2010). Some studies have estimated flows of specific materials contained in e-waste such as steel, aluminum, copper, lead, nickel, and zinc (Yamasue et al. 2007; Lam et al. 2013; Habuer et al. 2014) and parts such as lithium-ion batteries (Chang et al., 2009). Tracking international trade of second-hand e-products is another research objective of MFA to assess the potential negative impacts on the environment in importing countries (Kahhat and Williams 2009, 2012; Yoshida and Terazono 2010; Breivik et al. 2014).

In many countries, data on e-products and e-waste are limited. Further, consumers tend to store old e-products at home even if they are no longer used. Therefore, methodological aspects for estimating e-waste flows have been discussed (Leigh et al. 2007; Yoshida et al. 2009; Gutierrez et al., 2010; Araújo et al. 2012; Wang et al. 2013; Li et al. 2015). Moreover, Lam et al. (2013) combined MFA with ecological and human health impact assessment caused by heavy metals in e-waste. Metals in end-of-life vehicles and e-waste need to be quantified for planning recycling and assessing risks. Streicher-Porte et al. (2009) integrated MFA, life cycle assessment (LCA), and multiple attribute utility theory to evaluate scenarios for computer supplies to schools in Colombia. These methodological developments and integration of different tools are important next steps for MFAs.

2.4   Metals in Waste

There is a substantial body of research related to metal flows and stocks (Chen and Graedel 2012), inevitably including waste and recycling flows.

One of the motivations of metal flow studies is to estimate recycling rates of those metals. Graedel et al. (2011) provide an overview on the current knowledge of recycling rates for 60 metals and show that many end-of-life recycling ratios (EOL-RRs) are very low: only for 18 metals (silver, aluminum, gold, cobalt, chromium, copper, iron, manganese, niobium, nickel, lead, palladium, platinum, rhenium, rhodium, tin, titanium, and zinc) is the EOL-RR above 50 % at present. We need further research on recycling flows; this should be standardized and institutionalized in the compilation of statistics.

How many times materials are expected to be recycled is also an interesting and important question (see Chap. 7). Markov chain modeling has been applied to estimate average times of use of steel (Matsuno et al. 2007), stainless steel (Hashimoto et al. 2010), nickel (Eckelman et al. 2012), and copper (Eckelman and Daigo 2008). Results were, respectively, 2.7, 1.9–4.3, 3, and 1.9 times.

Some studies discuss alloying elements in metal recycling (Nakajima et al. 2011, 2013; Nakamura et al. 2012; Ohno et al. 2014). For example, Ohno et al. (2014) showed that considerable amounts of alloying elements, which correspond to 7–8 % of the annual consumption in electric arc furnace (EAF) steelmaking, are unintentionally introduced into EAFs. This type of analysis is an interesting application of MFA to help development of more appropriate recycling systems.

As metal stocks in a society are important sources of secondary resources, research related to the assessment of stocks has been increasing (Gerst and Graedel 2008). Using satellite night-time light observation data is an innovative and interesting methodological approach to stock estimation (Takahashi et al. 2009; Hsu et al. 2013). On the conceptual aspect of stocks as resources, classification of secondary resources in the anthroposphere was proposed (Hashimoto et al. 2008), based on the classification of primary resources in the lithosphere, i.e. so-called McKelvey diagram (McKelvey 1972). Analysis of stocks must be an important topic in coming decades.

2.5 Phosphorus in Waste

Phosphorus is as an essential nutrient for agriculture, but phosphate rock is a non-renewable resource and its deposits will be exhausted in a long term. Therefore, in addition to increasing phosphorus use efficiency in agricultural systems, phosphorus needs to be recovered from all current waste streams (Cordell et al. 2011). Especially, wastes rich in phosphorus can represent new sources. MFA can provide an effective tool to identify such new sources.

Sewage sludge is one candidate as a potential source of phosphorus and commonly utilized by spreading directly to farm land (Lederer and Rechberger 2010). However, by conducting MFA of phosphorous in Gothenburg, Sweden, Kalmykova et al. (2012) concluded that solid waste incineration residues represented a large underestimated sink of phosphorus and that focusing on wastewater as the sole source of recovered phosphorus was not sufficient. Further, Matsubae-Yokoyama et al. (2009) analyzed availability of phosphorus resources that remain untapped for Japan and found that the quantity of phosphorus in iron and steel-making slag was almost equivalent to that in imported phosphate ore in terms of both amount and concentration.

2.6 Waste Plastics

Attention has been paid to the management of waste plastics. MFAs have been carried out for different types of plastics in many countries, e.g. plastics streams in general in Austria and Poland (Bogucka et al. 2008), in Germany (Patel et al. 1998), in India (Mutha et al. 2006), in the Netherlands (Joosten et al. 2000) and in Serbia (Vujic et al. 2010); for polyethylene terephthalate flows in Colombia (Rochat et al. 2013) and in the US (Kuczenski and Geyer 2010); flows of polyvinyl chloride in

China (Zhou et al. 2013), in Japan (Nakamura et al. 2009), and in Sweden (Tukker et al. 1997; Kleijn et al. 2000); and flows of waste tires in China (Yang et al. 2010), Thailand (Jacob et al. 2014) and in small island developing states (Sarkar et al. 2011).

Implementing MFA of plastics is not easy because many plastics, such as polyethylene and polypropylene, are used in a variety of products, from construction to consumer products, whose lifetimes are different. Therefore, input-output analysis has been applied to estimate flows of plastics (Joosten et al. 2000; Nakamura et al. 2009; Yang et al. 2010).

2.7   Spatial System Boundaries

Amongst the body of studies targeting countries and cities are a number with interesting system boundaries.

For example, many islands face waste issues because of their limited availability of land for waste disposal as well as constrained availability of physical material resources. Therefore, research on waste flows has been performed for some small island states (Eckelman and Chertow 2009; Sarkar et al. 2011; Eckelman et al. 2014). An island is also a useful system for industrial ecology studies as it is a valuable unit of study for biological sciences (Deschenes and Chertow 2004).

Industrial ecosystems are another example, where one company's waste becomes another company's feedstock. Many studies have been undertaken in the field of industrial symbiosis research (see Chap. 5). For example, Lyons (2007) examined the issue of geographic scale and loop closing for heterogeneous wastes through an analysis of the location and materials flows of a set of recycling, remanufacturing, and waste treatment firms in Texas and showed that there was no preferable scale at which loop closing should be organized.

A process of waste treatment can also be the subject of MFA. For example, Chancerel et al. (2009) assessed precious metal flows during preprocessing of waste electrical and electronic equipment and showed that only 11.5 % of the silver and 25.6 % of the gold and of the palladium reach output fractions from which they can potentially be recovered.

International transfer of e-waste has been a major concern. Therefore, MFA has also been applied to international trade of secondhand products (Kahhat and Williams 2009, 2012; Yoshida and Terazono 2010; Breivik et al. 2014). Such trade can be seen as the trade of metals contained in the products. Fuse et al. (2009) estimated global flows of metal resources in the used automobile trade.

3   MFA-Based Policies and Concepts for Sustainable Resource and Waste Management

3.1   *Conceptual Progress for Sustainable Resource and Waste Management and Its Relevance to MFA – Cases in China and Japan*

According to Yuan et al. (2006), the circular economy (CE) was first proposed as a concept by scholars in China in the 1990s, and subsequently formally adopted in 2002 by the central government as a new development strategy. Yuan et al. (2006) state that the CE concept originates from the industrial ecology paradigm, building on the notion of loop-closing emphasized in German and Swedish environmental policy. According to another article on industrial ecology research in China (Shi et al. 2002), the first time that the term "industrial ecology" appeared in a Chinese academic publication was in 1990, published by Tsinghua University Press. Most recently, the CE was adopted as a keyword to promote resource efficiency policy in Europe. In summer 2014, the European Commission adopted the Communication "Towards a circular economy: a zero waste program for Europe" to establish an EU framework to promote the circular economy. Chap. 7 explores the different concepts brought together under the heading "circular economy."

Shi et al. (2002) also reviewed the state of research on and practice of core constituents of industrial ecology: LCA, DfE, MFA, EIPs, and closed-loop economies. They found that the wide gap between LCA application and policy making needed to be filled and, as compared to LCA, MFA was even less developed in China, as of early 2000s. Thus MFA was not explicitly linked to CE in China, at least in the early stage of the policy.

Earlier in 2000 in Japan, a new fundamental law towards "Jun-kan" (circular) society was adopted. The initial official translation of the circular society was "recycling-based society" but this was subsequently revised to "sound material-cycle (SMC) society." "Jun-kan" in Japanese and "XÚNHUÁN" in Chinese are synonymous, both of them meaning "circulation." Japan's preliminary economy-wide MFA can be found in a report by a committee organized by the Environment Agency in 1991 to examine the "Jun-kan" socio-system (Hashimoto 2009). Here, a similar phrase to "Jun-kan" society appeared for the first time in Japan's environmental administration. MFA and the circular society concept, therefore, have close relations for Japan.

A review by Takiguchi and Takemoto (2008) confirmed that the framework of 3Rs (reduce, reuse, and recycle) policies to establish a SMC Society was designed on the basis of research on MFA. A set of three economy-wide MF indicators was introduced into the Fundamental Plan for SMC Society in 2003, and numerical targets were set for each indicator. The concept of the 3Rs by Japan was also shared on a global scale through the Group of Eight (G8) process known as the 3R Initiative.

3.2   Initiatives in National and Intergovernmental Activities, Focusing on Policy Application of Economy-Wide MFA Indicators

Institutional aspects of progress in international MFA studies were reviewed by Moriguchi (2007), focusing on developments in Japan and in international fora such as within the Organisation for Economic Cooperation and Development (OECD). Experts in industrial ecology, including the authors of this chapter themselves, have played catalytic role in linking industrial ecology studies and their applications to national and international policies for sustainable resource and waste management. A typical example of the outcomes from these international activities is the OECD three-volume guidelines focusing on economy-wide MFA (OECD 2008), published when OECD and UNEP coorganized a Conference on Resource Efficiency in 2008.

Fischer-Kowalski et al. (2011) reviewed the state-of-the-art and reliability of MFA data across different sources. They concluded that the MFA framework and the data generated have reached a maturity that warrants MF indicators to complement traditional economic and demographic information in providing a sound basis for discussing national and international policies for sustainable resource use. International comparison of a few economy-wide MFA indicators (Bringezu et al. 2004), cross country comparisons of economy-wide MFA within the EU (Weisz et al. 2006) as well as country case studies such as in an EU accession country (Kovanda et al. 2010) have been conducted. There has been much criticism against using simplistic summation of different materials by mass as an indicator. Van der Voet et al. (2005) added a set of environmental weights to the flows of the materials to calculate indicators with differentiated environmental impact. Another key argument is accounting for indirect material requirement associated with processed materials and products. The "raw material equivalents" (RME) metric of material consumption addresses the issue of including the full supply chain (including imports) when calculating national or product level material impacts (Muñoz et al. 2009; Schor et al. 2013).

Aoki-Suzuki et al. (2012) conducted a study of how economy-wide MFA indicators are used in a number of developed countries, including analysis of the commonalities between countries that are actively using these indicators in policy, and a survey of the current capacity for economy-wide MFA in developing countries, including data availability and policy uptake.

4   Current and Future Developments

In parallel with the remarkable progress of empirical MFA studies in industrial ecology, there have been developments in statistical institutionalization of physical accounting as well as methodological elaboration employing Input-Output Tables. SEEA 2003 (System of integrated Environmental-Economic Accounting, 2003 edition) pays considerable attention to physical flow accounting, in addition to natural resource stock accounts and environmental protection expenditure accounts, and introduced the National Accounting Matrix including Environmental Accounts (NAMEA) as one of the main building blocks (Pedersen and de Haan 2006). Accounting framework was also discussed in one of the three volumes of OECD's MFA guidance in 2008. Under the Regulation (EU) 691/2011 on European environmental economic accounts, Eurostat collects economy-wide MFA from the national statistical institutes of the EU Member States.

Environmental extension of economic Input-Output Tables (IOT) has been undertaken both by experts in Input-Output analysis and by users of this approach (see Chap. 8). IOT describes intersectoral monetary flows, but this approach can provide a consistent framework to describe the physical balance of inflows and outflows. Wastes have conventionally not been included in monetary IOT (MIOT) because of their negative economic value. Attempts were made to include waste flows by physical IOT (PIOT) (Dietzenbacher 2005) and consistency between MIOT and PIOT was discussed (Weisz and Duchin 2006). A new framework designed specifically for the analysis of waste issues was developed by Nakamura and Kondo (2002), and they and their colleagues published a number of studies on framework and applications of Waste Input Output (WIO) tables (Nakamura et al. 2007, 2008, 2009). Lenzen and Reynolds (2010) applied a "supply and use" framework to WIO and confirmed the consistency between their Waste Supply-Use Tables (WSUT) and WIO. Recently, empirical studies of PIOT at local levels in China were reported by Liang and Zhang (2011).

Literally, MFA analyses flows of materials. However, there is no reason to restrict the scope of MFA within this definition. MFA is a general system approach that can be used to explore various interfaces between

- Flows and stocks
- Technosphere and ecosphere
- Upstream resource issues and downstream waste issue
- Valuable materials and toxic substances
- Energy (with GHGs) and materials (from non-renewables)
- Theoretical analysis/models and on-site practices
- Scientific studies and policy applications

so as to understand and improve socio-economic metabolism. Further, integration of other methodologies such as IO analysis, LCA, risk assessment, environmental impact assessment, and technology assessment are expected.

References

Alavi, N., Shirmardi, M., Babaei, A., Takdastan, A., & Bagheri, N. (2015). Waste electrical and electronic equipment (WEEE) estimation: A case study of Ahvaz City, Iran. *Journal of the Air and Waste Management Association, 65*(3), 298–305.

Andarani, P., & Goto, N. (2014). Potential e-waste generated from households in Indonesia using material flow analysis. *Journal of Material Cycles and Waste Management, 16*(2), 306–320.

Aoki-Suziki, C., Bengtsson, M., & Hotta, Y. (2012). Policy application of economy-wide material flow accounting – International comparison and suggestions for capacity development in industrializing countries. *Journal of Industrial Ecology, 16*(4), 467–480.

Araújo, M. G., Magrini, A., Mahler, C. F., & Bilitewski, B. (2012). A model for estimation of potential generation of waste electrical and electronic equipment in Brazil. *Waste Management, 32*(2), 335–342.

Ayres, R., & Ayres, L. (1998). *Accounting for resources 1* (245 pp). Cheltenham: Edward Elgar.

Ayres, R., & Ayres, L. (1999). *Accounting for resources 2* (371 pp). Cheltenham: Edward Elgar.

Ayres, R., & Ayres, L. (Eds.). (2002). *A handbook of industrial ecology* (680 pp). Cheltenham: Edward Elgar.

Ayres, R., & Kneese, A. (1969). Production, consumption and externalities. *American Economic Review, 59*(3), 282–297.

Bailey, R., Bras, B., & Allen, J. K. (2008). Measuring material cycling in industrial systems. *Resources, Conservation and Recycling, 52*(4), 643–652.

Bergsdal, H., Bohne, R. A., & Brattebo, H. (2007). Projection of construction and demolition waste in Norway. *Journal of Industrial Ecology, 11*(3), 27–39.

Bogucka, R., Kosinska, I., & Brunner, P. H. (2008). Setting priorities in plastic waste management – Lessons learned from material flow analysis in Austria and Poland. *Polimery, 53*(1), 55–59.

Breivik, K., Armitage, J. M., Wania, F., & Jones, K. C. (2014). Tracking the global generation and exports of e-waste. Do existing estimates add up? *Environmental Science and Technology, 48*(15), 8735–8743.

Bringezu, S., & Moriguchi, Y. (2002). Material flow analysis. In Ayres, R. & Ayres, L. (Eds.), *A handbook of industrial ecology* (pp. 79–90), Cheltenham: Edward Elgar.

Bringezu, S., Schqtz, H., Steger, S., & Baudisch, J. (2004). International comparison of resource use and its relation to economic growth – The development of total material requirement, direct material inputs and hidden flows and the structure of TMR. *Ecological Economics, 51*(1-2), 97–124.

Chancerel, P., Meskers, C. E. M., Hageluken, C., & Rotter, V. S. (2009). Assessment of precious metal flows during preprocessing of waste electrical and electronic equipment. *Journal of Industrial Ecology, 13*(5), 791–810.

Chang, T. C., You, S. J., Yu, B. S., & Yao, K. F. (2009). A material flow of lithium batteries in Taiwan. *Journal of Hazardous Materials, 163*, 910–915.

Cheah, L., Heywood, J., & Kirchain, R. (2009). Aluminum stock and flows in U.S. passenger vehicles and implications for energy use. *Journal of Industrial Ecology, 13*(5), 718–734.

Chen, W.-Q., & Graedel, T. E. (2012). Anthropogenic cycles of the elements: A critical review. *Environmental Science and Technology, 46*(16), 8574–8586.

Chung, S. S. (2012). Projection of waste quantities: The case of e-waste of the People's Republic of China. *Waste Management and Research, 30*(11), 1130–1137.

Chung, S. S., Lau, K. Y., & Zhang, C. (2011). Generation of, and control measures for, e-waste in Hong Kong. *Waste Management, 31*(3), 544–554.

Cordell, D., Rosemarin, A., Schroder, J.-J., & Smit, A. L. (2011). Towards global phosphorus security: A systematic framework for phosphorus recovery and reuse options. *Chemosphere, 84*, 747–758.

Daniels, P. (2002). Approaches for quantifying the metabolism of physical economies – Part II review of individual approaches. *Journal of Industrial Ecology, 6*(1), 65–88.

Daniels, P., & Moore, S. (2001). Approaches for quantifying the metabolism of physical economies – Part I methodological overview. *Journal of Industrial Ecology, 5*(4), 69–93.

Deschenes, P. J., & Chertow, M. (2004). An island approach to industrial ecology: Towards sustainability in the island context. *Journal of Environmental Planning and Management, 47*(2), 201–217.

Dietzenbacher, E. (2005). Waste treatment in physical input–output analysis. *Ecological Economics, 55*, 11–23.

Dwivedy, M., & Mittal, R. K. (2010a). Estimation of future outflows of e-waste in India. *Waste Management, 30*(3), 483–491.

Dwivedy, M., & Mittal, R. K. (2010b). Future trends in computer waste generation in India. *Waste Management, 30*(11), 2265–2277.

Eckelman, M. J., & Chertow, M. R. (2009). Using material flow analysis to illuminate long-term waste management solutions in Oahu, Hawaii. *Journal of Industrial Ecology, 13*(5), 758–774.

Eckelman, M., & Daigo, I. (2008). Markov chain modeling of the technological lifetime of copper. *Ecological Economics, 67*(2), 265–273.

Eckelman, M. J., Reck, B. K., & Graedel, T. E. (2012). Exploring the global journey of nickel with Markov chain models. *Journal of Industrial Ecology, 16*(3), 334–342.

Eckelman, M. J., Ashton, W., Arakaki, Y., Hanaki, K., Nagashima, S., & Malone-Lee, L. C. (2014). Island waste management systems – Statistics, challenges, and opportunities for applied industrial ecology. *Journal of Industrial Ecology, 18*(2), 306–317.

Eurostat. (2001). *Economy-wide material flow accounts and derived indicators – A methodological guide*. Luxembourg: European Communities.

Fisher-Kowalski, M., Krausmann, F., Giljum, S., Lutter, S., Mayer, A., Bringezu, S., Moriguchi, Y., Schütz, H., Schandl, H., & Weisz, H. (2011). Methodology and indicators of economy-wide material flow accounting. *Journal of Industrial Ecology, 15*(6), 855–876.

Fuse, M., Nakajima, K., & Yagita, H. (2009). Global flow of metal resources in the used automobile trade. *Materials Transactions, 50*(4), 703–710.

Gerst, M. D., & Graedel, T. E. (2008). In-use stocks of metals – Status and implications. *Environmental Science and Technology, 42*(19), 7038–7045.

Graedel, T. E., Allwood, J., Birat, J.-P., Buchert, M., Hagelueken, C., Reck, B. K., Sibley, S. F., & Sonnemann, G. (2011). What do we know about metal recycling rates? *Journal of Industrial Ecology, 15*(3), 355–366.

Gutiérrez, E., Adenso-Díaz, B., Lozano, S., & González-Torre, P. (2010). A competing risks approach for time estimation of household WEEE disposal. *Waste Management, 30*, 1643–1652.

Habuer, Nakatani, J., & Moriguchi, Y. (2014). Time-series product and substance flow analyses of end-of-life electrical and electronic equipment in China. *Waste Management, 34*(02), 489–497.

Hashimoto, S. (2009). A Junkan-Gata Society – Concept and progress in material flow analysis in Japan. *Journal of Industrial Ecology, 13*(5), 655–657.

Hashimoto, S., & Moriguchi, Y. (2004). Proposal of six indicators of material cycles for describing society's metabolism: From the viewpoint of material flow analysis. *Resources, Conservation and Recycling, 40*(3), 185–200.

Hashimoto, S., Tanikawa, H., & Moriguchi, Y. (2007). Where will the large amounts of materials accumulated within the economy go? – A material flow analysis of construction minerals for Japan. *Waste Management, 27*(12), 1725–1738.

Hashimoto, S., Daigo, I., Murakami, S., Matsubae-Yokoyama, K., Fuse, M., Nakajima, K., Oguchi, M., Tanikawa, H., Tasaki, T., Yamasue, E., & Umezawa, O. (2008). Framework of material stock accounts – Toward assessment of material accumulation in the economic sphere. In *Proceedings of the eighth international conference on ecobalance, C-08*, Tokyo.

Hashimoto, S., Tanikawa, H., & Moriguchi, Y. (2009). Framework for estimating potential wastes and secondary resources accumulated within an economy – A case study of construction minerals in Japan. *Waste Management, 29*(11), 2859–2866.

Hashimoto, S., Daigo, I., Eckelman, M., & Reck, B. (2010). Measuring the status of stainless steel use in the Japanese socio-economic system. *Resources, Conservation and Recycling, 54*(10), 737–743.

Hatayama, H., Daigo, I., Matsuno, Y., & Adachi, Y. (2012). Evolution of aluminum recycling initiated by the introduction of next-generation vehicles and scrap sorting technology. *Resources, Conservation and Recycling, 66*, 8–14.

Hsiao, T. Y., Huang, Y. T., Yu, Y. H., & Wernick, I. K. (2002). Modeling materials flow of waste concrete from construction and demolition wastes in Taiwan. *Resources Policy, 28*(1-2), 39–47.

Hsu, F.-C., Elvidge, C. D., & Matsuno, Y. (2013). Exploring and estimating in-use steel stocks in civil engineering and buildings from night-time lights. *International Journal of Remote Sensing, 34*(2), 490–504.

Hu, D., You, F., Zhao, Y., Yuan, Y., Liu, T., Cao, A., Wang, Z., & Zhang, J. (2010a). Input, stocks and output flows of urban residential building system in Beijing city, China from 1949 to 2008. *Resources, Conservation and Recycling, 54*(12), 1177–1188.

Hu, M., van der Voet, E., & Huppes, G. (2010b). Dynamic material flow analysis for strategic construction and demolition waste management in Beijing. *Journal of Industrial Ecology, 14*(3), 440–456.

Huang, T., Shi, F., Tanikawa, H., Fei, J., & Han, J. (2013). Materials demand and environmental impact of buildings construction and demolition in China based on dynamic material flow analysis. *Resources, Conservation and Recycling, 72*, 91–101.

Jacob, P., Kashyap, P., Suparat, T., & Visvanathan, C. (2014). Dealing with emerging waste streams – Used tyre assessment in Thailand using material flow analysis. *Waste Management and Research, 32*, 918–926.

Joosten, L. A. J., Hekkert, M. P., Worrell, E., & Turkenburg, W. C. (2000). Assessment of the plastic flows in The Netherlands using STREAMS. *Resources, Conservation and Recycling, 30*, 135–161.

Kahhat, R., & Williams, E. (2009). Product or waste? Importation and end-of-life processing of computers in Peru. *Environmental Science and Technology, 43*(15), 6010–6016.

Kahhat, R., & Williams, E. (2012). Materials flow analysis of e-waste – Domestic flows and exports of used computers from the United States. *Resources, Conservation and Recycling, 67*, 67–74.

Kalmykova, Y., Harder, R., Borgestedt, H., & Svanang, I. (2012). Pathways and management of phosphorus in urban areas. *Journal of Industrial Ecology, 16*(6), 928–939.

Kang, H. Y., & Schoenung, J. M. (2006). Estimation of future outflows and infrastructure needed to recycle personal computer systems in California. *Journal of Hazardous Materials, 137*(2), 1165–1174.

Kim, S., Oguchi, M., Yoshida, A., & Terazono, A. (2013). Estimating the amount of WEEE generated in South Korea by using the population balance model. *Waste Management, 33*(2), 474–483.

Kleijn, R., Huele, R., & van der Voet, E. (2000). Dynamic substance flow analysis – The delaying mechanism of stocks, with the case of PVC in Sweden. *Ecological Economics, 32*(2), 241–254.

Kovanda, J., Weinzettel, J., & Hák, T. (2010). Material flow indicators in the Czech Republic in light of the accession to the European Union. *Journal of Industrial Ecology, 14*(4), 650–665.

Kuczenski, B., & Geyer, R. (2010). Material flow analysis of polyethylene terephthalate in the US, 1996–2007. *Resources, Conservation and Recycling, 54*(12), 1161–1169.

Lam, C. W., Lim, S.-R., & Schoenung, J. M. (2013). Linking material flow analysis with environmental impact potential – Dynamic technology transition effects on projected e-waste in the United States. *Journal of Industrial Ecology, 17*(2), 299–309.

Lau, W. K.-Y., Chung, S.-S., & Zhang, C. (2013). A material flow analysis on current electrical and electronic waste disposal from Hong Kong households. *Waste Management, 33*(3), 714–721.

Lederer, J., & Rechberger, H. (2010). Comparative goal-oriented assessment of conventional and alternative sewage sludge treatment options. *Waste Management, 30*, 1043–1056.

Lee, J. C., Song, H. T., & Yoo, J. M. (2007). Present status of the recycling of waste electrical and electronic equipment in Korea. *Resources, Conservation and Recycling, 50*, 380–397.

Leigh, N. G., Realff, M. J., Ai, N., French, S. P., Ross, C. L., & Bras, B. (2007). Modeling obsolete computer stock under regional data constraints: An Atlanta case study. *Resources, Conservation and Recycling, 51*(4), 847–869.

Lenzen, M., & Reynolds, C. (2010). A supply-use approach to waste input-output analysis. *Journal of Industrial Ecology, 18*(2), 212–226.

Li, B., Yang, J., Lu, B., & Song, X. (2015). Estimation of retired mobile phones generation in China: A comparative study on methodology. *Waste Management, 35*, 247–254.

Liang, S., & Zhang, T. (2011). Data acquisition for applying physical input-output tables in Chinese cities. *Journal of Industrial Ecology, 15*(6), 825–835.

Liu, X., Tanaka, M., & Matsui, Y. (2006). Generation amount prediction and material flow analysis of electronic waste: A case study in Beijing, China. *Waste Management and Research, 24*(5), 434–445.

Lyons, D. (2007). A spatial analysis of loop closing among recycling, remanufacturing, and waste treatment firms in Texas. *Journal of Industrial Ecology, 11*(1), 43–54.

Mathieux, F., & Brissaud, D. (2010). End-of-life product-specific material flow analysis – Application to aluminum coming from end-of-life commercial vehicles in Europe. *Resources, Conservation and Recycling, 55*(2), 92–105.

Matsubae-Yokoyama, K., Kubo, H., Nakajima, K., & Nagasaka, T. (2009). A material flow analysis of phosphorus in Japan. *Journal of Industrial Ecology, 13*(5), 687–705.

Matsuno, Y., Daigo, I., & Adachi, Y. (2007). Application of Markov chain model to calculate the average number of times of use of a material in society. *International Journal of Life Cycle Assessment, 12*(1), 34–39.

McKelvey, V. E. (1972). Mineral resource estimates and public policy. *American Scientist, 60*(1), 32–40.

Modaresi, R., & Muller, D. B. (2012). The role of automobiles for the future of aluminum recycling. *Environmental Science and Technology, 46*(16), 8587–8594.

Moriguchi, Y. (2007). Material flow indicators to measure progress toward a sound material-cycle society. *Journal of Material Cycles and Waste Management, 9*(2), 112–120.

Müller, D. B. (2006). Stock dynamics for forecasting material flows – Case study for housing in The Netherlands. *Ecological Economics, 59*(1), 142–156.

Muñoz, P., Giljum, S., & Roca, J. (2009). The raw material equivalents of international trade – Empirical evidence for Latin America. *Journal of Industrial Ecology, 13*(6), 881–897.

Mutha, N. H., Patel, M., & Premnath, V. (2006). Plastics materials flow analysis for India. *Resources, Conservation and Recycling, 47*, 222–244.

Nakajima, K., Takeda, O., Miki, T., Matsubae, K., & Nagasaka, T. (2011). Thermodynamic analysis for the controllability of elements in the recycling process of metals. *Environmental Science and Technology, 45*(11), 4929–4936.

Nakajima, K., Ohno, H., Kondo, Y., Matsubae, K., Takeda, O., Miki, T., Nakamura, S., & Nagasaka, T. (2013). Simultaneous material flow analysis of nickel, chromium, and molybdenum used in alloy steel by means of input-output analysis. *Environmental Science and Technology, 47*(9), 4653–4660.

Nakamura, S., & Kondo, Y. (2002). Input-output analysis of waste management. *Journal of Industrial Ecology, 6*(1), 39–63.

Nakamura, S., Nakajima, K., Kondo, Y., & Nagasaka, T. (2007). The waste input-output approach to materials flow analysis – Concepts and application to base metals. *Journal of Industrial Ecology, 11*(4), 50–63.

Nakamura, S., Murakami, S., Nakajima, K., & Nagasaka, T. (2008). Hybrid input-output approach to metal production and its application to the introduction of lead-free solders. *Environmental Science and Technology, 42*(10), 3843–3848.

Nakamura, S., Nakajima, K., Yoshizawa, Y., Matsubae-Yokoyama, K., & Nagasaka, T. (2009). Analyzing polyvinyl chloride in Japan with the waste input-output material flow analysis model. *Journal of Industrial Ecology, 13*(5), 706–717.

Nakamura, S., Kondo, Y., Matsubae, K., Nakajima, K., Tasaki, T., & Nagasaka, T. (2012). Quality and dilution losses in the recycling of ferrous materials from end-of-life passenger cars – Input-output analysis under explicit consideration of scrap quality. *Environmental Science and Technology, 46*(17), 9266–9273.

Nnorom, I. C., & Osibanjo, O. (2008). Electronic waste (e-waste) – Material flows and management practices in Nigeria. *Waste Management, 28*(8), 1472–1479.

OECD. (2005). *OECD environmental data compendium 2004*. OECD, Paris.

OECD. (2008). *Measuring material flows and resource productivity*, Volume I The OECD Guide, Volume II The Accounting Framework, Volume III Inventory of Country Activities. OECD, Paris.

Oguchi, M., Kameya, T., Yagi, S., & Urano, K. (2008). Product flow analysis of various consumer durables in Japan. *Resources, Conservation and Recycling, 52*(3), 463–480.

Ohno, H., Matsubae, K., Nakajima, K., Nakamura, S., & Nagasaka, T. (2014). Unintentional flow of alloying elements in steel during recycling of end-of-life vehicles. *Journal of Industrial Ecology, 18*(2), 242–253.

Osibanjo, O., & Nnorom, I. C. (2008). Material flows of mobile phones and accessories in Nigeria – Environmental implications and sound end-of-life management options. *Environmental Impact Assessment Review, 28*(2-3), 198–213.

Patel, M. K., Jochem, E., Radgen, P., & Worrell, E. (1998). Plastics streams in Germany – An analysis of production, consumption and waste generation. *Resources, Conservation and Recycling, 24*, 191–215.

Pedersen, G., & de Haan, M. (2006). The system of environmental and economic accounts-2003 and the economic relevance of physical flow accounting. *Journal of Industrial Ecology, 10*(1-2), 19–42.

Polak, M., & Drapalova, L. (2012). Estimation of end of life mobile phones generation – The case study of the Czech Republic. *Waste Management, 32*(8), 1583–1591.

Rahmani, M., Nabizadeh, R., Yaghmaeian, K., Mahvi, A. H., & Yunesian, M. (2014). Estimation of waste from computers and mobile phones in Iran. *Resources, Conservation and Recycling, 87*, 21–29.

Richa, K., Babbitt, C. W., Gaustad, G., & Wang, X. (2014). A future perspective on lithium-ion battery waste flows from electric vehicles. *Resources, Conservation and Recycling, 83*, 63–76.

Rochat, D., Binder, C. R., Diaz, J., & Jolliet, O. (2013). Combining material flow analysis, life cycle assessment, and multiattribute utility theory – Assessment of end-of-life scenarios for polyethylene terephthalate in Tunja, Colombia. *Journal of Industrial Ecology, 17*(5), 642–655.

Sarkar, S., Chamberlain, J. F., & Miller, S. A. (2011). A comparison of two methods to conduct material flow analysis on waste tires in a small island developing state. *Journal of Industrial Ecology, 15*(2), 300–314.

Schor, K., Wood, R., Arto, I., & Weinzettel, J. (2013). Estimating raw material equivalents on a macro-level – Comparison of multi-regional input-output analysis and hybrid LCI-IO. *Environmental Science and Technology, 47*(24), 14282–14289.

Schumacher, K. A., Schumacher, T., & Agbemabiese, L. (2014). Quantification and probabilistic modeling of CRT obsolescence for the State of Delaware. *Waste Management, 34*(11), 2321–2326.

Shi, H., Moriguichi, Y., & Yang, J. (2002). Industrial ecology in China, part I: Research. *Journal of Industrial Ecology, 6*(3-4), 7–11.

Shi, F., Huang, T., Tanikawa, H., Han, J., Hashimoto, S., & Moriguchi, Y. (2012). Toward a low carbon-dematerialization society – Measuring the materials demand and CO2 emissions of building and transport infrastructure construction in China. *Journal of Industrial Ecology, 16*(4), 493–505.

Steubing, B., Boeni, H., Schluep, M., Silva, U., & Ludwig, C. (2010). Assessing computer waste generation in Chile using material flow analysis. *Waste Management, 30*(3), 473–482.

Streicher-Porte, M., Marthaler, C., Boni, H., Schluep, M., Camacho, A., & Hilty, L. M. (2009). One laptop per child, local refurbishment or overseas donations? Sustainability assessment of computer supply scenarios for schools in Colombia. *Journal of Environmental Management, 90*, 3498–3511.

Takahashi, K. I., Terakado, R., Nakamura, J., Daigo, I., Matsuno, Y., & Adachi, Y. (2009). In-use stock of copper analysis using satellite night-time light observation data. *Materials Transactions, 50*(7), 1871–1874.

Takiguchi, H., & Takemoto, K. (2008). Japanese 3R policies based on material flow analysis. *Journal of Industrial Ecology, 12*(5), 792–798.

Tanikawa, H., Managi, S., & Lwin, C. M. (2014). Estimates of lost material stock of buildings and roads due to the great east Japan earthquake and tsunami. *Journal of Industrial Ecology, 18*(3), 421–431.

Tukker, A., Kleijn, R., van Oers, L., & Smeets, E. (1997). Combining SFA and LCA – The Swedish PVC analysis. *Journal of Industrial Ecology, 1*(4), 93–116.

Van der Voet, E., van Oers, L., & Nikolic, I. (2005). Dematerialization – Not just a matter of weight. *Journal of Industrial Ecology, 8*(4), 121–137.

Vujic, G. V., Jovicic, N. M., Babic, M. J., Stanisavljevic, N. S., Batinic, B. J., & Pavlovic, A. R. (2010). Assessment of plastic flows and stocks in Serbia using material flow analysis. *Thermal Science, 14*(Suppl), S89–S95.

Walk, W. (2009). Forecasting quantities of disused household CRT appliances – A regional case study approach and its application to Baden-Wurttemberg. *Waste Management, 29*(2), 945–951.

Wang, F., Huisman, J., Stevels, A., & Balde, C. P. (2013). Enhancing e-waste estimates: Improving data quality by multivariate input-output analysis. *Waste Management, 33*(11), 2397–2407.

Weisz, H., & Duchin, H. (2006). Physical and monetary input-output analysis: What makes the difference? *Ecological Economics, 57*, 534–541.

Weisz, H., Krausmann, F., Amann, C., Eisenmenger, N., Erb, K.-H., Hubacek, K., & Fischer-Kowalski, M. (2006). The physical economy of the European Union: Cross-country comparison and determinants of material consumption. *Ecological Economics, 58*(4), 676–698.

Wen, Z., & Li, R. (2010). Materials metabolism analysis of China's highway traffic system (HTS) for promoting circular economy. *Journal of Industrial Ecology, 14*(4), 641–649.

Yamasue, E., Nakajima, K., Daigo, I., Hashimoto, S., Okumura, H., & Ishihara, K. N. (2007). Evaluation of the potential amounts of dissipated rare metals from WEEE in Japan. *Materials Transaction, 48*(9), 2353–2357.

Yang, J., Lu, B., & Xu, C. (2008). WEEE flow and mitigating measures in China. *Waste Management, 28*, 1589–1597.

Yang, N., Chen, D., Hu, S., Li, Y., & Jin, Y. (2010). Evaluation of the tire industry of China based on physical input-output analysis. *Journal of Industrial Ecology, 14*(3), 457–466.

Yano, J., Hirai, Y., Okamoto, K., & Sakai, S. (2014). Dynamic flow analysis of current and future end-of-life vehicles generation and lead content in automobile shredder residue. *Journal of Material Cycles and Waste Management, 16*, 52–61.

Yoshida, A., & Terazono, A. (2010). Reuse of secondhand TVs exported from Japan to the Philippines. *Waste Management, 30*(6), 1063–1072.

Yoshida, A., Tasaki, T., & Terazono, A. (2009). Material flow analysis of used personal computers in Japan. *Waste Management, 29*(5), 1602–1614.

Yu, J., Williams, E., Ju, M., & Yang, Y. (2010). Forecasting global generation of obsolete personal computers. *Environmental Science and Technology, 44*(9), 3232–3237.

Yuan, Z., Bi, J., & Moriguchi, Y. (2006). The circular economy – A new development strategy in China. *Journal of Industrial Ecology, 10*(1-2), 4–8.

Zhang, L., Yuan, Z., Bi, J., & Huang, L. (2012). Estimating future generation of obsolete household appliances in China. *Waste Management and Research, 30*(11), 1160–1168.

Zhou, Y., Yang, N., & Hu, S. (2013). Industrial metabolism of PVC in China: A dynamic material flow analysis. *Resources, Conservation and Recycling, 73*, 33–40.

Part II
Case Studies and Examples of the Application of Industrial Ecology Approaches

Chapter 13
Circular Economy and the Policy Landscape in the UK

Julie Hill

Abstract This chapter sets out the European policy origins of 'circular economy' thinking in the UK and discusses the extent to which the waste prevention plans written by the four countries of the UK (to fulfill the EU requirement) start to move the UK in the direction of more circular approaches. This is important for an understanding of what has driven UK action on this agenda. I argue that the 'circular economy' has become an increasingly vigorous topic of debate in the UK. This has been manifested mainly through interest and use of the language by leading companies, but more recently also through political interest in Scotland and Wales, resulting in diverging policies in the countries of the UK. Heightened political interest in some parts of the UK has coincided with uncertainty about activity in the European Commission. The chapter discusses some of the difficulties in turning the concept into policy prescriptions.

Keywords Circular economy • Waste prevention • Resource efficiency • Resource security • Resilience

1 Introduction

The circular economy debate in the UK has evolved over the last 3–4 decades from a number of converging strands of thinking and activity, with their origins chiefly in Europe (Hill 2014). European Commission policy development on waste has been one of the key strands. Academic institutions and think tanks, with support from some leading businesses, have built on the foundations provided by European policy to raise awareness of the circular economy concept, but translating the aspirations into more progressive policies is a mixed picture among the four countries comprising the UK.

J. Hill (✉)
Senior Visiting Fellow, Centre for Environmental Strategy, University of Surrey and Associate of Green Alliance, 36, Buckingham Palace Road, SW1W 0RE London, UK
e-mail: jhill@green-alliance.org.uk

© The Author(s) 2016
R. Clift, A. Druckman (eds.), *Taking Stock of Industrial Ecology*,
DOI 10.1007/978-3-319-20571-7_13

What links the various strands of 'circular economy' discourse in the UK is 'systems thinking' – that keeping resources in productive use is not just a matter for individual firms on the one hand, or consumers on the other, but part of the whole economic system. This holistic view distinguishes these initiatives from much of waste management policy in the UK through the 1980s, 1990s and early twenty-first century, which has taken a predominantly 'end of pipe' view of the problems of waste. It is also a step on from political discourse concerning 'resource efficiency', which has often focused on industrial process efficiency rather than the whole life cycle of products, and is often unspecific as to which resources it is considering and what kind of efficiencies count most.

2 The European Union's Development of Waste Policy and Resource Efficiency Initiatives

European policy and legislation provides the overarching framework for the development of circular economy thinking in Europe and the UK. Without the development of legislation to limit landfill, reduce carbon emissions, improve recycling of key materials, introduce producer responsibility for end of life products, restrict toxic substances in the environment, influence the design of products, develop 'integrated product policy' and, crucially, sign up leading businesses to the ambition of greater resource efficiency, circular economy thinking would have far less traction. These actions have themselves been influenced by the more environmentally progressive European nations, in particular the Netherlands, Germany, Denmark and Sweden. The EU-promulgated notion of the 'waste hierarchy' is an important precursor to the ideas set out by many of those espousing the circular economy. The waste hierarchy has legal force through the Waste Framework Directive (EC 2008) and indicates the order of preference for waste options: prevention before reuse, reuse before recycling, recycling before energy recovery and lastly disposal. The circular economy approach can be seen as a more systematic (rather than incremental and material-based) application of this thinking.

The requirement of the 2008 Waste Framework Directive, for all member states to produce Waste Prevention Programmes by the end of 2013 (EC 2008), forced the issue further and gave policy makers an opportunity to set out circular economy ambitions. Much of the thinking behind this requirement had been done through the Commission's development of 'Integrated Product Policy (IPP)' – a series of initiatives starting in 2003 to understand which products accounted for the greatest environmental impact. Work under the IPP banner promoted 'life-cycle thinking' and advanced policy instruments, including the Ecodesign of Energy Using Products Directive, which aims to reduce the impact of products (EC 2014b).

Integrated Product Policy also underlies another relevant development – the process begun by the publication of the Roadmap to a Resource Efficient Europe championed by Environment Commissioner Janez Potocnik, and adopted by the full

Commission in September 2011 (EC 2011b). The Resource Efficiency Roadmap is part of the Resource Efficiency Flagship of the Europe 2020 Strategy, which the European Commission describes as 'the European Union's growth strategy for the next decade and aimed at establishing a smart, sustainable and inclusive economy with high levels of employment, productivity and social cohesion' (EC 2011a).

The roadmap can be seen as a move to advance from the position reached by virtue of decades of waste legislation, and guide member states and businesses towards a consolidated and comprehensive vision of a resource efficient, circular economy. The roadmap does not place any legal obligations on member states, but it does set a number of benchmarks and milestones (notional targets), and has an ambition (vigorously debated) to agree indicators to measure resource efficiency. It has also given rise to the European Resource Efficiency Platform (EREP), a multi-stakeholder group of influential politicians, business people, NGOs and academics (EREP 2013). In addition, Horizon 2020, which is the EU's framework programme for research and innovation, is beginning to feature 'Circular Economy' and 'Industrial Symbiosis' as recognized terms, while OECD has started to refer to 'Industrial Ecology'.

A parallel initiative has been the development of the EU Raw Materials Initiative (EC 2014a), which has examined the future prospects for the availability of raw materials crucial to the economies of the EU. Fourteen materials, mainly metals, have been identified as critical and recommendations have been developed to secure future supplies. These recommendations fall into three categories, or 'pillars': (1) ensuring a 'level playing field' for access to resources in third countries (often referred to as 'resource diplomacy'); (2) securing supplies within Europe, such as reopening historic mines; and (3) improving resource efficiency and recycling, not least by highlighting how poor our current recovery of key metals is at present. It is the third 'pillar' that has contributed to the growing circular economy debate.

In 2014, in an attempt to consolidate and extend the progress made to date, Environment Commissioner Janez Potocnik proposed a Circular Economy package of new policy initiatives. This appeared at the very end of the 2010–2014 Commission in July 2014 (EC 2014c). Its principle measures were:

- A target of 70 % recycling for municipal waste by 2030
- A target of 80 % recycling of packaging waste by 2030
- Landfill bans from 2025 for plastics, metals, glass, paper, card and biodegradable waste.

The policy package also included two non-binding targets: the Commission wanted member states to adopt national strategies to reduce food waste by 30 % by 2025 and proposed a target of a 30 % increase in EU resource productivity by 2030.

The package was greeted with disappointment from environmental groups who wanted more measures to stimulate activities such as re-use and remanufacturing (the 'inner' or 'tighter' loops, as described by the Ellen MacArthur Foundation) (EMF 2013). However, it was not welcomed by some key member states and leading business groups who felt that the 70 % target was unachievable, and that the resource productivity target was not implementable, given the absence of good baseline data on which to assess progress against the target. The package did not survive the formation of the new Commission and was formally withdrawn in February 2015, as part of a drive to cut 'red tape' (Euroactive 2015). At the same time, the EU Commission pledged to propose a 'new and more ambitious' package by the end of 2015. The uncertainty thus generated has been widely criticized by NGOs and some businesses.

3 UK Policy Responses to Circular Economy Objectives

Businesses are ahead of the policy debate on circular economy in the UK, in terms of their promulgation of the ideas and their understanding of the opportunities and barriers. This is evidenced by think tank/business partnerships such as the Ellen Macarthur Foundation (EMF 2013), the RSA's Great Recovery Project (RSA 2013) and the Green Alliance's Circular Economy Task Force (2015).

In the UK, the last decade of policy developments in Europe have worked through into policy into four main ways:

- Efforts to implement the 50 % recycling target.
- Efforts to divert waste from landfill into recycling and energy from waste, particularly biodegradable wastes.
- The implementation of producer responsibility schemes for packaging, end of life vehicles, electronics and batteries.
- Discussions in the four countries comprising the UK of how to move to the 'inner loops' of re-use and remanufacturing, as well as greater product longevity.

The first three have been relatively successful, but focus on the lower parts of the waste hierarchy or the 'outer loops' of the circular economy. The last has been most evident in (1) the publication of a Resource Security Action Plan jointly by the UK Departments for Environment and for Business, and (2) the development of 'Waste Prevention Plans' by the four countries of the UK.

4 The Resource Security Action Plan

The UK's Resource Security Action Plan (RSAP) (Defra 2012) was a joint initiative of the UK Department for Environment, Food and Rural Affairs (Defra) and the Department for Business Innovation and Skills (BIS) to examine strategies for addressing resource security in the UK. The RSAP put more emphasis on recovery (i.e. circular approaches) than on opening up new sources of materials as a means to provide greater resource security. It also encouraged the environmental think tank Green Alliance to establish the Circular Economy Task Force as a means of engaging businesses in the solutions. The task force's first report, Resource Resilient UK, was published in July 2013 (Benton and Hazel 2013) and provided a new account of material security, as related to the environmental impacts and reputational threats of raw materials as much as to access. It also made recommendations for how UK policy could support the development of more circular approaches in pursuit of greater resource security. The report was well received by Government Ministers and by businesses.

5   Waste Prevention Plans

The Article 29 requirement of the revised EU Waste Framework Directive (2008/98/EC) that every member state should produce a Waste Prevention Programme by the end of 2013 provided a vehicle for countries that were so minded to produce something close to circular economy plans. Of the UK's devolved administrations, the most engagement has been seen in Scotland and Wales, with Zero Waste initiatives in Scotland developing rapidly before and during the independence referendum debates in 2014.

Looking at the Waste Prevention Programmes and the targets (Table 13.1) of the four countries of the UK side by side, some observations can be made:

Table 13.1   Waste prevention targets set by the four countries of the UK

Country and document	Waste prevention targets
'Safeguarding Scotland's Resources: Blueprint for a more resource efficient and circular economy' (Scottish Government 2013)	Reduce Scotland's waste by 7 % by 2017 from 2011 levels and achieve a 15 % reduction by 2025
'Towards Zero Waste: One Wales, One Planet' (Welsh Government 2013)	Overall goal of achieving zero (non-recyclable) waste by 2050 (67 % less than 2007 levels) and an interim goal of 27 % less by 2025
	For household waste, a reduction of 1.2 % every year to 2050 based on 2006/7 baseline
	A general reduction of 1.4 % every year to 2050 based on 2006/7 baseline for industrial waste, with specific targets for individual priority sectors: metals, paper, chemicals and food
	A reduction of 1.2 % every year to 2050 based on 2006/7 baseline for commercial waste
'Prevention is better than cure: the role of waste prevention in moving to a more resource efficiency economy' HM Government 2013 (but only covers England). (HMG 2013)	No national waste prevention target
	The Greening Government Commitment aims, by 2015, to deliver a reduction in the amount of waste generated from the Government Estate by 25 % from a 2009/10 baseline and ensure redundant ICT equipment is reused or responsibly recycled
'The Road to Zero Waste' (The waste prevention programme for Northern Ireland 2014) (DOENI 2014)	No targets proposed

– The economic advantages of waste prevention and circular economy are emphasised far more strongly by the Scottish and Welsh than by England and Northern Ireland.
– Resource security forms a central part of the rationale for waste prevention in all the plans.
– There is a strong emphasis on business taking up the challenge itself, rather than waiting for further policy initiatives, particularly from England and Northern Ireland.
– There is little money available from government to facilitate action.
– The Scottish and Welsh have set waste reduction targets (additionally to recycling targets), although the actions suggested are not explicitly linked to progress towards the targets.
– Targets are for waste reduction, rather than for the value to be recouped through circular economy approaches.

Other than the targets, when scrutinised in detail, the proposed actions of the four countries of the UK are not greatly divergent, but there are significant differences in language and tone. The Scottish Plan is most firmly aligned with the language of the circular economy, has strong political backing for the idea and has some of the most practical actions. These include the establishment of the Scottish Institute of Remanufacture at Strathclyde University, and the Resource Efficient Scotland service, which integrates advice to businesses on water, energy and materials, the first of its kind in the UK. Scottish Government is also considering Resource Utilisation Assessments which could develop into 'mass balance' exercises (helping businesses to understand inputs, outputs and consequent resource efficiency). Scotland has the advantage of building on legislation put in place in 2012, to require sorting of recyclable wastes by businesses. This was designed to create more certainty around supply of recyclates which, in turn, is hoped will generate demand for secondary materials, and there is a target 70 % recycling and maximum 5 % to landfill by 2025 for all Scotland's waste.

A test of the economic policy relevance is that circular economy is part of Scottish Enterprise's business strategy. The Scottish Government also sponsored a 2015 report by Green Alliance in partnership with the Scottish Council for Development and Industry detailing circular economy opportunities for three key sectors – Oil and Gas, Food and Drink, and Finance (Benton 2015), demonstrating the extent to which circular economy is seen as potential contributor to growth.

In Wales, the circular economy is tied up with a political commitment to future generations in a way that is probably unique in the world. The Welsh goal of 'One Planet Living' drives intervention and the key metric is ecological footprint reduction (NAWRS 2011). Separate recovery of materials from source is considered the best way to reduce the footprint (particularly the carbon component): this differs from the preference of many local authorities in England to collect recyclables together, or 'co-mingled'. To foster re-use, electronics are being kept whole wherever possible through source segregation.

As well as the waste reduction targets, Welsh measures also promote source segregation of materials for recycling, and there is a target of 70 % recycling of municipal waste by 2025. The existing Sustainable Development Duty is being given greater force through the forthcoming Environment Bill, with a requirement to make it an 'organising principle' which is given force through reporting requirements. All these measures are felt to contribute towards meeting the waste reduction targets.

The most significant action highlighted in the English Waste Prevention Plan is the Electrical and Electronic Equipment Sustainability Action Plan (ESAP). This work is being led by the charity the Waste and Resources Action Programme (WRAP) on behalf of UK governments and will help organisations that design, manufacture, sell, repair, re-use and recycle electrical and electronic products to work collaboratively across the product life-cycle. ESAP will stimulate action across five themes: extending product durability through design and customer information; minimising product returns; understanding and influencing consumer behaviour on product durability and reparability; implementing profitable, resilient and resource efficient business models and gaining greater value from re-use and recycling. By November 2014, over 50 organisations from across the UK electrical sector had signed up, including Argos, Beko, Dell, Ifixit, LG, Microsoft, Oxfam and Panasonic (WRAP 2015). This work forms part of WRAP's wider plan for assisting the transition of the UK to a more circular economy (WRAP 2014).

Overall, however, it is fair to say that policy makers in all the UK countries are only just getting to grips with what a circular economy might mean, and the opportunities it presents.

It is not easy to make national policy in this field. As well as trying to influence individual firms and householders to view their waste differently, and struggling to measure flows of materials, implementing a more circular economy might imply seeking to condition the entire economic system in three main ways:

– More centralised encouragement, even direction, of what infrastructure is needed (including infrastructure for collection, sorting and repair/reprocessing of products and materials). Without such an overview and framework to provide some certainties about the likely scale of the circular economy, it may continue to be hard to mobilise private sector investment in the large number of new facilities needed by a genuinely circular economy. None of the country plans suggest going down this route.
– Making secondary materials a more cost effective option than primary. Sometimes market conditions deliver this on their own, but where they don't (often due to the costs of reprocessing or other constraints on supply of secondary materials) there is a case for policy intervention. The 2015 collapse of recovered plastics companies after the fall in the price of oil illustrates this problem. The answer could be through taxes on primary resources, or on non-recyclable products. Alternatively, the incentive could be tax breaks (lowered VAT being the most often mentioned) on products and materials that are more durable or recyclable.

All these options have been debated by policy makers, but tend to end up in the 'too difficult' box. Plastic carrier bag charges are as far as any of the UK nations have got so far.

– Mandating certain aspects of product design, so that all products are designed for longevity and recovery from the outset. This should change the economics of remanufactured versus new goods, by lowering the costs of remanufacturing. This should also create a new generation of consumer products and, with it, it might be hoped, a change in consumer preferences away from pursuit of newness, towards embracing the durable, the 'pre-loved' and the reclaimed. It would create both supply push and demand pull for reused products and recycled materials. Here, the national plans for Scotland and England signal a cognisance of this possibility in relation to the EU's Ecodesign Directive but in fairly tentative language.

Another complexity for policy makers is that the circular economy debate includes recognition that products and materials have impacts at their point of sourcing, which may create risks for companies and which should be taken into account in their supply chain policies. This is challenging but feasible for companies. For a national government, however, this wider dimension is much harder to handle, as policy instruments that reach beyond national boundaries need careful justification, particularly if they seek to go beyond the internal trading arrangements of the EU. Even EU supranational law making is increasingly under pressure to justify itself to UK, and particularly English, political discourse.

6   Conclusion

We have seen that the idea of the circular economy has developed slowly over a period of at least three decades. It builds on the concepts of waste prevention and resource efficiency by showing where the greatest benefits are to be realised, and by emphasising the need to consider the sustainability of the sources of raw materials, as well as their fate. It adds to the development of EU waste and resources policy over the same period by emphasising the economic as well as environmental benefits of durability and recyclability, the need to more firmly link different actors in the economy to achieve comprehensive recovery, and the need to condition product design as the underpinning of the whole system. Progressive businesses are ahead of government in their use of the term, and they have tools available to drive it through their supply chains, but most smaller businesses have yet to fully understand the implications. Leading businesses also draw attention to the framework conditions they need from governments, which include help with collaborative approaches and better data, but also economic intervention in the form of more strategic direction to provide investment certainties, with industries (UK Govt 2013) being the obvious vehicle for this.

Beyond this, progressive companies often privately signal the need for fiscal intervention, and ways to make secondary resources more cost effective than primary resources, but rarely make these calls in public. The circular economy debate needs to move from being one that augments and energises the recycling debate of the past 30 years, to being about how national economies should be constructed to meet the challenges of the next 30 years. This means involving finance in all its forms: the finance industry (which will provide the capital for circular economy infrastructure); Chief Financial Officers of companies, who will be persuaded of the monetary benefits and new opportunities presented by circular business models; and, most crucially, those in finance ministries who set the framework for businesses through fiscal policy.

References

Benton, D. (2015). *Circular economy Scotland*. London: Green Alliance.

Benton, D., & Hazell, J. (2013). *Resource resilient UK* (Circular economy task force). London: Green Alliance.

Defra. (2012). *Resource Security Action plan*. London: HM Government.

DOENI (2014). *The road to zero waste – The waste prevention programme for Northern Ireland*. Belfast, Ireland.

EC. (2008). *Directive 2008/98/EC of the European Parliament and of the Council of 19th November 2008 on waste and repealing certain directives. E. Commission*. Brussels: European Commission.

EC. (2011a). *Communication from the Commission to the European Parliament, the Council, the European Economic and Social Committee and the Committee of the Regions. A resource-efficient Europe – Flagship initiative under the Europe 2020 strategy*. Brussels: European Commission.

EC. (2011b). *The roadmap to a resource efficient Europe*. Brussels: European Commission.

EC. (2014a, October 8, 2013). *Defining 'critical' raw materials*. Retrieved February 24, 2015, from http://ec.europa.eu/enterprise/policies/raw-materials/critical/index_en.htm

EC. (2014b). *Integrated product policy*. Retrieved February 24, 2015, from http://ec.europa.eu/environment/ipp/home.htm

EC. (2014c). *Communication from the Commission to the European Parliament, the Council, The European Economic and Social Committee and the Committee of the Regions: Towards a circular economy: A zero waste programme for Europe*. Brussels: European Commission.

EMF. (2013). *Towards the circular economy*. Ellen MacArthur Foundation Report No.1.

EREP. (2013). *Action for a resource efficient Europe*. Brussels: European Commission. Retrieved February 24, 2015, from http://ec.europa.eu/environment/resource_efficiency/re_platform/index_en.htm

Euroactive. (2015). *Ministers want circular economy saved but back commission's better regulation push*. Retrieved February 24, 2015, from http://www.euractiv.com/sections/sustainable-dev/ministers-want-circular-economy-saved-backcommissions- better-regulation

Green Alliance. (2015). *Circular economy task force*. Retrieved February 25, 2015, from http://www.green-alliance.org.uk/CETF.php

Hill, J. (2014, October). The circular economy: From waste to resource stewardship, part I. *Proceedings of the ICE – Waste and Resource Management, 168*(1), 3–13.

HMG. (2013). *Prevention is better than cure: The role of waste prevention in moving to a more resource efficiency economy*. London: Her Majesty's Government.

NAWRS. (2011). *National assembly for Wales Research Service, key issues for the fourth assembly* (p. 86). Cardiff.

RSA. (2013). *The great recovery – Investigating the role of design in the circular economy* (Vol. 1). London: Royal Society of Arts.

Scottish Government. (2013). *Safeguarding Scotland's Resources: Blueprint for a more resource efficient and circular economy*. Edinburgh: Scottish Government.

UK Govt. (2013). *Collection. Industrial Strategy: Government and industry in partnership*. Retrieved March 5, 2015, from https://www.gov.uk/government/collections/industrial-strategy-government-and-industry-in-partnership

Welsh Government. (2013). *Towards Zero Waste: One wales, one planet*. Cardiff: Welsh Government.

WRAP. (2014). *WRAP's vision for the UK circular economy to 2020*. Retrieved February 24, 2015, from http://www.wrap.org.uk/content/wraps-vision-uk-circular-economy-2020

WRAP. (2015). *Electronic and Electricals Sustainability Action Plan (ESAP)*. Retrieved February 24, 2015, from http://www.wrap.org.uk/content/esap-generating-value-business-through-sustainability

Chapter 14
Industrial Ecology and Portugal's National Waste Plans

Paulo Ferrão, António Lorena, and Paulo Ribeiro

Abstract This chapter explores how industrial ecology concepts and tools were used to support the design of waste management systems and policies in Portugal. The focus is on a set of case studies that illustrate the results of a successful cooperation between government, private institutions, and academia to transform waste into a useful resource for socio-economic development.

The "Relvão Eco Industrial Park", an industrial symbiosis case study, is analyzed, showing that it was possible to build from scratch a large number of synergies between companies, creating over 300 local jobs and attracting an investment of over €19 million to a region which was industrially undeveloped.

The partnership between the Portuguese Environment Agency and IST to develop the National Waste Management Plan enabled design of a policy instrument that explicitly identified the need for a life-cycle approach to underpin waste management policies and that supported a circular economy to contribute to increasing resource efficiency.

The recent national strategy for urban waste management (PERSU 2020), developed in 2014, is the latest case study of cooperation between academia and the government to develop a public policy whose results show that the proposed changes will lead to a major qualitative leap in the environmental and economic performance of the sector by 2020. It is estimated that the net GHG emissions will be reduced by 47 %, as demonstrated by an LCA study promoted to support policy development. These benefits are due not only to reduction of the quantity sent to landfill, especially the biodegradable fractions, but also to the expected increase in MSW recycling resulting from the increase of selective collection and more efficient treatment and recovery of mixed wastes. An hybrid input-output model, i.e. with both monetary and mass flows, that explicitly considers seven types of waste, showed that the new policy will allow for increasing the economic added value of the urban waste management system by 26 % to €451million and that the number of direct and indirect jobs will increase to 13,000 and 5,500, respectively.

P. Ferrão (✉)
IST – Instituto Superior Técnico, University of Lisbon, Lisbon, Portugal
e-mail: ferrao@tecnico.ulisboa.pt

A. Lorena • P. Ribeiro
3 Drivers, Engineering, Innovation and Environment, Lisbon, Portugal

© The Author(s) 2016   275
R. Clift, A. Druckman (eds.), *Taking Stock of Industrial Ecology*,
DOI 10.1007/978-3-319-20571-7_14

The evidence reported in this chapter shows that the cooperation between government, academia and private sector in Portugal, based on industrial ecology principles and tools, has been able to significantly improve waste management performance in Portugal since the late 1990s, making the sector an important actor of the green economy, by combining better environmental performance with economic growth and job creation, critical dimensions for enabling sustainable development.

Keywords Waste • Industrial symbiosis • LCA • Hybrid input-output analysis • Waste management policy

1   Introduction

Earlier works of industrial ecology placed strong emphasis in resource efficiency and waste minimization. The seminal paper by Frosch and Gallopoulos (1989) summarized the industrial ecosystem as one in which "the consumption of energy and materials is optimized, waste generation is minimized and the effluents of one process serve as the raw material for another process. Manufacturing processes in an industrial ecosystem simply transform circulating stocks of materials from one shape to another; the circulating stock decreases when some material or energy is unavoidably lost, and it increases to meet the needs of a growing population." The authors argue that the traditional model of industrial activity, in which individual manufacturing processes take in raw materials and generate products to be sold plus waste to be disposed of, should be transformed into an industrial ecosystem model, analogous to a biological ecosystem.

Industrial ecology is rooted in this ecological metaphor that provides a conceptual model for industrial development, based on improving the material interactions in and between firms, within a certain geographical distribution or industrial cluster, emulating ecosystems. Industrial systems would adopt principles such as cooperation, symbiosis (Ehrenfeld 2007), loop-closing (Graedel and Allenby 1995), diversity and systemic approaches (Lifset and Graedel 2002). Industrial ecology has long evolved from a metaphorical approach to a broad transdisciplinary field (Ehrenfeld and Gertler 1997, Ehrenfeld 2004) that encompasses both conceptual approaches (e.g. industrial symbiosis, loop closing, design for recycling) and tools (e.g. life-cycle assessment, materials and substance flow accounting, economic and physical input-output based techniques), making it a rich discipline and ripe for inter-disciplinary knowledge transfer.

This chapter explores how industrial ecology related concepts and tools were used to support the design of waste management systems and policies in Portugal, through a set of challenges in which the authors have been involved at Instituto Superior Técnico (IST) of the University of Lisbon during their academic careers and PhD studies and subsequently, for Lorena and Ribeiro, as consultants.

Industrial ecology's systemic approach to waste management extends the analysis beyond the waste and its collection and treatment processes and includes the production and use phase of the products that are transformed into waste at the end of their life. Outcomes of this approach include the Industrial Symbiosis concept, in which waste flows from one industry are used as resource inputs for another, and loop-closing, which emphasizes the reintroduction of used material and energy resources at the production phase in order to reduce extraction and extend the lifetime of stocks.

This chapter is organized in four sections, including this introduction. The next section provides a historic perspective of the Portuguese waste management policies and highlights key developments in terms of Industrial Ecology principles. Section 3 is focused on the most recent municipal solid waste management plan recently developed in Portugal, and Sect. 4 quantifies its impact assessment making use of different industrial ecology tools.

2   Portuguese Waste Management Policy 1990–2014: The Contribution of Industrial Ecology

Modern Portuguese waste management policies date from the 1990s, when urban waste was first considered to be a priority for the national environmental policies. In this context, in 1996, the first national municipal solid waste (MSW) management plan (PERSU) was approved for the period 1997–2007. Its main goals were focused on banning waste dumping in inadequate sites, which represented 73 % of the destination of the MSW generated in Portugal, and structuring a set of regional entities that would be responsible for local waste management and would have access to funding for new infrastructures.

In 2006, the revision of the PERSU was launched; it resulted in a new strategy, PERSU II (2007–2016), and provided additional funding for installing a set of new infrastructures that the regional management systems considered necessary to fulfill the national waste management targets, as derived from EU directives, in terms of separating the different waste streams during waste collection, avoiding deposition in landfill and promoting the valorization of the waste collected, particularly wastes from packaging and those which are biodegradable.

In parallel, industrial ecology principles emanating from academia, with major involvement of the authors at Instituto Superior Técnico (IST) – University of Lisbon, were fundamental to establish common development strategies shared by governmental, private and regional institutions, and to create mechanisms to foster the closing of material cycles through integrated management systems for end of life (EOL) products and industrial symbiosis in eco-parks.

The industrial ecology academic contributions to reshape the Portuguese waste management activities were particularly relevant in the design of the systems that managed the Extended Producer Responsibility (EPR) of different products such as automobiles, tires, electric and electronic products, lubricants and packaging. They were focused on improving the environmental performance of products and services through their life cycles, and, in particular, their end-of-life (EOL) disposal and subsequent processing. This was particularly successful in Portugal due to the excellent articulation between governmental authorities, academia and the private sector. More recently, IST has contributed to assessing the benefits of EPR schemes with the use of IE tools, as will be described in the third section.

Another significant contribution of the IST team consisted in supporting the development of an eco-industrial park to promote industrial symbiosis. In January 2004, the Portuguese government issued the Law Decree 3/2004, with the objective of calling for a new approach to hazardous waste management. To prevent fragmented treatments and minimize movements and transfers, the Government called for the creation of an integrated recovery, treatment and elimination center for hazardous wastes (CIRVER). As a result, two companies were to be established in the municipality of Chamusca. Soon after, several companies approached the municipality with projects and requests to locate their business in the municipality.

IST was invited by the mayor to be a strategic advisor to establish and disseminate information about IE and IS principles to local agents and to analyze how IE and IS could be better integrated into the municipality's strategy. These efforts resulted in the "Relvão Eco Industrial Park" (REIP) project, in 2006, under the leadership of the municipal government. A middle-out approach was adopted (Costa and Ferrão 2010), that supported the successive governmental, university and business interventions in the municipality of Chamusca. National government introduced the objectives and policy instruments, and businesses were free to provide solutions to those challenges. Local government was able to influence the social context, at the local level, by promoting a variety of events in which agents interacted (e.g. community, industry, university). By establishing an eco-industrial park to anchor important waste management infrastructures, industries were attracted to the region and synergies began to emerge. Local government and the university monitored the process evolution and disseminated information to agents at all levels, in order to foster IS development.

As a result, between 2006 and 2009, a large number of synergies were established between companies installed in the eco-park and beyond it, and local jobs increased from 50 to about 350 in 14 companies, attracting an investment of over €19 million, to a region which was industrially undeveloped. The eco-park was viewed as very successful model for combining environmental and economic development, or the green economy in practice.

All this work has resulted in various academic contributions covering different aspects, set out in a number of scientific papers, including:

1. The contribution of industrial ecology principles to environmental policy and business practices: Ferrão and Heitor (2000); Eherenfeld et al. (2002); Thore and Ferrão (2002); Ferrão and Nhambiu (2006); Dijkema et al. (2006); Ferrão (2007); Costa and Ferrão (2010); Niza et al. (2014).

2. The contribution of industrial ecology principles and tools to the design of specific EOL managing schemes for products: Freire et al. (2000, 2001); Giacommucci et al. (2002); Ferrão et al. (2002, 2006, 2008); Ferrão and Amaral (2006a, b); Amaral et al. (2006a, b); Ferrão and Lorena (2014).

In Portugal, the successful cooperation between academia, the Portuguese government and the private sector was also reflected in different partnerships, notably between IST and the authorities responsible for waste management, to perform a variety of tasks such as improving the information management system in Portugal, technical support on waste management systems and support for policy development.

An example was the partnership between the Portuguese Environment Agency and IST to develop a proposal for the Portuguese National Waste Management Plan (PNGR), made available for public discussion in 2011, which represented the first policy instrument in Portugal that explicitly identified the need for a life-cycle approach to underpin waste management policies and that supported a circular economy to contribute to increasing resource efficiency. The goal of the PNGR was to "promote waste prevention and management as a part of the life-cycle of products, support the circular economy and enable a greater resource efficiency." This vision or goal is clearly aligned with central concepts of industrial ecology – life-cycle approach and its systemic approach.

The 2011 version of the PNGR was revised in 2014 in order to reflect the most recent European strategic and legislative framework and waste generation and management data. However, as a testament to its conceptual relevance and currency, the vision, goals and objectives of the 2011 version were not significantly altered.

In 2014, the municipal solid waste management strategy, a component of PNGR, was also reviewed to enable more ambitious targets. Paulo Ferrão was invited to coordinate the working group responsible for the task and the authors of this chapter were involved in developing the models that facilitated it. As the most recent example of the use of IE principles in establishing advanced waste management policies, the next section will detail the use of IE principles and tools in the design and impact assessment of the new Portuguese National Plan for MSW – PERSU 2020.

3   PERSU 2020

The Portuguese National Plan for Municipal Solid Waste 2014–2020 – PERSU 2020, was commissioned by the Secretary of State for the Environment to be developed by a Working Group that included the participation of the national and local public entities, government and academia. This section summarizes the methodology adopted and main results of the work developed.

In 2010 and 2011, the European strategic framework emphasized resource efficiency as a pillar of environmental policy. While this was not new – for instance, the 6th Environmental Action Program had already pointed in this direction – there was a set of strategic and legislative proposals that marked a clear shift toward the intersection of waste and resource efficiency policies (e.g. Commission Communication "Roadmap for a Resource Efficient Europe"). Waste was to be seen as a resource that should be reintroduced into productive processes, avoiding the extraction and imports of resources.

PERSU 2020 would need to provide a response to the MSW management targets for 2020, particularly the target of reaching 50 % of preparation for reuse and recycling of MSW (as defined by EU Directive 2008/98/EC). The ambition was also for the MSW sector to open up tangible opportunities for economic growth and job creation. Constraints on the growth of the sector were to be identified and avoided so that key players could take advantage of these opportunities and contribute simultaneously to the reduction of environmental impacts of waste management and increase of wealth. The impact of the new plan in these dimensions was to be quantified making use of IE tools.

In this context, the PERSU 2020 approved by the law: "Portaria n.º 187-A/2014, publicada em DR (I Série) n.º 179, de 17 de setembro de 2014", proposes objectives for the MSW sector that stem from the PNGR and are aligned with the European strategic framework: "The PERSU 2020 maintains the objective of protecting the environment and human health through the correct use of processes, technologies and infrastructures, but goes further by promoting waste prevention and the reintroduction of waste in the productive processes, thus reducing the need for resource extraction and ensuring a secure supply to the Portuguese economy, while creating new jobs and value added".

In designing the PERSU 2020 vision, several principles were established, including:

• Protection of the environment and of human health by preventing or reducing the adverse impacts of waste generation and management.
• Full compliance with the national and European legislation.
• PERSU 2020 should not impose technological solutions and allow regional waste management institutions to determine their own approaches to reach the defined targets.
• Performance targets should be established at the regional level.
• Efficient use of infrastructures – regional infrastructures are encouraged to increase process efficiency and to share available treatment capacity.
• MSW treatment should support principles of regional self-sufficiency and proximity.
• Emphasize actions targeting the upstream stages, namely waste prevention and the collection of sorted MSW.

The IE principle of loop closing stands at the base of PERSU 2020, as clearly stated in the vision: "PERSU 2020 (…) goes further by promoting (…) the reintroduction of waste in the productive processes, thus reducing the need for resource extraction." It is a central objective that is materialized in the national targets of preparation for reuse and recycling and recycling of packaging waste. Regarding policy options, circularity is promoted through two major vectors: the first is by limiting new investments in end-of-the-line technologies – new landfills and incineration plants – to cases where these are necessary as operational buffers (e.g. when an adjacent treatment plant is shut down for maintenance) or for waste streams where recovery is not economically or technically viable (e.g. waste from road sweeping). This policy is further supported by a suggested increase in landfill and incineration taxes, historically lower than other Member-states (Watkins et al. 2012). The second vector aims to upgrade the materials recovered from MSW in order to increase the uptake, in quantity and value, by upstream industries. From the several actions proposed to achieve this, the priority was to strengthen collection schemes to increase separation at source, thus reducing contamination and overall unsorted MSW, to promote ecodesign through voluntary agreements with the packaging industry, and to review the regulatory framework concerning secondary streams from the waste sector (Residue derived fuels – RDF, compost, biogas).

The concepts of cooperation and symbiosis in industrial systems are based on the premise that a firm can divert unwanted materials from the waste system (or ecosystem) by having them used as resource in a different firm. The same reasoning can be extended to the secondary wastes of waste treatment. For instance, the bottom ash from MSW incineration can be used as secondary material for cement production and the stabilized organic material from mechanical and biological treatment facilities (MBT) can be used to increase organic content in forest soil. Waste management systems can also be considered as necessary intermediary processes between industries, i.e. enablers of industrial symbiosis at a larger scale, if some constraint prevents the transformation process from occurring at the point of generation of the waste (e.g. limitations by environmental protection legislation).

Other forms of collaboration occur when firms of the same sector share infrastructures. For instance, a MSW management firm can use a neighboring RDF preparation facility to avoid the deposition of residual streams from material recovery facilities in landfill. A positive corollary of this approach is that it allows greater economies of scale.

The principle of efficient use of infrastructures is supported by the concepts of symbiosis and cooperation in the Portuguese MSW sector. By restricting new infrastructures to situations where these are absolutely necessary, PERSU 2020 induces cooperation between regional companies and the establishment of agreements with firms from other sectors. Practical examples include the use of RDF in energy intensive industries or deposition of compost in vineyards, forest and other soils with low organic content. The plan suggests that financial support from available funding to the MSW sector should be focused on processes that add value to MSW streams (e.g. MSW separation at source, RFD preparation facilities).

To realize the vision, principles and objectives, it was necessary to develop a full material flows model of the MSW management system from 2012 to 2020 in order to evaluate the distance-to-target and identify challenges and opportunities in the sector.

All 23 regional institutions responsible for local MSW management – MSW RMS (MSW regional management systems) – were consulted in order to establish a Business-as-Usual (BaU) scenario for MSW generation and management. The material flows expected to occur in 2020 are represented in Fig. 14.1. It is clear that there is a strong reliance on treatment operations for MSW from unsorted collection. This

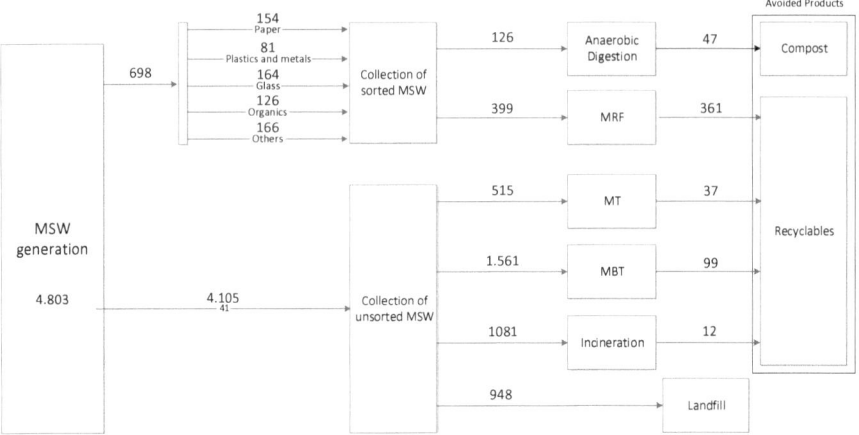

Fig. 14.1 Most important material flows in the Portuguese MSW sector in the BaU 2020 scenario. Values in thousand tons (*MRF* material recycling facility, *MT* mechanical treatment facility)

results in significantly lower recycling rates when compared to other EU Member-States, even if significant progress is made in the diversion of MSW from landfill. These data provided a strong argument for the policy option of emphasizing collection of sorted waste. Despite the increase in MBT and MT capacity compared to 2012, the BaU pathway was not sufficient to reach the 2020 proposed targets, specifically in:

- Preparation for reuse and recycling: to reach a minimum of 50 % preparation for reuse and recycling of MSW
- Recycling of packaging waste: to reach a minimum of 70 % recycling of packaging waste
- Diversion of biodegradable MSW from landfill: to reach a maximum of 35 % of deposition of biodegradable MSW in landfill (measured relative to the amount of biodegradable MSW in 1995)

PERSU 2020 assumed the innovative challenge of defining specific targets for each of the MSW RMS, in accordance with the overall goals established and taking into account the existing differences between regions and infrastructures. Each of the 23 MSW RMS is responsible for the treatment and deposition of waste and, in most cases, they are also responsible for the collection of sorted MSW (waste that is sorted by households and deposited in specific containers). Exceptions occur in the most densely populated areas of Lisboa and Porto, where municipalities are responsible for the collection of sorted waste. Finally, unsorted waste always falls under municipality responsibility. The involvement of private companies has been mostly restricted to service contracting by municipalities or minor participation in the capital of regional companies. There are also major differences in terms of territory, population and socio-economic profile, which translate to different waste generation patterns and management options (collection schemes, infrastructures,

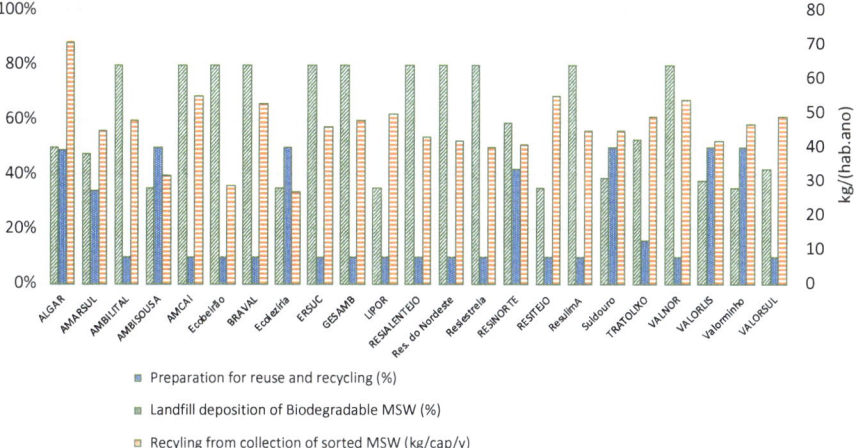

Fig. 14.2 PERSU 2020 targets for each MSW regional management system

technology). Target setting for each regional company was a major innovation in the planning process in Portugal, as in the past the national target did not result in specific targets for each of the MSW RMS. The targets for each system were based not only in their development stage and technology profiles but also on exogenous factors such as population density, in order to ensure proportionality of the effort by each MSW RMS.

The resulting targets for each of the MSW RMS are represented in Fig. 14.2. It shows that there is room to attend to the specificities of each system while promoting a fair incentive for each of them to improve their operation in order for the national targets to be fulfilled with optimal economic and environmental performance.

4 Impact Assessment of the Portuguese National Plan for Municipal Solid Waste 2014–2020

The assessment of the environmental, economic and employment impacts of packaging waste and MSW was commissioned by Sociedade Ponto Verde (SPV), the Portuguese Producer Responsibility Organization (PRO) for packaging waste, and resulted in two studies whose results are summarized here.

4.1 Environmental Impacts

The environmental assessment of the MSW management system in 2012 and 2020 was performed using attributional life cycle assessment (LCA). The functional unit selected as the basis for comparison was 1 ton of average MSW managed in Portugal in 2012. Coproducts were taken into account by the "substitution by system expansion" or "avoided burden method" considering the average primary route market consumption mix. The "zero burden assumption" was also used, which means that it was considered that the waste carries none of the upstream burdens into the waste management system (Clift et al. 2000).

For modeling purposes, the various activities were organized as groups of associated processes, as follows: (1) waste generation, (2) collection of sorted MSW (material and biodegradable waste); (3) collection of unsorted MSW; (4) organic recovery; (5) sorting at MRF; (6) mechanical treatment; (7) mechanical-biological treatment; (8) energy recovery; (9) recycling; (10) landfill. The secondary wastes from waste treatment, as also the materials and energy recovery processes, were also taken into account.

The detailed analysis of the most important foreground processes constitutes the backbone of the life cycle inventory and was based on primary data obtained from the companies that handle the waste (e.g. SPV) and on secondary data from LCA databases and scientific bibliography which were adapted to better reflect the Portuguese reality.

The results obtained show that for 7 of the 11 environmental impact categories considered in the study, the reference MSW system configuration leads to a positive or neutral environmental balance. In these cases, the benefits due to the recovery of materials and energy obtained by waste recovery processes (avoided impacts), with special focus on recycling, are bigger or at least equal to the negative impacts generated by the various waste management activities, like collection, sorting, transport, treatment and recovery. In the remaining four impact categories, the benefits obtained from waste recovery are not sufficient to mitigate the impacts generated by its management activities. This is due mainly to a still significant MSW fraction that is not recovered and is mostly eliminated in landfills (in 2012, 5 4 % of the Portuguese MSW was directly put in landfill). This is the case, for example, in the climate change category, where it was estimated that the net GHG emissions reached 1.1 Mt CO_2eq in 2012, equivalent to 1.6 % of the annual GHG emissions in Portugal.

Comparing the MSW management as a whole with the specific management of packaging materials, it was found that, on a unit basis, the environmental balance of the packaging materials is more favorable than the remaining fractions. This is due in part to the intrinsic characteristics of packaging materials, but also to higher recycling rates achieved and the consequently lower amount sent to landfill, particularly for glass, paper/cardboard and ferrous metals.

Comparing the 2012 performance with the target for the year 2020 with the strategy defined in PERSU 2020, the results show that the proposed changes will lead to a major qualitative leap in the environmental performance of the sector. For example, it is estimated that the net GHG emissions will be reduced by 47 %, which translates into a emission savings of 522 kt CO2 eq and that the benefits obtained from the recovery of mineral resources, both fossil and renewable, will increase by 61 % compared to 2012. These benefits are due not only to reduction of the quantity sent to landfill, especially the biodegradable fractions, but also to the expected increase

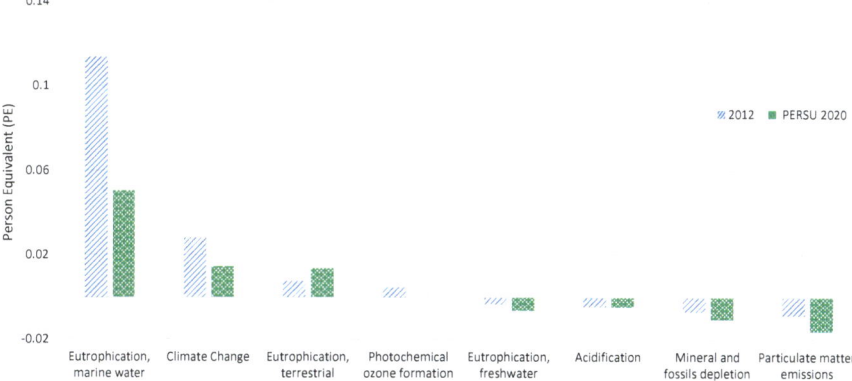

Fig. 14.3 Environmental impacts of MSW management per ton of MSW in 2012 and 2020, according to the objectives of PERSU 2020. Three categories were omitted since in both cases the values were approximately null

in MSW recycling resulting from the increase of selective collection and more efficient treatment and recovery of mixed wastes (Fig. 14.3).

4.2 Economic Impacts

To assess the economic impacts of the PERSU 2020, it was assumed that in 2020 all MSW RMS would meet each of their specific targets. The 2020 scenario (based on these assumptions) was then compared to the reference scenario (based on data from 2012) to determine the net contribution of the PERSU 2020.

Two variables were covered by the assessment of the economic impacts: the gross value added and the number of jobs. The methodology used is based on the input-output (IO) model presented in Ferrão et al. (2014). The economic assessment considers that in the course of its economic activity, a firm uses resources, such as labor or imports, and that these are direct impacts of the firm. A firm also purchases goods and services from other companies, which themselves are going to use resources (these are the first order indirect impacts of the first firm). These firms in turn make purchases to other firms, leading to second order indirect effects of the first firm, and so on (Ferrão et al. 2014). Finally, substitution impacts arise from, as the name implies, the materials and energy recovered in the sector studied which replace output from other sectors (e.g. electricity production from the MSW management sector reduces the output of thermal power plants).

The IO model developed to study the economic impact of the MSW management system was a hybrid, i.e. with both monetary and mass flows, that explicitly considers seven types of waste: unsorted MSW, paper, plastic and metal, glass, refuse (secondary wastes from MT, MBT and MRF) and ashes. It also considered eight

MSW sectors: collection of unsorted waste, collection of sorted waste, material recovery facilities, incineration, mechanical biological treatment, incineration, landfill, packaging PRO, besides the rest of the economy (ROE). The sector ROE was modeled using the 2008 basic prices symmetric product-by-product Input–Output tables of the Portuguese Department of Prospective Planning and International Relations (Dias and Domingos 2011), with an aggregation of 85 sectors using the official Portuguese NACE Rev.3 two digit classification. Data for the MSW management itself was obtained mostly from publicly available accounting reports from regional companies, annual reports from the Portuguese Environment Agency (APA references), and, for the 2020 scenario the PERSU 2020 was used.

PERSU 2020 requires the Portuguese MSW management system to increase the collection rates of sorted waste and to avoid direct deposition in landfill by increasing MT and MBT capacity. Intuitively, the shift from collection of unsorted waste and landfill deposition to these solutions will lead to higher value added; first, because more materials with positive economic value are being returned to the economy (e.g. scrap metal, plastic film); second, because the shift is toward higher labor and capital intensive processes. The obvious increase in the total cost of the MSW management system will be partially supported by the producers' responsibility organizations (PRO) for packaging waste, which are mandated to compensate the higher collection cost of sorted waste. The remaining part should be transferred to the population.

The results obtained for 2012 show that the GVA of the MSW management activities is €357 million, 55 % of which is related to the collection of unsorted waste. The activities related to material recycling, which include collection of sorted MSW and sorting at material recovery facilities, represent 22 % of the total direct contribution in terms of GVA. Indirect impacts, i.e. the economic activity on other sectors due to the operational expenses of MSW management system, were estimated to be €114 million, mostly concentrated in construction activities (contracts for new infrastructure), installation and repair of machinery and specialized services (consultancy, architecture, engineering, etc.). In the reference year, the number of jobs in the MSW management system was 11.700, of which nearly 84 % was attributed to collection activities, especially of unsorted waste. In the rest of the economy, the number of jobs due to the MSW management system amounts to 3.400.

If we look at the GVA and number of jobs per ton, there is a clear indication that the collection of sorted waste and sorting at MRF has higher economic impact than end-of-the-line solutions such as incineration or landfill.

The 2020 results are thus a consequence of the higher unitary impact of these activities. In relation to 2012, the GVA in MSW management system is expected to increase by 26 % to €451 million. The increase in the rest of the economy due to these activities is significantly higher at 55 %, reflecting the higher operational costs, i.e. inter-sectorial transactions, of the growing activities (collection of sorted waste, sorting at MRF, MBT and MT). The number of direct and indirect jobs will also increase to 13,000 and 5,500, respectively (Fig. 14.4).

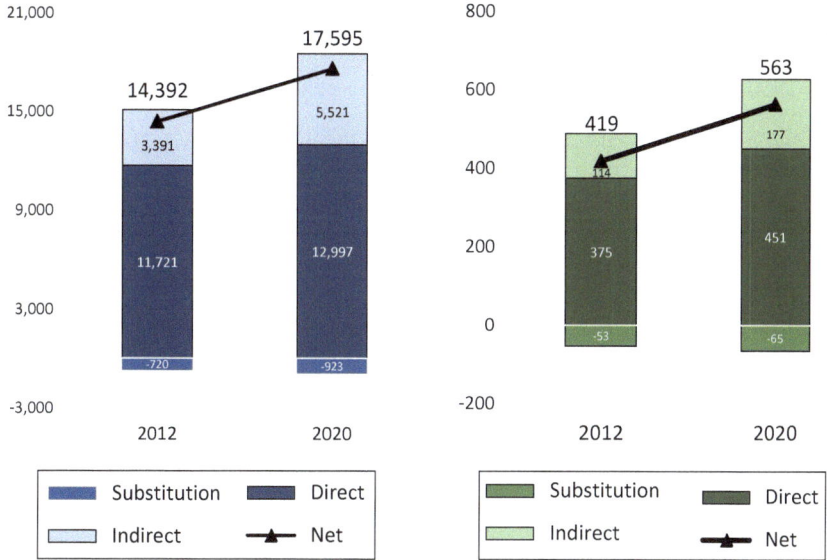

Fig. 14.4 Estimated economic impacts and job creation of the MSW sector in 2012 and 2020 (PERSU 2020 scenario)

The evidence reported in this chapter shows that the cooperation between government, academia and private sector in Portugal, based on industrial ecology principles and tools, has been able to significantly improve waste management performance in Portugal, since the late 1990s, making it an important actor of the green economy, by combining better environmental performance with economic growth and job creation, critical dimensions for enabling sustainable development.

References

Amaral, J., Ferrão, P., & Rosas, C. (2006). Is recycling technology innovation a major driver for technology shift in the automobile industry under an EU context? *International Journal of Technology Policy and Management, 6*, 385–398.

Clift, R., Doig, A., & Finnveden, G. (2000). The application of life cycle assessment to integrated solid waste management: Part 1 – Methodology. *Trans IChemE (Process Safety and Environmental Protection), Special Issue: Sustainable Development, 78*, 279–287.

Costa, I., & Ferrão, P. (2010). A case study of industrial symbiosis development using a middle-out approach. *Journal of Cleaner Production, 18*(10), 984–992.

Dias, A. M., & Domingos, E. (2011). *Avaliação do impacto macroeconómico do Quadro de Referência Estratégica Nacional 2007–2013 (QREN)*. Lisboa: Departamento de Prospectiva e Planeamento de Relações Internacionais.

Dijkema, G. P. J., Ferrão, P., Herder, P. M., & Heitor, M. V. (2006). Trends and opportunities framing innovation for sustainability in the learning society. *Technological Forecasting and Social Change, 73*, 215–227.

Ehrenfeld, J. (2004). Industrial ecology: A new field or only a metaphor? *Journal of Cleaner Production, 12*(8), 825–831.

Ehrenfeld, J. (2007). Would industrial ecology exist without sustainability in the background? *Journal of Industrial Ecology, 11*(1), 73–84.

Ehrenfeld, J., & Gertler, N. (1997). Industrial ecology in practice: The evolution of interdependence at Kalundborg. *Journal of Industrial Ecology, 1*(1), 67–79.

Ehrenfeld, J., Ferrão, P., & Reis, I. (2002). Tools to support innovation of sustainable product systems. In D. Gibson et al. (Eds.), *Knowledge for the inclusive development* (pp. 417–434). Westport: Quorum Books.

Ferrão, P. (2007). Industrial ecology: A step towards sustainable development. In M. Seabra Pereira (Ed.), *A portrait of state-of-the-art research at the Technical University of Lisbon* (pp. 357–383). Dordrecht: Springer.

Ferrão, P., & Amaral, J. (2006a). Design for recycling in the auto industry: New approaches and new tools. *Journal of Engineering Design, 17*(5), 447–462.

Ferrão, P., & Amaral, J. (2006b). Assessing the economics of auto recycling activities in relation to European Union Directive on End of Life Vehicles. *Technological Forecasting and Social Change, 73*, 277–289.

Ferrão, P., & Heitor, M. V. (2000). Integrating environmental policy and business strategies: The need for innovative management in industry. In D. Gibson et al. (Eds.), *Science technology and innovation policy: Opportunities and challenges for the knowledge economy* (pp. 503–518). Westport: Quorum Books.

Ferrão, P., & Lorena, A. (2014). PERSU 2020 – Plano Estratégico para os Resíduos Urbanos, um contributo para a Produtividade dos Recursos. *Indústria e Ambiente, 88*, 20–22.

Ferrão, P., & Nhambiu, J. (2006). The use of EIO-LCA in assessing national environmental polices under the Kyoto protocol: The Portuguese economy. *International Journal of Technology Policy and Management, 6*, 361–371.

Ferrão, P., Reis, I., & Amaral, J. (2002). The industrial ecology of the automobile: A Portuguese perspective. *International Journal of Ecology and Environmental Sciences, 28*, 27–34.

Ferrão, P., Nazareth, P., & Amaral, J. (2006). Strategies for meeting EU end-of-life vehicles re-use/recovery targets. *Journal of Industrial Ecology, 10*(4), 77–93.

Ferrão, P., Ribeiro, P., & Silva, P. (2008). A management system for end-of-life tyres: The Portuguese case study. *Waste Management, 28*(3), 604–614.

Ferrão, P., Ribeiro, P., Rodrigues, J., Marques, A., Preto, M., Amaral, M., Domingos, T., Lopes, A., & Costa, I. (2014). Environmental, economic and social costs and benefits of a packaging waste management system: A Portuguese case study. *Resources, Conservation and Recycling, 85*, 67–78.

Freire, F., Ferrão, P., Reis, C., & Thore, S. (2000). *Life cycle activity analysis applied to the Portuguese used tire market*, SAE Technical Paper 2000-01-1507.

Freire, F., Thore, S., & Ferrão, P. (2001). Life cycle activity analysis: Logistics and environmental policies for bottled water in Portugal. *OR Spektrum, 23*(1), 159–182.

Frosch, R. A., & Gallopoulos, N. E. (1989). Strategies for manufacturing. *Scientific American, 261*(3), 144–152.

Giacommucci, M., Graziolo, P., Ferrão, & Caldeira Pires, A. (2002). Environmental assessment in the electromechanical industry. In D. Gibson et al. (Eds.), *Knowledge for the inclusive development* (pp. 465–476). Westport: Quorum Books.

Graedel, T. E., & Allenby, B. R. (1995). *Industrial ecology* (1st ed.). Upper Saddle River: Prentice-Hall.

Lifset, R., & Graedel, T. E. (2002). Industrial ecology: Goals and definitions. In R. Ayres & L. Ayres (Eds.), *A handbook of industrial ecology* (pp. 3–15). Cheltenham/Northampton: Edward Elgar.

Niza, S., Santos, E., Costa, I., Ribeiro, P., & Ferrão, P. (2014). Extended producer responsibility policy in Portugal: A strategy towards improving waste management performance. *Journal of Cleaner Production, 64*, 277–287.

Thore, S., & Ferrão, P. (2002). The environmental impact of new products. In S. Thore (Ed.), *Technology commercialization: DEA and related analytical methods for evaluating the use and implementation of technical innovation* (pp. 277–290). Norwell: Kluwer Academic Publishers.

Watkins, E., Hogg, D., Mitsios, A., Mudgal, S., Neubauer, A., Reisinger, H., Troelzsch, J., & Van Acoleyen, M. (2012). *Use of economic instruments and waste management performances.* Final report prepared for the European Commission – DG Environment.

Chapter 15
The Role of Science in Shaping Sustainable Business: Unilever Case Study

Sarah Sim, Henry King, and Edward Price

Abstract Unilever is a leading example of a multinational company in the Fast-Moving Consumer Goods (FMCG) sector. Unilever has long been an advocate of sustainable business, using scientific assessment as the basis for its strategy and initiatives. Given its business, Life Cycle Assessment (LCA) is established within the company and there is a current focus on improving the methodology and scope of LCA. Recent developments include new approaches to fill data gaps for agricultural ingredients and new impact assessment methods for assessing land use change. We have also adapted LCA approaches to inform corporate strategy and to engage a broad range of stakeholders both within the company and outside. The most recent and significant example of this has been the use of product footprinting as an integral element of Unilever's Sustainable Living Plan (USLP); currently over 2000 products are footprinted annually across 14 countries.

LCA approaches will continue to play an important role in Unilever's strategy. However, there is an urgent need to develop more predictive, regional/global level approaches that take into account the limited availability of many earth resources, the non-linearity of certain impacts and the absolute limits of sustainability. Several conceptual systems-level frameworks and theories already exist, but the Planetary Boundary (PB) approach has been selected as the most promising for developments in data, modelling and contextualization of environmental assessment. We have identified the need for developments in informatics to exploit new data gathering approaches as well as new modelling initiatives utilizing Geographical Information Systems (GIS) mapping and 'big data' approaches. In particular, we see real value in developing a distinct and novel, 'PB-enabled' normative LCA approach to support product/service/sectorial decision-making.

Keywords Unilever sustainable living plan • Fast-moving consumer goods sector • FMCG • Life cycle assessment • Environmental footprinting • Earth systems science • Resilience science • Planetary boundaries • Ecosystem services

S. Sim (✉) • H. King • E. Price
Safety and Environmental Assurance Centre (SEAC), Unilever,
Colworth Park, Sharnbrook, Bedford MK44 1LQ, UK
e-mail: sarah.sim@unilever.com

1   Introduction

Unilever is a Fast Moving Consumer Goods (FMCG) company with over 400 brands, operations in nearly 100 countries and sales in nearly every country in the world. Our products (foods, beverages, ice cream, home and personal care) are used 2 billion times a day in over half the households on the planet. Our strategy for sustainable growth sets out a vision for leveraging this global reach to improve health and well-being, reduce environmental impacts and enhance livelihoods. Unilever has long been a pioneer of sustainability, and throughout this journey, science has been the foundation for the company's sustainability strategy and initiatives. As such, the focus of this chapter is not the business case for sustainability in Unilever (which can be found in Bell (2013a, b); Lingard (2012)), but an illustration of how science has helped inform and shape the company's thinking on sustainability so far and also how we will continue to use science to help us think about future challenges.

To date, Unilever has relied on scientific methods based around Life Cycle Assessment (LCA) to guide and inform our sustainability decision-making and such approaches will continue to have an important role to play. However, LCA does not address all relevant environmental impacts; nor does it deal with each impact category in an equally robust way. Therefore, there is an urgent need to develop more predictive approaches that take into account the limited availability of many resources, the non-linearity of certain impacts and the absolute nature of sustainability as articulated in the planetary boundary concept.

2   The Journey So Far

Life cycle assessment (LCA) is the preferred tool for many organizations to help them understand the impacts and performance of their products and services in a rigorous and scientific manner (Baitz et al. 2013). The LCA methodology was codified in the late 1980s/early 1990s through the work of SETAC and ISO. However, LCA is not a static tool and it is continually being evolved and improved to reflect new science and to expand the robustness and scope of the environmental impact assessment.

The flexibility of the LCA concept is one of its strengths; in Unilever, we have exploited this feature and in doing so helped inform and shape our environmental strategy over the past 20 years (Fig. 15.1). In the mid-1990s, Unilever developed the Overall Business Impact Assessment (OBIA) approach (Clift and Wright 2000) as a means to scientifically identify our priority environmental impacts and to inform the selection of key sustainability programs. Unlike traditional LCA which focuses on single systems, usually products, the OBIA approach can be used to assess the effects of all the individual products produced by a business for a given period of time, typically 1 year. The potential environmental impacts of the business are

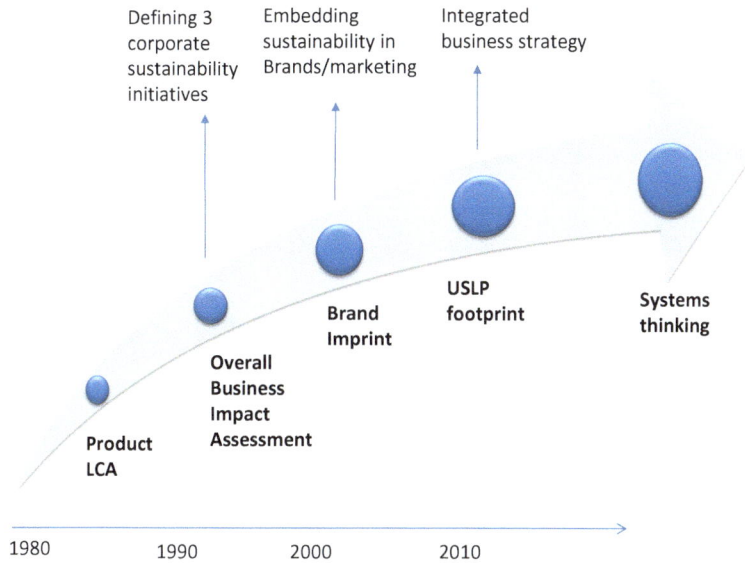

Fig. 15.1 Evolving development and application of environmental sustainability science within Unilever

presented as a profile or imprint and contributions are scaled (normalized) against annual global totals for each of the indicators and the economic size of the business (Clift and Wright 2000). By comparing the potential environmental impacts of a business to the economic size of the business, the OBIA approach helped to identify those areas where most benefit to the global environment can be achieved. The insights arising from the application of OBIA led to Unilever focusing on three sustainability themes in the late 1990s and early 2000s, namely sustainable fisheries, including the co-creation of the Marine Stewardship Council together with WWF (Constance and Bonanno 2000), the Sustainable Agriculture Initiative (Unilever 2010) and the creation of institutions such as the Roundtable for Sustainable Palm Oil and the Sustainable Water Initiative.

Unilever continues to innovatively develop and apply LCA and life cycle thinking to help inform strategy and to engage a broader range of stakeholders both within the company and outside. At a product level, recent method developments and applications have included new approaches: to fill data gaps in agricultural data (Milà i Canals et al. 2011; Roches et al. 2010; Nemecek et al. 2012); to better represent the impacts arising from land use change (Flynn et al. 2012; Milà i Canals et al. 2013); to better represent the impacts arising from product disposal (Muñoz et al. 2013); and the application of new methods for assessing impacts related to water (Jefferies et al. 2012 and Van Hoof et al. 2011). In addition, we have developed approaches to conduct brand/portfolio footprints with examples of Knorr and Ben & Jerry's (Garcia Suarez et al. 2008; Milà i Canals et al. 2010). LCA also formed one of the elements of the innovative Brand Imprint process that Unilever

Fig. 15.2 Unilever
sustainable living plan

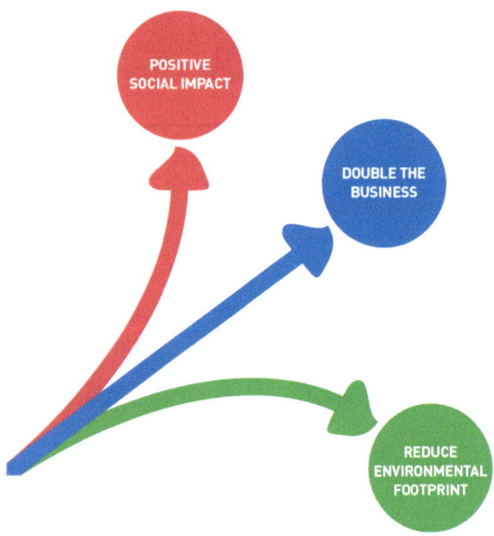

deployed with its Brand teams to help them understand the sustainability impacts and issues of their brands and to identify sustainability marketing opportunities in the late 2000s (Gowland 2009).

The most recent and significant example of the use of the LCA approach to guide sustainability strategy has been the use of product footprinting as an integral element of Unilever's Sustainable Living Plan (USLP) (Rigarlsford et al. 2010; UNEP 2015). The USLP footprint, which is an example of an organizational LCA (UNEP 2015) is based on a streamlined process that involves the definition of representative countries and products. Fourteen countries were selected on the basis of business parameters (e.g. annual sales and consumer habits) and environmental ones (e.g. carbon intensity of the electricity grid, waste management infrastructure and water scarcity). The business and sales in each country are then described by a series of representative products. Currently over 2000 products are footprinted annually and one of the key USLP objectives is to decouple business growth in sales from an increase in the footprint (Fig. 15.2).

3   Looking to the Future

As outlined in the previous section, to date our ability to think about sustainability has mostly focused around classical LCA, but we recognize the existence of global environmental limits; the earth's systems are not behaving as we have designed our economic and social systems. 'What is becoming apparent is that earlier assumptions about the stability, linearity, and reversibility of changes in ecosystems and the Earth systems fell short of what actually happens' (Whiteman et al. 2013). Assessment of corporate decisions in this context cannot be wholly facilitated by

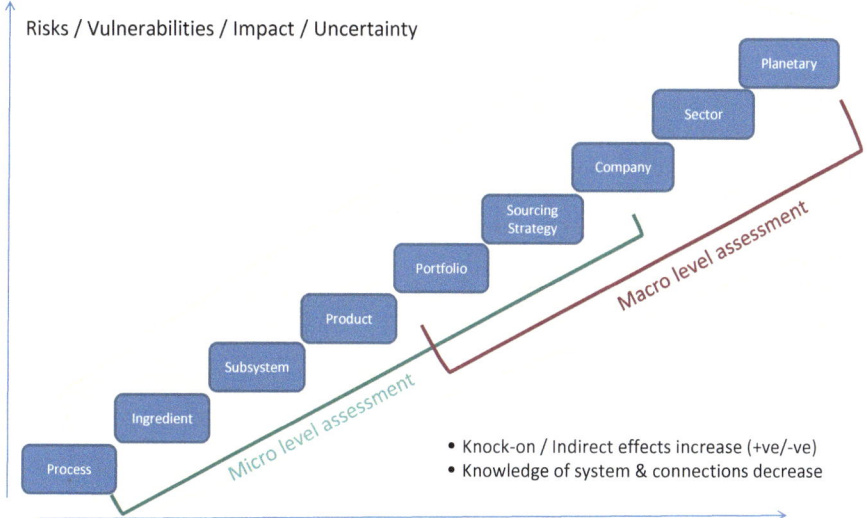

Fig. 15.3 Systems thinking – implications for method development and data

traditional LCA approaches. Though these remain important, their focus on eco-efficiency and relative sustainability will not be sufficient.

The refresh of Unilever's Sustainable Living Plan (USLP) in 2014 (Unilever 2014) articulates a desire to drive transformational change. There are three particular aspects to this: (1) helping to eliminate deforestation; (2) championing sustainable agriculture and smallholder farmers and (3) improving water, sanitation and hygiene (WaSH). This marks the start of a move towards more holistic, systems-level analysis and action to achieve a decoupling of growth and impact. Systems-thinking requires new concepts, approaches, methods and tools. These need to be predictive, point to early warning signals or non-linear relationships between growth and environmental impacts (Biggs et al. 2009; Barnosky et al. 2012; Sheffer et al. 2009, 2012; Wang et al. 2012), be spatially resolved and operational at different scales of decision-making, but particularly portfolio, company and sectorial decisions (Fig. 15.3). Whilst clearly a 'tall order' when considering also the urgency with which we must act, Unilever has already begun efforts in this direction, for example, working with the Natural Capital Project at Stanford University to develop methodologies to help us understand the spatial dependency of Biodiversity and Ecosystem Services (BES) impacts associated with large scale agricultural expansion (as implied by the large and converging demand on agricultural raw materials from the foods, chemicals, energy and textile sectors). The research was set to explore possible nonlinear changes in environmental impacts which may arise from different amounts and spatial configurations of land use change, and which may be affected by corporate strategic sourcing decisions (Chaplin-Kramer et al. 2015).

The need for such systems thinking and macro-level approaches, enabling us to predict and avoid unintended environmental consequences of growth and strategic choices (as discussed in Chap. 2), is amply illustrated by the example of biofuels. 'IPCC(2007) highlighted the large potential for biofuels to meet the growing energy needs as well as contributing to GHG emissions reduction, especially in the transportation sector. Escalating oil prices and the uncertainty about sustained oil supplies further added to the growing interest on biofuels' (Ravindranath et al. 2009). As such, a number of governments developed policies and financial incentives to encourage the production and use of biofuels. Unfortunately, large scale expansion of biofuel crops has led to land use change (LUC), notably the conversion of natural lands such as peatlands, forests and grasslands to the production of biofuel crops. 'Studies have shown that the possible GHG emissions from the induced LUC can substantially influence the climate benefit of biofuels production and use (Leemans et al. 1996; Schlamadinger et al. 2001; Fargione et al. 2008; Searchinger et al. 2008; Gibbs et al. 2008) [...] Fargione et al. (2008) shows that land-use conversion from native land-uses to biofuel crops leads consistently to significant GHG emissions and a negative carbon balance, or carbon-debt, for many years' (Ravindranath et al. 2009).

3.1   Conceptual Basis for Developing Scientific Approaches

Several conceptual systems-level frameworks and theories already exist, such as Ecological Footprint (Wackernagel and Rees 1996), Carrying Capacity (Rees and Wackernagal 1994) and 'Limits to Growth' (Meadows et al. 1972), but by far the most promising as a guiding concept for developments in data, modelling and contextualisation of environmental assessment is the Planetary Boundaries (PB) concept (Rockström et al. 2009a, b; Steffen et al. 2015). This concept stands out for a number of reasons:

1. *The positive framing* of a 'safe operating space' or an 'earth system stability domain' is helpful in a corporate innovation context. Planetary Boundaries firmly establishes the principle of 'absolute sustainability', attempting to set limits on how much impact or change can be tolerated in various PB categories before boundaries are transgressed and the Earth system moves outside of the set of parameters that are deemed 'safe' for humanity (into a 'danger zone') and beyond which global change is likely to have profound negative consequences for us. However, unlike other concepts such as 'Limits to Growth', it does not make assumptions about human ingenuity in terms of technology. As noted by Steffen et al. (2015), 'the PB approach is embedded in this emerging social context [rapid increase in human pressures on the planet], but it does not suggest *how* to manoeuvre within the safe operating space in the quest for global sustainability.' Rather we can view PB as presenting the context for transformative innovation.

2. For Unilever specifically, there is a strong *connection between this framing and the USLP transformational change agenda*, particularly zero net deforestation (principally aligned to the land system change boundary) and improving water, sanitation and hygiene (WaSH) (particularly aligned to the biogeochemical flow and freshwater boundaries) (see Hague 2014).

3. The PB focus and framing also *encourages predictive assessment*, as opposed to descriptive or retrospective assessment, since the idea is to recognise the approach to boundaries so as to find 'risk-reducing interventions' (Steffen et al. 2015) and avoid transgressing the boundaries.

4. Planetary boundaries has received strong interest (with more than 60 peer-reviewed papers on the subject since the seminal Rockström et al. paper in 2009a) and reasonably widespread acceptance, due to its *robust empirical base* which draws on both Earth System and Resilience Science. Clearly there is potential to improve quantification for all nine PB categories and efforts are on-going in this regard (Carpenter and Bennett 2011; de Vries et al. 2013; Gerten et al. 2013; Mace et al. 2014), but 'the approach guarantee[s] a higher degree of consistency and meaningful aggregation (commensurability) than composite indices' (Whiteman et al. 2013) such as the 'Ecological Footprint [...] which fail to fulfil fundamental scientific requirements of validity and reliability (i.e. normalization, weighting, and aggregation), and reveal a high degree of arbitrariness' (Böhringer and Jochem 2007, in Whiteman et al. 2013).

5. Finally, the concept aims to hold focus on these nine categories simultaneously, recognising the inter-dependencies between them: this imposes *limits to trade-offs* in that temporal and spatial trade-offs could be considered within a PB category but not between them (Murphy et al. in prep). Clearly this implies considerable space for the development of multi-disciplinary approaches.

The planetary boundary (PB) concept is increasingly being accepted as a science basis for understanding sustainability in business and government policy contexts (e.g. EU Sustainable Foods Policy development, WBCSD Action 2020), although measurement and analysis of the actions advocated is required to provide assurance that actions will and indeed are leading to the right outcomes; or, otherwise stated, 'to quantitatively measure the role of companies within the decline [or maintenance] of Earth systems' (Whiteman et al. 2013). 'We therefore need more studies that analyse how the micro role of firms and industries interacts with a macro-view of the world informed by system dynamics in order to better address environmental externalities (Whiteman et al. 2013). Indeed this is the challenge: PBs are planetary-scale and "conceptual" and we need to find ways in which they can be made operational at various geographical scales (local, regional and global) (see for example, Nykvist et al. 2013; Cole et al. 2014; Dearing et al. 2014) but particularly decision-making scales (product, portfolio, company and industry sector). This is where scientific advance aligned to the PB concept is required. An early attempt can be seen at the sectorial scale with the 'Mind the Science, Mind the Gap project' (CDP et al. 2014), which proposes guidance on methodology to set science based GHG emissions reduction targets in line with a 2 °C decarbonisation pathway.

3.2 Applying the Planetary Boundaries Approach for Business Decision-Making

There are, however, few such examples since the operationalization of PB is still immature. We argue for further methodological development to build on existing data gathering and modelling initiatives using new GIS-based mapping and 'big data' approaches e.g. Future Earth (www.futureearth.org) and the UK Centre for Agricultural Innovation in the area of agri-informatics and sustainability metrics (White 2015) in order to improve our ability to measure the PBs appropriately, but to do so in a 'solution-focused' frame. That is to say that even when risks, impacts or transgression of the PBs are uncertain, we still need approaches that will help to move the PB approach into day-to-day decisions as a matter of priority. For this reason, towards the end of 2014, Unilever's Safety and Environmental Assurance Centre (SEAC) and the Centre for Environmental Strategy, University of Surrey, co-hosted an expert workshop on this topic. The outcome is a 'roadmap' for how operationalizing the PB concept can be approached in practical decision-making. This is a 'forward looking' agenda for research and implementation in which we will propose a possible framework for progress (Murphy et al. in prep). In summary, the agenda includes the need for: (1) additional science to underpin each boundary, but particularly the biodiversity (now revised to 'biosphere integrity' in Steffen et al. (2015)), chemical pollution (now 'novel entities' in Steffen et al. (2015)) and freshwater boundaries[1]; (2) development or identification of 'rules of thumb' that can be implemented rapidly; (3) normative debate for example around equity, human/societal values, governance, land use contests, lifestyle changes, etc. and; (4) tools for integrating the science in decision-making contexts. In particular, we see real value in developing a distinct and novel, 'PB-enabled' normative LCA approach to support the adoption of the PB concept in product/service/sectorial decision-making; this would be a complement to, and not a replacement of, existing LCA uses.

4   Conclusion

It is evident that Unilever's pioneering position in regards to sustainability is firmly rooted in our innovative application of, and commitment to, environmental sustainability science. This has helped inform and shape the company's thinking on

---

[1] These boundary categories were chosen as the focus for the workshop as they are of particular relevance to Unilever because of the types of products the company designs and markets (acknowledging of course that due to inter-linkages between PB categories, all are relevant in some way or another to companies such as Unilever). In addition, we believe that (better) definition and consensus of the planetary boundary for these three is particularly important since the boundaries are either missing entirely (chemical pollution/novel entities), challenging on the basis of scale (regional vs global) (all three), or indeed now considered to be 'core PBs' in a 'two level hierarchy of boundaries' (biodiversity/biosphere integrity) (Steffen et al. 2015).

sustainability and we will continue to use science to help us address future challenges. Currently, systems thinking remains in the margins of scientific development, but increasingly this needs to be brought to the fore so that large companies such as Unilever, as well as governments, are better equipped to make choices that can drive transformational change. This implies even more intensive cross-disciplinary activity, merging elements from Environmental Sustainability Science, LCA, Earth Systems Science, Resilience Science and Economics for more holistic insights and business/policy relevant assessment tools. The emerging developments in big data, informatics, and the deployment of imaging technologies and information systems to visualize earth systems will facilitate the move towards a more holistic understanding of environmental impacts and the sustainable management of Earth's resources.

References

Baitz, M., Albrecht, S., Brauner, E., Broadbent, C., Castellan, G., Conrath, P., Fava, J., Finkbeiner, M., Fischer, M., Fullanai Palmer, P., Krinke, S., Leroy, C., Loebel, O., McKeown, P., Mersiowsky, I., Möginger, B., Pfaadt, M., Rebitzer, G., Rother, E., Ruhland, K., Schanssema, A., & Tikana, L. (2013). LCAs theory and practice: Like ebony and ivory living in perfect harmony. *International Journal of Life Cycle Assessment, 18*, 5–13.

Barnosky, A. D., Hadly, E. A., Bascompte, J., Berlow, E. L., Brown, J. H., Fortelius, M., Getz, W. M., Harte, J., Hastings, A., Marquet, P. A., Martinez, N. D., Mooers, A., Roopnarine, P., Vermeij, G., Williams, J. W., Gillespie, R., Kitzes, J., Marshall, C., Matzke, N., Mindell, D. P., Revilla, E., & Smith, A. B. (2012). Approaching a state shift in Earth's biosphere. *Nature, 486*(7401), 52–58.

Bell, G. (Interview by) (2013a). Doing well by doing good. *Strategic Direction, 29*(4), 38–40.

Bell, G. (Interview by) (2013b). Want to change the world? Think differently. *Strategic Direction, 29*(5), 36–39.

Biggs, R., Carpenter, S. R., & Brock, W. A. (2009). Turning back from the brink: Detecting an impending regime shift in time to avert it. *Proceedings of the National academy of Sciences of the United States of America, 106*, 826–831. doi:10.1073/pnas.0811729106.

Böhringer, C., & Jochem, P. E. P. (2007). Measuring the immeasurable – A survey of sustainability indices. *Ecological Economics, 63*, 1–8.

Carpenter, S. R., & Bennett, E. M. (2011). Reconsideration of the planetary boundary for phosphorus. *Environmental Research Letters, 6*, 014009. doi:10.1088/1748-9326/6/1/014009.

CDP, GHGP, & WWF. (2014, March). *Mind the science, mind the gap.* Concept Note. Retrieved January 30, 2015, from http://www.ghgprotocol.org/files/ghgp/Concept%20Note%20-%20MScienceMGap.pdf

Chaplin-Kramer, R., Sharp, R., Mandle, L., Sim, S., Johnson, J., Butnar, I., Milà i Canals, L., Eichelberger, B., Ramler, I., Mueller, C., McLachlan, N., Yousefi, A., King, H., & Kareiva, P. (2015). Spatial patterns of agricultural expansion determine impacts on biodiversity and carbon storage. *Proceedings of the National academy of Sciences of the United States of America, 112*(24), 7402–7407. doi:10.1073/pnas.1406485112.

Clift, R., & Wright, L. (2000). Relationships between environmental impacts and added value along the supply chain. *Technological Forecasting and Social Change, 65*, 281–295.

Cole, M. J., Bailey, R. M., & New, M. G. (2014). Tracking sustainable development with a national barometer for South Africa using downscaled "safe and just space" framework. *Proceedings of the National academy of Sciences of the United States of America, 111*(42), E4399–E4408. doi:10.1073/pnas.1400985111.

Constance, D. H., & Bonanno, A. (2000, June). Regulating the global fisheries: The World Wildlife Fund, Unilever, and the Marine Stewardship Council. *Agriculture and Human Values, 17*(2), 125–139.

De Vries, W., Kros, J., Kroeze, C., & Seitzinger, S. P. (2013). Assessing planetary and regional nitrogen boundaries related to food security and adverse environmental impacts. *Current Opinion in Environment. Sustainability, 5*, 392–402. doi:10.1016/j.cosust.2013.07.004.

Dearing, J. A., Wang, R., Zhang, K., Dyke, J. G., Haberl, H., Hossain, M. S., Langdon, P. G., Lenton, T. M., Raworth, K., Brown, S., Carstensen, J., Cole, M. J., Cornell, S. E., Dawson, T. P., Doncaster, C. P., Eigenbrod, F., Flörke, M., Jeffers, E., Mackay, A. W., Nykvist, B., & Poppy, G. M. (2014). Safe and just operating spaces for regional social-ecological systems. *Global Environmental Change, 28*, 227–238.

Fargione, J., Hill, J., Tilman, D., Polasky, S., & Hawthorne, P. (2008). Land clearing and the bio-fuel carbon debt. *Science 319*, 1235. American Association for the Advancement of Science, New York/Washington DC.

Flynn, H. C., Milà i Canals, L., Keller, E., King, H., Sim, S., Hastings, A., & Smith, P. (2012). Quantifying global greenhouse gas emissions from land-use change for crop production. *Global Change Biology, 18*(5), 1622–1635. doi:10.1111/j.1365-2486.2011.02618.x.

Garcia-Suarez, T., Sim, S., Mauser, A., & Marshall, P. (2008, November 12–14). Greenhouse gas assessment of Ben & Jerry's ice cream: Communicating their 'Climate Hoofprint.' *Paper in the proceedings of the 6th international conference on life cycle assessment in the Agri-Food sector: Towards a sustainable management of the food chain*. Zurich, Switzerland.

Gerten, D., Hoff, H., Rockstrom, J., Jagermeyr, J., Kummu, M., & Pastor, A. V. (2013). Towards a revised planetary boundary for consumptive freshwater use: Role of environmental flow requirements. *Current Opinion on Environmental Sustainability, 5*, 551–558.

Gibbs, H. K., Johnston, M., Foley, J., Holloway, T., Monfreda, C., Ramankutty, N., & Zaks, D. (2008). Carbon payback times for crop-based biofuel expansion in the tropics: The effects of changing yield and technology. *Environmental Research Letters, 3*, 034001 (10pp).

Gowland, S. (2009). *Unilever's sustainable brand and business strategy. Melodies in marketing.* Retrieved January 30, 2015, from http://www.melodiesinmarketing.com/2009/04/11/unilever-sustainable-brand-lipton-knorr-dove

Hague, J. (2014). *Building future sanitation models: A report prepared for the toilet board coalition, sponsored by Unilever.* Retrieved January 30, 2015, from http://www.toiletboard.org/media/downloads/6.pdf

IPCC. (2007). *Mitigation of climate change: Technical summary.* Geneva: Intergovernmental Panel on Climate Change.

Jefferies, D., Muñoz, I., Hodges, J., King, V. J., Aldaya, M., Ercin, A. E., Milà i Canals, L., & Hoekstra, A. Y. (2012, September). Water footprint and life cycle assessment as approaches to assess potential impacts of products on water consumption. Key learning points from pilot studies on tea and margarine. *Journal of Cleaner Production, 33*, 155–166.

Leemans, R., van Amstel, A., Battjes, C., Kreilman, E., & Toet, S. (1996). The land cover and carbon cycle consequences of large-scale utilizations of biomass as an energy source. *Global Environmental Change, 4*, 335–357.

Lingard, T. (2012). Unilever's strategic response to sustainable development and its implications for public affairs professionals. *Journal of Public Affairs, 12*, 224–229. doi:10.1002/pa.1436.

Mace, G. M., Reyers, B., Alkemade, R., Biggs, R., Chapin, F. S., III, Cornell, S. E., Díaz, S., Jennings, S., Leadley, P., Mumby, P. J., Purvis, A., Scholes, R. J., Seddon, A. W. R., Solan, M., Steffen, W., & Woodward, G. (2014). Approaches to defining a planetary boundary for biodiversity. *Global Environmental Change, 28*, 289–297. doi:10.1016/j.gloenvcha.2014.07.009.

Meadows, D. H., Meadows, D. L., Randers, J., & Behrens, W. W. (1972). *Limits to growth.* New York: New American Library.

Milà i Canals, L., Sim, S., Garcia-Suarez, T., Neuer, G., Herstein, K., Kerr, C., Rigarlsford, G., & King, H. (2010). Estimating the greenhouse gas footprint of Knorr. *The International Journal of Life Cycle Assessment, 16*(1), 150–158. doi:10.1007/s11367-010-0239-5.

Milà i Canals, L., Azapagic, A., Doka, G., Jefferies, D., King, H., Mutel, C., Nemecek, T., Roches, A., Sim, S., Stichnothe, H., Thoma, G., & Williams, A. (2011). Approaches for addressing life cycle assessment data gaps for bio-based products. *Journal of Industrial Ecology, 15*(5), 707–725.

Milà i Canals, L., Rigarlsford, G., & Sim, S. (2013). Land use impact assessment of margarine. *Journal of Life Cycle Assessment, 18*(6), 1265–1277.

Muñoz, I., Rigarlsford, G., Mila, I., Canals, L., & King, H. (2013). Accounting for greenhouse gas emissions from the degradation of chemicals in the environment. *International Journal of Life Cycle Assessment, 18*(1), 252–262.

Murphy, R. J. et al. (in prep). *Planetary boundaries as a basis for decision-making.*

Nemecek, T., Weiler, K., Plassmann, K., Schnetzer, J., Gaillard, G., Jefferies, D., García-Suárez, T., King, H., & Milà I Canals, L. (2012, August). Estimation of the variability in global warming potential of worldwide crop production using a modular extrapolation approach. *Journal of Cleaner Production, 31*, 106–117.

Nykvist, B., Persson, Å., Moberg, F., Persson, L., Cornell, S., & Rockström, J. (2013, June). National Environmental Performance on Planetary Boundaries – A study for the Swedish Environmental Protection Agency. Report 6573.

Ravindranath, N. H., Manuvie, R., Fargione, J., Canadell, P., Berndes, G., Woods, J., Watson, H., & Sathaye, J. (2009). SCOPE Biofuel Report. Chapter 4, GHG implications of land use and land conversion to biofuel crops. Retrieved January 30, 2015, from http://www.globalbioenergy.org/uploads/media/0810_Ravindranath_et_al_-_GHG_implications_of_land_use_and_land_conversion_to_biofuel_crops.pdf

Rees, W., & Wackernagel, M. (1994). Ecological footprints and appropriated carrying capacity: Measuring the natural capacity requirements of the human economy. In A. Jansson, M. Hammer, C. Folke, & R. Costanza (Eds.), *Investing in natural capital.* Washington, DC: Island Press.

Rigarlsford, G., Garcia-Suarez, T., Milà i Canals, L., David, H., Sim, S., Unger, N., & King, H. (2010, September 22–24). *Estimating the greenhouse gas footprint of Unilever's food business.* Presented at the 7th international conference on LCA in the agri-foods sector. Bari, Italy.

Roches, A., Nemecek, T., Gaillard, G., Plassmann, K., Sim, S., King, H., & Milà i Canals, L. (2010). MEXALCA: A modular method for the extrapolation of crop LCA. *The International Journal of Life Cycle Assessment, 15*(8), 842–854. doi:10.1007/s11367-010-0209-y.

Rockström, J., Steffen, W., Noone, K., Persson, Å., Chapin, F. S., III, Lambin, E., Lenton, T. M., Scheffer, M., Folke, C., Schellnhuber, H., Nykvist, B., De Wit, C. A., Hughes, T., van der Leeuw, S., Rodhe, H., Sörlin, S., Snyder, P. K., Costanza, R., Svedin, U., Falkenmark, M., Karlberg, L., Corell, R. W., Fabry, V. J., Hansen, J., Walker, B., Liverman, D., Richardson, K., Crutzen, P., & Foley, J. (2009a). A safe operating space for humanity. *Nature, 461*, 472–475.

Rockström, J., Steffen, W., Noone, K., Persson, Å., Chapin, F. S., III, Lambin, E., Lenton, T. M., Scheffer, M., Folke, C., Schellnhuber, H., Nykvist, B., De Wit, C. A., Hughes, T., van der Leeuw, S., Rodhe, H., Sörlin, S., Snyder, P. K., Costanza, R., Svedin, U., Falkenmark, M., Karlberg, L., Corell, R. W., Fabry, V. J., Hansen, J., Walker, B., Liverman, D., Richardson, K., Crutzen, P., & Foley, J. (2009b). Planetary boundaries: Exploring the safe operating space for humanity. *Ecology and Society, 14*(2), 32.

Scheffer, M., Bascompte, J., Brock, W. A., Brovkin, V., Carpenter, S. R., Dakos, V., Held, H., van Nes, E. H., Rietkerk, M., & Sugihara, G. (2009). Early-warning signals for critical transitions. *Nature, 461*, 53–59. doi:10.1038/nature08227.

Scheffer, M., Carpenter, S. R., Lenton, T. M., Bascompte, J., Brock, W., Dakos, V., van de Koppel, J., van de Leemput, I. A., Levin, S. A., van Nes, E. H., Pascual, M., & Vandermeer, J. (2012). Anticipating critical transitions. *Science, 338*, 344–34. doi:10.1126/science.1225244.

Schlamadinger, B., Grubb, M., Azar, C., Bauen, A., & Berndes, G. (2001). *Carbon sinks and biomass energy production: A study of linkages, options and implications.* Climate Strategies: International Network for Climate Policy Analysis. London.

Searchinger, T., Heimlich, R., Houghton, R. A., Dong, F., Elobeid, A., Fabiosa, J., Tokgoz, S., Hayes, D., & Yu, T. H. (2008). Use of U.S. croplands for biofuels increases greenhouse gases through emissions from land-use change. *Science, 319*(5867), 1238–1240.

Steffen, W., Richardson, K., Rockström, J., Cornell, S. E., Fetzer, I., Bennett, E. M., Biggs, R., Carpenter, S. R., de Vries, W., de Wit, C. A., Folke, C., Gerten, D., Heinke, J., Mace, G. M., Persson, L. M., Ramanathan, V., Reyers, B., & Sörlin, S. (2015). Planetary boundaries: Guiding human development on a changing planet. *Science, 347*(6223), 1259855. doi:10.1126/science.1259855.

UNEP. (2015) Guidance on Organisational Life Cycle Assessment. Available at: http://www.life-cycleinitiative.org/wp-content/uploads/2015/04/o-lca_24.4.15-web.pdf

Unilever. (2014). *Unilever sustainable living plan.* Retrieved January 30, 2015, from http://www.unilever.co.uk/sustainable-living-2014/our-approach

Unilever Sustainable Agriculture Code. (2010). Retrieved January 30, 2015, from http://unilever.com/images/sd_Unilever_Sustainable_Agriculture_Code_2010_tcm13-216557.pdf

Van Hoof, G., Schowanek, D., Franceschini, H., & Muñoz, I. (2011). Ecotoxicity impact assessment of laundry products: A comparison of USEtox and critical dilution volume approaches. *International Journal of Life Cycle Assessment, 16*(8), 803–818.

Wackernagel, M., & Rees, W. (1996). *Our ecological footprint: Reducing human impact on the earth.* Gabriola Island: New Society Publishers. ISBN 0-86571-312-X.

Wang, R., Dearing, J. A., Langdon, P. G., Zhang, E., Yang, X., Dakos, V., & Scheffer, M. (2012). Flickering gives early warning signals of a critical transition to a eutrophic lake state. *Nature, 492*, 419–422. doi:10.1038/nature11655.

White, S. (2015, January 02). *Preferred Bidder selected for first centre for Agricultural Innovation Centre.* Retrieved January 30, 2015, from https://agritech.blog.gov.uk/feed/atom/

Whiteman, G., Walker, B., & Perego, P. (2013). Planetary boundaries: Ecological foundations for corporate sustainability. *Journal of Management Studies, 50*(2), 307–336.

Chapter 16
Practical Implications of Product-Based Environmental Legislation

Kieren Mayers

Abstract A number of approaches to industrial ecology are now employed within environmental legislation, targeting products at various stages of their life-cycle. These require producers to reduce the hazardous substances content of their products during production, increase product energy efficiency during use, and organise and finance improved recycling and treatment of their products at end of life (Extended Producer Responsibility, or 'EPR'). Such requirements are increasingly commonplace in the Americas, Eurasia, and Pacific Rim countries and have substantial impact. If companies can't comply, then they can't sell their products. There appears to be little research on the practical steps producers have taken to manage compliance with this new-wave of product-based requirements, as compared to the more established areas of environmental management addressing site-based air and water emissions, resource and energy use, and waste management. Based on a number of case studies, this chapter explains how such product-based legislation operates in practice.

Keywords Environmental legislation • Environmental management • Extended producer responsibility • Industrial ecology • Product-based legislation • Risk management • Supplier commitment

1 Introduction

Environmental legislation is increasingly targeting various stages of the product life-cycle. There are a widening range of restrictions on hazardous substances in different types of products globally, covering an increasing number of substances in their scope. To ensure their products comply, producers can ask the suppliers to commit to and sign declarations of compliance, audit production facilities, and

K. Mayers (✉)
INSEAD Social Innovation Center, Sony Computer Entertainment Europe,
10 Great Marlborough Street, London W1F 7LP, UK
e-mail: kieren_mayers@scee.net

© The Author(s) 2016
R. Clift, A. Druckman (eds.), *Taking Stock of Industrial Ecology*,
DOI 10.1007/978-3-319-20571-7_16

303

undertake chemical testing of products. If these measures do not succeed, and producers are found to be non-compliant, corrective steps are needed to isolate and clear affected products from distribution.

Energy efficiency standards, including power caps, allowances, and power management, and information requirements, focus on reducing the energy used by a device to perform a particular defined task. For example, energy allowances are higher for categories of PCs with more powerful processors. Such standards do not necessarily result in an overall reduction in energy use of a product, as new products with increasing performance over time may use more energy overall, even if energy used per unit output falls.

Extended Producer Responsibility (EPR) legislation, requiring producers to finance and organise collection, treatment, and recycling of their products at end of life, has been enacted for a range of different types of products and in many different jurisdictions internationally. The main intention of such regulations is to ensure producers have financial incentives to design their products to be easier to treat and recycle at end-of-life, and also therefore to improve recycling and standards of environmental protection at end of life. Typically such regulations covers batteries, packaging, waste electrical and electronic products, tyres, household hazardous wastes, and automobiles in many areas of the world.[1]

This chapter draws on two decades of the author's personal experience working as an environmental/sustainability professional in the electronics and recycling sector within Europe, and also a number of associated case studies (Martin 2008; Webb 2014; Mayers 2007a, b; Mayers and Butler 2013) as an example of how producers respond to this new wave of product-based regulations. Content has also been taken from a presentation by the author prepared for a taught module on Life Cycle Assessment at the Centre for Environmental Strategy at the University of Surrey (Mayers 2011).

2   Dealing with Hazardous Substance Restrictions in Products

Regulations can target substance use in specific sectors, such as with the European Union (EU) Restriction of Hazardous Substances Directive covering electrical and electronic equipment (2011/65/EU) and the Packaging and Packaging Waste Directive (94/62/EC), or may focus on specific chemicals across product classes, such as with Danish lead restrictions (Danish Statutory Order No. 1012). The EU REACH Regulation (EC/1907/2006) takes a combined approach by restricting chemicals according to their specific applications. Overall, applicable substance limits depend on the product and material concerned. For example, as of January 2015, there are around 17 different hazardous substance regulations in Europe affecting Sony Computer Entertainment products, restricting 46 different substances, and including 78 different limit values for different types of products and materials.

---

[1] See special feature volume on EPR: Journal of Industrial Ecology, Vol. 17, Issue 2, 2013.

Typically, producers must consider at least five different factors when assessing the applicability of substance restrictions to their products:

- Which categories of products are in scope e.g. electronic, batteries, packaging?
- What is the limit value (allowable concentration per homogenous material or part)?
- When will restrictions be implemented?
- Which countries does it apply to?
- Which materials are affected e.g. plastics, solders, etc.?

The impacts of non-compliance, if caught, are severe. In the last year alone over 370 products were withdrawn from the market due to enforcement actions on hazardous substances within the EU (European Commission 2015). As well as halting sales, there is a reputational impact on brands, with governments and nongovernmental organisations (NGOs) generating exposure by naming and shaming companies in the media. To ensure they comply with such regulations, producers must (Martin 2008):

- Monitor and track continually evolving legislation
- Ensure products can meet prescribed limits
- Check that suppliers and factory management understand applicable requirements
- Plan phase-out of substances ahead of regulatory deadlines
- Check and approve parts and materials before shipping
- Retain technical documentation as evidence of compliance
- Correct non-compliance incidents
- Determine preventative and proactive measures to minimise risk of non-compliance

Supply chains can be extremely complex, involving networks of tens or hundreds of actors. Manufacturing activities are most commonly outsourced by brand holders to third parties. As a result, specific production processes may not be even known to or under the control of producers and brand-holders. Even if suppliers make stated commitments to phase out hazardous substances, and include requirements in documented specifications for new products, they may struggle to track and correctly interpret the plethora of different regulations globally. It may be challenging to find substitute materials, and so some companies may simply manufacture products that only comply with local or selected legislation. Upstream changes in suppliers or materials can result in unexpected changes to substance concentrations. Producers may also be unaware of chemical contaminants not deliberately added to their products e.g. black pigments may contain soot, which may contain a wide variety of heavy metals. Also, different laboratories may sample products differently with differing results. Finally, national authorities often take differing approaches to compliance and enforcement (Martin et al. 2007). In some countries, enforcement agencies test products selected from retail outlets to detect non-compliance; in others, a lack of technical documentation available from producers is considered to be an offence.

To protect against these risks and manage elimination of hazardous substances in their products, producers can use a combination of approaches. As a basic underpinning and assurance, producers can ask their third party manufacturers and suppliers to complete and sign declarations that they will ensure all materials, components or products comply with all hazardous substance restrictions applicable in the jurisdictions where the product will be distributed. First and foremost, this alerts suppliers to the necessary requirements. It also may provide some assurance to producers in terms of liability for any resulting financial losses. Signed supplier declarations, however, do not provide guarantees that the products themselves will be in compliance. Government authorities often find quite a high proportion of products investigated through market surveillance do not actually comply with substance restrictions. For an example, in 2014, the Swedish Chemicals Agency found that over 40 % of the plastic articles such as handbags, wallets, pencil boxes and cases for mobile phones that they tested contained short-chain chlorinated paraffins which are prohibited under the EU Persistent Organic Pollutants Regulation (KEMI 2014).

To minimise risk further, producers can submit samples of products for testing at laboratories, undertake chemical testing themselves or ask suppliers to provide test reports that show their products meet legal limits. Testing provides a robust check; for example, test reports can be used to detect any substances suppliers are unaware of from sources 'upstream' in their supply chain. Alongside signed supplier declarations, test reports can also be used to show regulatory enforcement agencies documented evidence of due diligence. Testing, however, also has its limitations: it only provides a 'snap-shot' of one or perhaps a few products at one point in time, so multiple samples may be needed as well as retesting on periodic basis e.g. monthly, quarterly, or annually. As testing is expensive and samples are usually destroyed in the testing process, testing statistically representative samples of products can be infeasible.

To gain an overall perspective and level of assurance, producers can also audit their supplier's manufacturing facilities. Auditing can be used to ascertain the level of competency and understanding of staff working in manufacturing, ensure the necessary controls are actually in place, check the effectiveness of procedures used to control hazardous substances, and assess unforeseen risks in the process. For example, if manufacturers do not have a process to isolate any products suspected to have compliance issues from compliant stock, then there is a risk the products may enter the supply chain. The limitation of such audits is that they too only provide a 'snap shot' in time of how any supplier may be operating. Audits should, therefore, be repeated every year or so, but this may involve a substantial amount of time and resources as global supply chains typically have many different suppliers involved.

For any new substance compliance requirements, producers must ensure their products comply by time they come into force. There are several different approaches to stock control that can be used to clear-out older stocks of non-compliant products (Fig. 16.1).

If preventative measures fail and non-compliant products enter distribution (where, in the worst-case, producers may face fines and sales prohibitions), then corrective measures are needed both to identify any current non-compliant stocks of

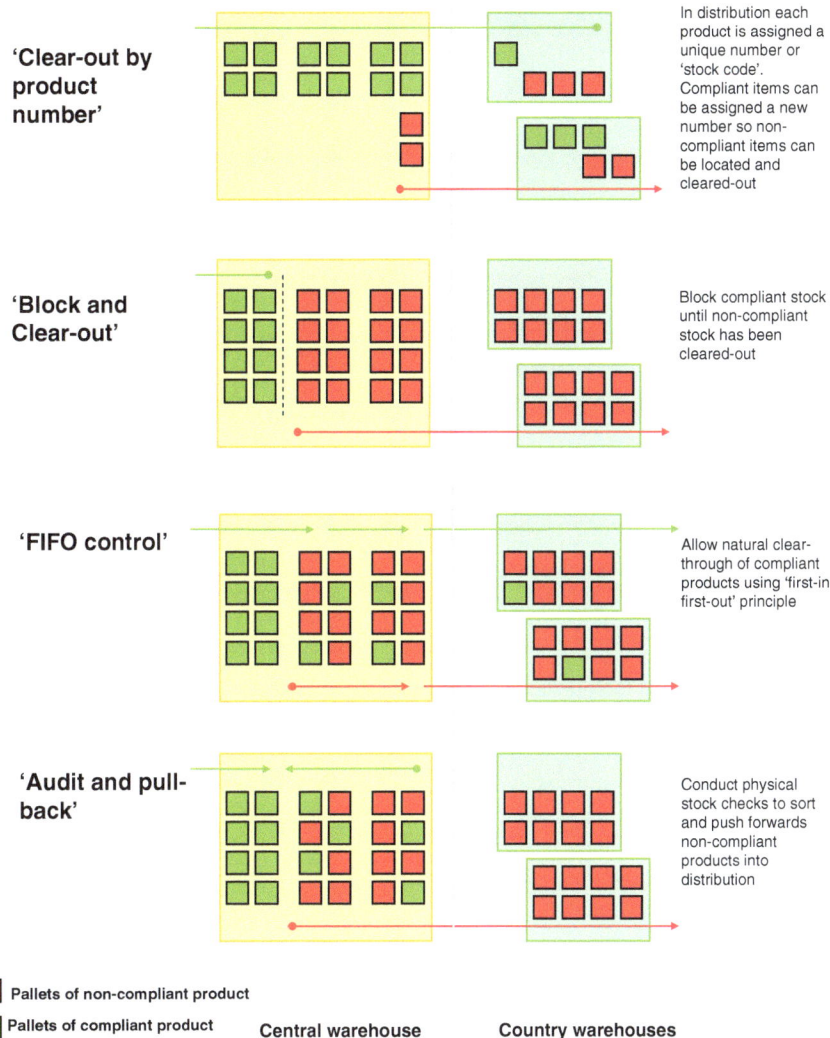

Fig. 16.1 Stock management approaches to clear-out non-compliant products (Adapted from Mayers 2011)

products, and also to find and fix whatever oversight or issue is at cause. A producer's supply chain may be made up of both pan-regional, and state or national warehouses, filled with millions of units of potentially hundreds of different product lines. Each product line may be supplied by several different manufacturers. Producers must ascertain which products are affected; whether it is just products from one supplier in particular, all products, or just those manufactured in a certain time period. The compliance issue may be related only to specific components or materials e.g. lead in plastic could be from either the pigment or plasticiser used (Fig. 16.2).

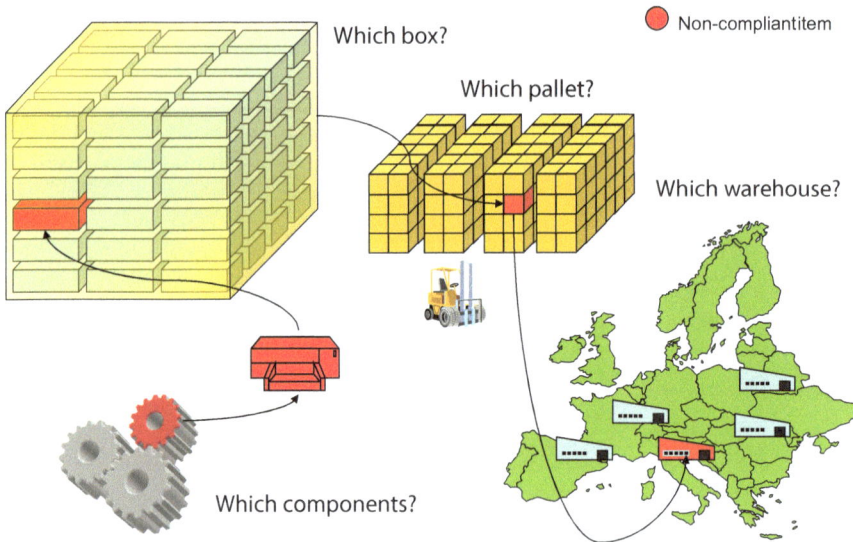

Fig. 16.2 Distinguishing non-compliant products in a supply chain (Adapted from Mayers (2007a, b). Lifecycle environmental management in the electronics sector. Lecture, University of Surrey, 12 January)

Not all of these factors are easy to assess or understand, and it may not be straight forward to distinguish stocks of compliant from non-compliant products across many warehouses. As a consequence, producers may have to send many samples of product for testing to try to determine which products in their supply chain are affected. Once identified, producers may have to take a number of actions: they may have to rework or replace non-compliant items e.g. power cables, they may dispose of the items if there is no other option feasible, or they may return the products to the original manufacturer for a refund (which is where the signed declaration can be of use).

To solve the problem at source, suppliers must identify and address (a) what went wrong, (b) how it can be fixed, and (c) what can be done to preventive recurrence.

3 Ensuring Energy Efficiency

Energy using products are covered by a range of different types of energy standards worldwide. Focusing on electrical and electronic products these include:

- Requirements for power management, such that a device switches automatically into the lowest power mode possible for the function required.
- Reductions in power consumption, where the power consumed for a given task is reduced, such as modal power caps or allowances.
- Requirements to inform consumers and users and label devices on their energy use, and how to use them in a way to minimise energy use.

Standards may involve voluntary commitments, such as US Energy Star (US Environmental Protection Agency 2015), or sector voluntary agreements used in the EU (2009/125/EC), US (US Department of Energy 2015a), Australia and New Zealand (E3 2015a); regulatory requirements, such as EU Energy Using Products (2009/125/EC) and Energy Labelling Directives (2010/30/EU), and Australia and New Zealand Minimum Energy Performance Standards (E3 2015b), and US Federal energy efficiency standards (US Department of Energy 2015b); or regulatory benchmarks, such as Japan's 'top runner' programme to ensure all best technologies are adopted over time according to leading products on the market (Energy Conservation Centre Japan 2015).

As technological development and innovation occurs relatively rapidly, energy efficiency requirements are typically organised into a succession of chronological 'tiers' over which requirements are ratcheted up. To comply producers may either need to retain technical documentation and test reports showing that their products meet each of the applicable criteria (as in the EU and Australia), or in some cases may be required to submit their products themselves for testing and certification (as under US Energy Star requirements). Non-compliant products may either lose their certification status and must withdraw or change energy labels used, or ultimately, if mandatory standards are involved, producers may face fines and sales blocks.

The most complex challenge for producers is to understand and keep pace with the future energy implications of technological development. With the pace of technological advancement, energy efficiency standards are updated every few years to ensure improvement vs. 'business as usual'. Such assessments consider and compare estimates of total energy use for any potential improvements, considering power consumed, usage time, and number of units of a product in use to calculate estimates of total electricity consumption (TEC).

To engage in this process, producers need in-depth understanding and available research on their consumer usage behaviour, and the energy implications of different technology scenarios, to consider energy implications at the early stages of product development, and also to engage with and gain the understanding of stakeholders such as environmental and consumer NGOs. Predicting power consumption of future technology 3–4 years in advance, considering the timescales for developing new regulations, involves large amounts of risk for producer. Where it may not be clear the extent to which a new energy efficient technology may be suitable, or what implications it may have, further research and development may be needed. Unless producers engage in continuous dialogue with policy makers and NGO stakeholders at an early stage, regulations and standards may be developed based on only rudimentary understanding of their products and services, which may not result in optimal solutions to energy efficiency and may impede innovation.

Compliance with energy standards appears relatively straight forward in comparison to substance compliance (discussed above) and Extended Producer Responsibility (EPR) (discussed below). This is because standards are uniformly applied and relatively easy to assess. The challenge for producers is to anticipate and even influence the direction of future energy policy and standards. If producers are unable to keep pace and comply with these evolving standards, they may be forced to withdraw many of their products, as recently observed for vacuum cleaners that could not meet 1,800 W power cap within the EU (BBC 2014).

4   Managing Products at End of Life

There are many thousands of different producers, and usually many hundreds or thousands of waste collection points within any country. As a consequence, under EPR it is absolutely infeasible for each producer to set-up an individual system to collect their own branded products from all possible municipal collection centres and households. Conversely, it would also be an overwhelming task for each house-holder, or municipal waste collection point, to sort all their waste by hundreds of brands on a daily basis and try work out which producer to contact to arrange col-lection (assuming even that the original producer still exists).

This insurmountable 'economy of scope' limits the effectiveness of EPR in prac-tice; producers have little option other than to co-operate and manage waste collec-tively within each country. To this end producers have established Producer Recycling Organisations (PROs) to manage and administer waste arrangements on their behalf (Fig. 16.3). Based on 2014 data, over 400 EPR systems (most of which include *at least* one PRO) have been established worldwide (Lifset 2014).

Setting-up and running PROs can be quite an involved process (Mayers and Butler 2013). PROs must organise sufficient management, expert, and administra-tive staff to organise their activities. Management overheads can, therefore, consti-tute up to 20 % of overall PRO expenditure. There can be considered to be three stages to PRO implementation: 'design', 'build', and 'operate':

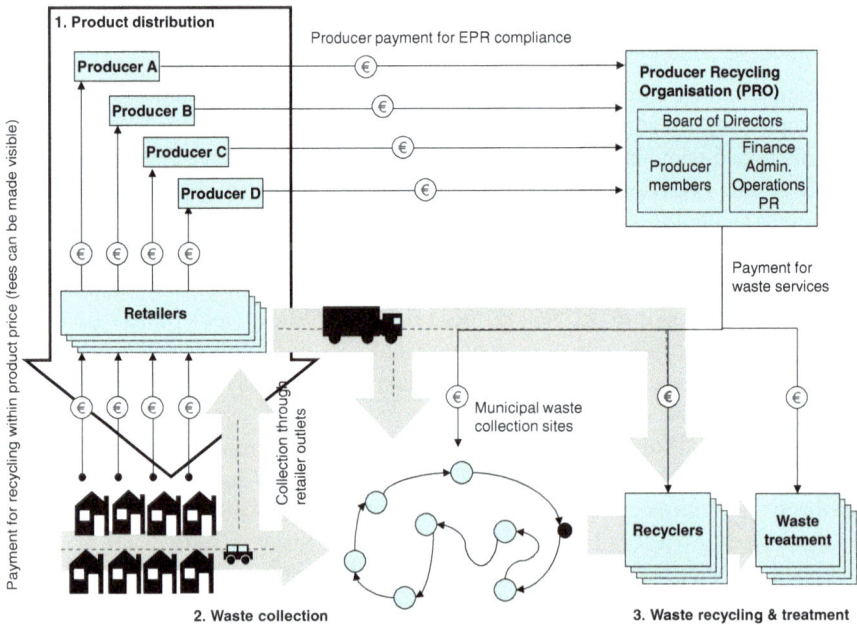

Fig. 16.3 Producer recycling organisation set-up and operation (Mayers 2007a)

- PRO planning and system design:

  - Investigating existing waste infrastructure and legislation
  - Determining PRO license conditions and requirements
  - Identifying where waste materials arise, and how the PRO will access them
  - Assessing which recycling services are available, and at what price
  - Agreeing how competing PROs will balance their share of waste

- Building-up PRO operations and processes:

  - Appointing PRO staff, and setting-up administration and reporting procedures
  - Auditing and approval of recycling companies against required standards
  - Deciding on containers and transport needed for each collection point
  - Selecting and appointing recyclers and collection companies

- Running ongoing operations:

  - Receiving and responding to requests for collection
  - Organising day-to-day collection, treatment, and recycling
  - Collating and submitting reports of quantities collected and processed
  - Accounting and payment for collection, treatment, and recycling services
  - Trouble-shooting operational and service problems
  - Optimising activities to meet cost and key performance indicators

To comply with EPR legislation, producers must firstly understand specific regulatory requirements within each country for each type of product. Different regulations may specify different requirements relevant to each producer, for example reporting and registration procedures. Once requirements have been checked, producers must choose which PROs are most appropriate for their products. It may be possible to choose between several competing PRO services. Some key considerations for producers to take into account before joining a PRO include:

- Does the PRO charge a membership fee? How often and how much?
- Are recycling fees fixed per product sold in advance, or will the amount charged vary according to the producer's share of treatment and recycling costs each month?
- Does the PRO accrue any financial reserves from the payments producers make? Who owns those funds and for what purpose?
- Does the PRO have the necessary permits and authorisations to operate and fulfil the producer's EPR obligations? Will it ensure the necessary environment standards?
- How long is the contract period for? Under which conditions can the PRO terminate the contract?
- Once the contract is agreed, are fees fixed or can they be changed by the PRO from time to time? Under what conditions can the PRO increase or lower their fees?

- Is the exact wording of the contract fixed, or can it be varied according to each member producer's policies?
- How often does the PRO require reporting? Annually, quarterly, or monthly?
- Are there any other useful services the PRO can provide e.g. collection of WEEE from offices, or pan-European EPR compliance services?

Once producers have signed-up to their selected PRO, they must then start reporting the amount of products sold and pay any PRO charges due to finance collection, treatment, and recycling as well as any management and administrative overhead. Reporting requirements between different countries, types of waste, and PROs can differ widely e.g. units vs. weight sold, reporting by different sub-categories of products and materials, monthly, quarterly, or annual reporting, or reporting to PROs, national enforcement agencies, or special registration or 'clearing house' bodies. Typically producers will have a list of components used in their products (bill of materials) available, but will not be necessarily aware of the weight of different packaging materials, or the weight of electronic products with cables but without batteries, etc. Collecting and then reporting such data, combining it with sales reports, and completing different formats of reporting forms for different PROs and different waste streams takes time for operations and environmental managers.

Unfortunately, financial incentives for improved design of products from EPR are limited to non-existent. Recycling fees tend to be higher for plastics than for card and paper packaging, as plastics are more complex to recycle. This provides some incentives towards use of paper packaging, but not for the packaging to be designed in a way to ensure it can be easily recycled e.g. easily separated into different material types rather than being glued together. In addition, all EPR fees are charged per unit or weight sold, which will only reward producers if they sell fewer products or sell smaller products (in the case of weight sold). As products are recycled collectively, and costs shared equally among producers, incentives for each individual producer for their own products are removed or diminished substantially.

The economies of scope explained above means that it is impractical for producers to collect and recycle only their own products in an individual EPR system. Separation of products into thousands of brands at municipal collection points with only enough space and staff to provide a few waste containers is logistically impossible. At the household level, putting aside the sheer impracticality of sorting for 'kerbside' or 'doorstep' collection, it would not be environmentally beneficial to arrange individual collections of waste by brand due to the need for dedicated transport for small volumes. In addition, collection costs would outweigh any recycling value or cost by several orders of magnitude (for example, consider the costs of mailing individual parcels).

Collective PRO systems appear to be a necessary component of EPR. Nevertheless, individual producers can still be made financially responsible for waste costs attributable to their products. PROs can allocate costs to producers more accurately and proportionately based upon the treatment and recycling costs of different types of products (Mayers et al. 2014). For example, display screens with mercury backlights must be treated before recycling to remove mercury, which is an expensive process, whereas mercury-free LCDs can be more conventionally recycled, potentially with a net value. This approach is already in place in France for packaging and WEEE, using a system of differentiated fees.

While EPR was developed with good motives, in practice its implementation is both administratively and logistically complex, and to date the main purpose to incentivise design is largely unfulfilled.

5   Discussion and Conclusion

Looking back, site-based approaches to environmental management, developed since mid-seventies, focus on treating environmental problems at 'end of pipe'. Education in environmental management and sciences from this time focussed mainly on environmental problems and their causes, environmental protection legislation, environmental management systems, air and water monitoring, water treatment and air pollution control, waste management, and resource and energy use. Such knowledge is important in managing and reducing the environmental impact of any particular industrial operation.

Looking forward, managing environmental impacts of the life-cycle of products from raw material extraction to end-of-life involves altogether different issues, requiring additional skills from environmental and 'sustainability' managers. In addition to a basis of understanding of current environmental issues, sociology, ethics, economics, and sustainable development, risk assessment and management, and life-cycle assessment methods, further expertise is essential as listed below:

- Managing hazardous substances in supply chains:

  - Substance compliance regulations
  - Materials usage and engineering
  - Product testing and chemical analysis
  - Supply chain management
  - Stock control and auditing

- Ensuring energy efficiency of products:

  - Energy efficiency legislation
  - Product development and engineering
  - Government relations and stakeholder engagement
  - International standardisation processes
  - Consumer behaviour and market research
  - Product testing and power measurement

- Implementing EPR for wastes:

  - EPR legislation
  - Contract management for PROs
  - Recycling standards and technologies
  - Company accounting systems such as SAP to report sales volumes

Managing flows of hazardous substances, recyclable materials, and energy use throughout the life-cycle of products and considering entire industrial supply chains with the aim of reducing their environmental impact are administratively and logistically complex. Producers often do not have complete knowledge or direct control of complex supply chains and cannot always accurately predict the future outcomes of technology development. Environmental managers must find new ways to address gaps in understanding and implement procedures to ensure producers can comply with this new wave of product-based environmental legislation, and ultimately to solve and prevent environmental problems at source.

There appears to be very little corresponding information or knowledge of contemporary approaches to environmental management in the available literature, and furthermore, students may miss out if further education only equips them with knowledge to understand and assess the implications from a life-cycle or sustainability perspective. For example, imagine a surgeon only taught how to diagnose heart disease, but not how to conduct heart surgery. From a policy perspective, lessons from practical experience reinforce the need for harmonised product-based requirements including applicable standards, product categorisation, reporting, and proof of compliance. For industrial ecology to move forwards, the practical challenges and approaches, the administrative procedures, and the management methods required to solve environmental problems at various stages of a product's lifecycle are surely worth further consideration.

Acknowledgements The author would like to thank Dr. Joachim Zietlow, and Reid Lifset for their assistance.

References

BBC, British Broadcasting Corporation. (2014). *Ten days left to vacuum up a powerful cleaner.* Retrieved February 25, 2015, from http://www.bbc.co.uk/news/business-28878432

E3, Equipment Energy Efficiency. (2015a). *Minimum Energy Performance Standards (MEPS).* Retrieved February 25, 2015, from http://www.energyrating.gov.au/about/other-programs/meps

E3, Equipment Energy Efficiency. (2015b). *Voluntary measures for reducing standby power in appliances and equipment.* Retrieved February 25, 2015, from http://www.energyrating.gov.au/wp-content/uploads/Energy_Rating_Documents/Library/Standby_Power/Standby_Power/2002sbpres6-harrington.pdf

Energy Conservation Centre Japan. (2015). *Final reports on the top runner target product standards.* Tokyo, Japan. Retrieved February 25, 2015, from http://www.eccj.or.jp/top_runner

European Commission. (2015). *RAPEX – Weekly notification reports.* Brussels, Belgium. Retrieved February 25, 2015, from http://ec.europa.eu/consumers/safety/rapex/alerts/main/index.cfm?event=main.listNotifications

KEMI, Kemikalieinspektionen. (2014). *Hälften av plastprodukterna innehöll farliga ämnen* [Half of the plastic products contained hazardous substances]. Stockholm, Sweden. Retrieved February 25, 2015, from http://www.kemi.se/en/Content/News/Half-of-the-plastic-products-contained--hazadous-substances

Lifset, R. (2014). Extended producer responsibility: Insights from the academic literature. In *Global forum on the environment: Promoting sustainable materials management through EPR*. Tokyo: Organization for Economic Cooperation and Development (OECD).

Martin, A. (2008). *Risk assessment and the management of environmentally hazardous substances in electrical and electronic equipment*. Doctoral dissertation, University of Surrey, Guildford.

Martin, A., Mayers, K., & France, C. (2007). The EU restriction of hazardous substances directive: Problems arising from implementation differences between member states and proposed solutions. *Review of European Community & International Environmental Law, 16*(2), 217–229.

Mayers, K. (2007a). Strategic, financial, and design implications of extended producer responsibility in Europe: A producer case study. *Journal of Industrial Ecology, 11*(3), 113–131.

Mayers, K. (2007b, January 12). *Lifecycle environmental management in the electronics sector*. Lecture, University of Surrey.

Mayers, K. (2011). *Product life-cycle management in the electronics industry. Presentation to life cycle assessment module*. Guildford: Centre for Environmental Strategy, University of Surrey.

Mayers, K., & Butler, S. (2013). Producer responsibility organizations development and operations. A case study. *Journal of Industrial Ecology, 17*(2), 277–289.

Mayers, K., Lifset, R., Bodenhoefer, K., & Van Wassenhowe, L. N. (2014). Implementing individual producer responsibility for waste electrical and electronic equipment through improved financing. *Journal of Industrial Ecology, 17*(2), 186–198.

US Department of Energy. (2015a). U.S. Energy Department, Pay-Television Industry and Energy Efficiency Groups Announce Set-Top Box Energy Conservation Agreement; Will Cut Energy Use for 90 Million U.S. Households, Save Consumers Billions. Washington, DC, USA. Retrieved February 25, 2015, from http://energy.gov/articles/us-energy-department-pay-television-industry-and-energy-efficiency-groups-announce-set-top

US Department of Energy. (2015b). *Regulatory processes*. Washington, DC, USA. Retrieved February 25, 2015, from http://energy.gov/eere/buildings/regulatory-processes-0

US Environmental Protection Agency. (2015). *Energy efficiency*. Washington, DC, USA. Retrieved February 25, 2015, from http://www.energystar.gov

Webb, A. (2014). *Evaluating games console electricity use: Technologies and policy options to improve energy efficiency*. Doctoral dissertation, University of Surrey, Guildford.

Dr. **Kieren Mayers** is Head of Environment and Technology Compliance at Sony Computer Entertainment Europe, London, and also Executive in Residence at INSEAD Social Innovation Centre in Fontainebleau.

Chapter 17
Multinational Corporations and the Circular Economy: How Hewlett Packard Scales Innovation and Technology in Its Global Supply Chain

Kirstie McIntyre and John A. Ortiz

Abstract Hewlett Packard discusses how companies can move from the conceptual ambiguity of the circular economy to operational reality. The development of the circular economy concept is described, in particular the extension from resource efficiency: the importance of moving from the idea of 'consumers' to 'users'. Transitioning from a linear economy to a circular one will require disruptive innovation. For more than 30 years, HP technologies have led large scale changes in a wide range of markets. We describe how HP is designing products and services which meet and enable circular economy applications. The examples demonstrate how a major multinational company like HP can build on its long-held resource efficiency principles to profitably drive industry forward in the circular economy. It is clear that the 'new style of IT' enables many future and current circular economy initiatives, from car sharing; community garden/power tool sharing and developing further connections between networks – i.e. the 'sharing economy'. The 'internet of things' has huge potential to retain and grow control over dispersed resources. Through collaborative technologies and partnerships, and by engaging the innovation potential of others, HP looks to lead the proliferation of full system solutions that can allow inventors and communities to design and innovate surpassing what can be imagined today.

Keywords Hewlett Packard • Circular economy • Up-cycling • Closed loop plastic manufacturing • Recycled content plastic • Servicization • 3D printing

K. McIntyre (✉) • J.A. Ortiz
Hewlett Packard, Bracknell RG121HN, UK
e-mail: kirstie.mcintyre@hp.com

© The Author(s) 2016
R. Clift, A. Druckman (eds.), *Taking Stock of Industrial Ecology*,
DOI 10.1007/978-3-319-20571-7_17

1   Circular Economy Introduction

The circular economy is a new buzzword that has caught the attention of a wide variety of actors within the public and private sector – particularly in the last 4 years or so. The circular economy has developed as a result of the contributions from various schools of thought (some inter-related). These include 'biomimicry', 'cradle to cradle', 'industrial ecology', 'resource efficiency' and the 'performance economy' (Masuda 2014).

Although definitions of circular economy seem to be converging, particularly towards the definition put forward by the Ellen MacArthur Foundation (Ellen MacArthur Foundation & McKinsey Co. 2014), a uniformly accepted definition is yet to be reached. In the past, public and private sector, actors have focused on the idea of resource efficiency – doing more with less. Eco-efficiency measures increase the ratio of units/value of products and services to environmental impact (positive or negative). According to the World Business Council for Sustainable Development (Schmidheiny 1992), 'eco efficiency is achieved by the delivery of competitively priced goods and services that satisfy human needs and bring quality of life, while progressively reducing ecological impacts and resource intensity throughout the life cycle to a level at least in line with the Earth's estimated carrying capacity. In short, it is concerned with creating more value with less impact'.

Potentially the starting point of the concept of circular economy comes from the 1966 paper, 'The economics of the coming spaceship earth' (Boulding 1966), which put forward the idea of circular material flows as a model for the economy. Boulding called for the need to consider the earth as a closed economic system where natural resources are limited. Following this industrial ecology made studies of the material and energy flows in industrial systems, using natural eco-systems as a guide to create sustainable schemes that operate in accordance with local and global ecological boundaries (Allenby 2006). Cradle to cradle thinking seeks to make the further distinction from eco-efficiency to 'eco-effectiveness'; the creation of 'cyclical' flows which allow materials to maintain their quality and status as a resource (up cycling) instead of minimizing cradle to grave material flows (Braungart et al. 2007).

There is some discrepancy between the older version of resource efficiency and the newer concept of circular economy. The understanding of resource efficiency often precludes the important idea of moving from 'consumers' to 'users' of durable goods in the economy. Conversely, circular economy can miss the elements of 'cleaner production' or 'eco-efficiency' measures. Currently it appears that the circular economy concept is better defined at the higher conceptual level, but not at the practical, operational level. Circular economy must make the shift from conceptual ambiguity to operational clarity if it is to be widely and uniformly incorporated by industry and governments (Masuda 2014).

What is clear to Hewlett Packard (HP) is the need to innovate in a resource constrained world. The traditional linear economy of 'take, make, consume, discard' will not long be viable where planetary resources are being 'overshot' earlier and earlier each year (considered to be August 19th in 2014) (Global Footprint Network 2014). For HP, the Circular Economy encompasses a system that is restorative or regenerative by intention with design that eliminates waste. As an alternative to the linear approach (take, make, use, discard), HP believes that connecting circular economy principles to resource efficiency is the route to success.

Resource efficiency is important to a global scale manufacturer like HP. We cannot abandon our 'design for environment' principles of:

1. Energy efficiency – reducing the energy needed to manufacture and use products
2. Materials innovation – decreasing the amount of materials used and selecting materials with lower environmental impact
3. Design for recyclability – designing equipment that has more value at end-of-life, is easier to upgrade and/or recycle.

(HP 2014a).

However, HP also understands the need for disruptive innovation to break through the limits of linear resource consumption models, irrespective of how efficient we become with the materials, energy, water and other resources we use.

2 Why Innovation in Circular Economy Is Important

Three main areas have emerged which have brought this opportunity into sharp focus:

1. Trends furthering the case for resource efficiency include:

    (a) Risk factors around resource availability and price volatility
    (b) Increased public opinion and government regulation concerning environmental and social issues

2. The opinion that traditional resource efficiency measures are insufficient to address current resource challenges
3. The alignment of conditions which will allow for the rapid diffusion of the circular economy such as:

    (a) The introduction of policies and regulations of governments around the world that support and promote the circular economy
    (b) Changing customer/consumer attitudes and other societal shifts that are essential for the diffusion of circular economy
    (c) The advancement of information technologies and other technologies which drastically increase the feasibility of circular economy.

Each of these drivers will be discussed in turn:

2.1 Resource Availability

As of 2015, sharp price increases in commodities since 2000 have erased the real price declines of the twentieth century. At the same time, price volatility levels for metals, food and non-food agricultural output in the first decade of the twenty-first century were higher than in any single decade in the twentieth century (McKinsey Global Institute 2013a). Extended producer responsibility is ever more important to public sector purchasing and individual consumption of durable and food goods. The Guardian Sustainable Business pages list climate change, supply chain responsibility, conflict minerals and factory workers' rights in their top 10 issues of 2014 (Buckingham 2014).

2.2 Resource Efficiency

In modern manufacturing processes, opportunities to increase efficiency still exist, but the gains are largely incremental and unlikely to generate real competitive advantage or differentiation. The latest IPCC report determined that the global emissions of greenhouse gases have risen to unprecedented levels, despite a growing number of policies to reduce climate change. Emissions grew more quickly between 2000 and 2010 than in each of the three previous decades (IPCC 2014). With three billion new middle class consumers expected to enter the market by 2030, current efficiency measures will not be enough to meet this demand.

2.3 Alignment of Conditions

Conditions for rapid diffusion are aligning. The IT sector and many other industries see increasing regulation being developed that moves beyond eco-efficiency into new forms of producer responsibility. The early adopters will be the ones to find competitive advantage and exponential growth. The millennial generation are more likely to demand access to services over ownership of products, for example through subscription services like Netflix or mobility access through car leasing (Ross 2014). Big data and data analytics helps companies to drive business growth by moving from 'transactions' to 'relationships' with their customers. This in turn drives increased brand loyalty, a concept which is well understood as valuable to industry via models such as Net Promoter Score (Reichheld 2003).

Having described circular economy principles and why HP thinks it is important for future business success, the remainder of this chapter will examine the real, at-scale programs currently underway at HP. These will demonstrate how a major multinational company like HP can build on its long-held resource efficiency principles to profitably drive industry forward in the circular economy.

3 The Shift from Conceptual Ambiguity to Operational Clarity

Scaling circular economy applications to a global level does not come without its complexities. In this section, three current programs at HP are described to demonstrate how circular economy principles are being applied now within a $130 billion business (Fig. 17.1).

At its core, a circular economy aims to design out waste; products are designed and optimized for a cycle of disassembly and reuse. For technical, durable products (like computers and printers), the circular economy largely replaces the concept of 'consumer' with 'user'. Unlike in today's buy-and-consume economy, durable products are leased, rented or shared wherever possible (Ellen MacArthur Foundation & McKinsey Co. 2014). In addition, new innovative technologies are required to fundamentally disrupt traditional manufacturing and supply chain models. New technological applications will enable greater collaboration in developing business models, job creation and further innovation.

HP's first example concerns closed-loop plastic manufacturing. While this may be considered a resource efficient measure, for many companies it is the first step to circular thinking. A report by Ellen MacArthur Foundation & McKinsey Co. for The World Economic Forum (Ellen MacArthur Foundation & McKinsey Co. 2015) has identified polypropylene plastic as a high potential material to demonstrate real change across supply chains. Our example shows the challenges and opportunities with successful scaling at a global level:

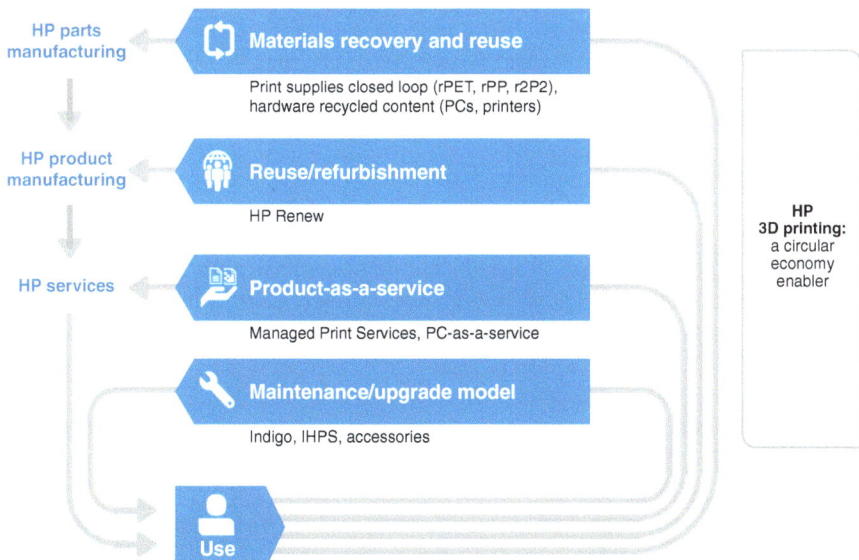

Fig. 17.1 HP circular economy diagram (Hewlett Packard 2015)

4   HP R2P2 program

This program highlights Circular Economy innovation of Inkjet Printer cartridges and demonstrates that by leveraging HP's scale, partnerships, customer relationships, materials knowledge and process innovation, we have closed material loops in technical grade polypropylene (PP).

1. Closed loop recycling with cascaded PP streams: HP inkjet cartridges returned by customers are collected, plastic separated, recovered and cleaned. They are combined with different cascaded PP streams (from other post-consumer applications) and plastic additives to create a ready-to-mold, 85 % recycled content plastic (RCP) replacement for virgin plastic resin.
2. Fully closed loop recycling with HP-only streams: HP inkjet cartridges returned by customers are collected, plastic separated, recovered, cleaned, re-pelletized and mixed at 20–30 % with virgin resin. This process allowed HP to accelerate the development cycle, providing a very solid example of developing "pure" material streams and then incorporating these streams into fully scaled manufacturing processes. With this most recent project, HP applied the knowledge and experience of the past 9 years of closed-loop recycling, reverse logistics, materials development, product design and separation technology to reduce the learning cycle from nearly 7 years, to approximately 9 months from project start to manufacturing ramp. This project began its worldwide production ramp on 27 October, 2014.

With the two PP projects fully implemented in manufacturing, HP expects to use nearly 5 million kilos of recycled PP in 2015. Combined with the multiyear effort on Polyethylene Terephthalate (PET), HP now uses approximately 10 million kilos of recycled plastic in inkjet cartridges. By 2014 year end, over 90 % of HP's inkjet cartridges shipped contain recycled plastic from closed loop and cascaded streams (Fig. 17.2).

HP has achieved this through collaborative effort with several key supply chain partners:

• Reverse logistics: HP is using multiple reverse logistics routes (channel partners, bulk enterprise customer shipments and logistics providers) to collect, process and recycle 10 million kgs of plastic material every year.
• Partner development: A collaborative recycling eco-system has been created to utilize the strengths of multiple partner suppliers including recycling, material refining, material development and plastics compounding.
• Recycling process and equipment development: This is a very critical piece of innovation which enables pure material streams to be used without requiring performance trade-offs.
• HP is driving recycling process innovation and recycling equipment development including patented recycling equipment and $Ms invested in product disassembly tooling.

Fig. 17.2 Design changes to high volume ink cartridges to allow for increased recycled plastic content (Hewlett Packard 2015)

- Cascading streams and materials development (Up cycling post-consumer materials from other products and industries). Developing up cycling solutions for low value materials (materials from garment industry used in products).
- Together with its suppliers, HP has developed analytical and functional quality tests for incoming RCP (recycled content plastic) feed streams to ensure product quality is not compromised. This is particularly important when considering for cascaded streams to be up-cycled into HP products.

  Certain challenges of using RCP have been overcome in the course of this work:

- Impurities – potential impact on ink quality. The ink inside cartridges is a complex chemical formulation designed to perform under specific conditions. HP doesn't consider that it sells ink or cartridges, customers want the printed page. Customers also expect that the output is perfect from the first page to the last.
- Molding and performance – negative impact to production molds and tools can cost millions of dollars in damage and lost manufacturing time.
- Product component dimensional tolerances (reduced flexing in large sheet components), manufacturability (ease of manufacturing), yield (reducing material waste) and supply chain implications.
- The assembly and performance in the customer's hands.
- The qualification of new resin formulations for manufacturing applications can be a slow process. HP has worked to reduce time for internal certification.
- Design teams have been involved in decisions around material properties and color and cosmetics of finished products. These technical discussions have debunked the commonly held belief that recovered/recycled materials are inferior through product and process performance testing and quality analysis. RCP materials have not resulted in any documented customer quality issue/concern.

- There have been significant market challenges to overcome: the main one has been assurance of supply – particularly with regard to consistency and quantity.
- Pricing stability has been found to be key. HP commodity purchasers are closely involved with comparisons to virgin materials to ensure there is no financial impact from using RCP. A real-time example comes from the late 2014 oil price slump which has created new challenges for the adoption of recycled plastic. Virgin resin prices have dropped as much as 30 % per month (HP internal data, 2015).

The economics of recycled plastic are interesting and merit further study. Recycled plastic providers who 'float' their prices with the virgin resin demand will most likely feel a 'pinch' when petroleum pricing drops. However, HP believes that recycled plastic can be priced independently of virgin resin when operating in a circular model. This makes most sense when long-term collaborative relationships are established within the supply chain.

HP intends to continue this path of increasing recycled content in products and components, wherever possible using its own closed-loop material streams. The next step in this program is to extend into further plastic types and more product families, thus disrupting the perception that closed-loop solutions can only be possible in small volume, localized situations. HP demonstrates that a world-wide solution involving millions of kilograms of material every year is indeed possible.

The next example from HP concerns an inner circle of the circular economy diagram: that of service models. This is where products are leased or rented to customers, either with a service contract or not. The products remain within HP's control which makes closing material loops, refurbishment, maintenance and repair much simpler and cost-effective.

5   HP Device-as-a-Service Program

HP is currently developing business models in all levels of circularity. Market forces and new customer norms are growing and shaping the company's portfolio of device-as-a-service product offerings to become a larger part of its business portfolio. Current programs are offered to all types of customer, from the home user to the large enterprise such as Government departments or companies like banks or manufacturers.

- *HP Renew* – This program currently remanufactures server, storage and networking products, offering the same reliability, functionality and warranty as new HP products. In 2013, 3.7 million units were processed through HP's 5 facilities across the world. Returned, loaned or trial units are completely remanufactured so they can be fully utilized and are not unnecessarily wasted. HP Renew is a full-scale worldwide product example of remanufactured and redeployed servers, storage and networking solutions.

- *MPS – Managed Print Services (printers)* – Printing is offered to enterprise and small businesses as a service. Through a variety of financing or lend/lease business offerings, customers can choose their level of involvement with the management of devices. HP today has nearly 1 million devices managed as a service, and the average age of the 'printer in management' is 5–7 years. At printer end-of-service, approximately 75 % of those printers are refurbished and remarketed to other customers. The remainder are determined unusable in their whole state, therefore go for recycling by certified providers, thus providing potential feed-stock into other closed loop material streams.
- *HP subscription services* (computers) – Launched in June 2014, this product offering meets the needs of small and medium businesses that need to scale their computing capabilities very quickly and have less desire to own products. A formal program has started now with several HP channel partners (resellers) who will sell bundles of hardware and services under a lease from HP Financial Services. Started in the UK, France and Spain, it is too early to show detailed results, however early feedback from customers and reseller partners is very positive. This is a good example of where business benefits from a circular product service model, customers are satisfied and the profit margins secure the long-term sustainability of the business model.
- *Instant Ink* – Inkjet printer ink supplies are managed for home users to address the recognised 'pain points' of (1) Ink availability (running out at the crucial moment), and (2) Ink cost. Through internet-connected printers, HP's Instant ink service allows customers to pay for only the ink they need to print, on a monthly basis. Ink is re-supplied automatically to the home via the national postal service. There is no contract and no get-out fee. Cartridge recycling is offered as part of the service. There are further opportunities to close material loops from consumer printers, HP will be exploring this in the future. By enabling this business model for users, HP's product designers can reduce material consumption by 45–65 % over traditional supplies purchase models.

The above programs briefly describe some of the device-as-a-service models that HP currently offers to the market. The success factors in these programs are as follows:

- The service must be what the customer wants.
- Products offered must not be perceived to be of lower value than outright purchased goods.
- The service must be easy (and fun) to use.
- Closing material/product loops to prolong product lives is key.
- The business model must be profitable for the company in order to facilitate further innovation to eliminate waste.

Connected technology is also key to the ease of use and rapid dissemination of such service models for durable goods such as IT products. HP service and repair businesses are using such technology to manage capital assets in customer premises. With 'Visual Remote Guidance – Integrated with Google Glass ™', HP has pioneered remote support on its IHPS (Inkjet High-Speed Print Systems) equipment. Customers can diagnose and service their printing systems with the virtual assistance of an HP technician viewing their problems in real time. This technology is intended for capital industrial printing systems now, but has far-reaching implications in the service economy.

In the last example described below, it is shown how innovative technology can be used to disrupt and transform accepted norms in manufacturing and supply chains.

6   HP Multijet Fusion 3D Printing

The HP-led digitization of printing revolutionized the industry, turning printing upside down, reducing waste and inefficiency (Hewlett Packard 2014b). It created the ability to print completely unique material with variable data, instead of being constrained by one master with many copies, transforming supply chains and industries as a result. Allowing printing to become personalized, localized, customized, targeted and unique, the results are valued and not 'thrown away', reducing waste. One example is grocery store circulars. By enabling one client to customize their magazines by local customer base, their circulars went from being 32 pages to just four, while conversion rates rose dramatically. Digitization has also led the print-on-demand revolution. In publishing, 30–40 % books are unsold. With digital printing, copies can mirror demand, whilst the digital print process prints pristine pages first time round, instead of creating large volumes of waste whilst presses are set up.

3D printing presents HP with another tremendous opportunity to transform supply chains and industries. Identified as one of '12 disruptive technologies that will transform life, business and the global economy' (McKinsey Global Institute 2013b), 3D printing offers the ability to produce – both rapidly and inexpensively – short runs or one-of-a-kind parts. In contrast to traditional manufacturing which typically cuts, grinds or molds raw materials into shape, 3D printing builds to shape.

In addition, 3D printing will revolutionize part manufacturing and the part distribution supply chain by offering local, on-demand production. It is easy to envisage the local car repair garage printing the replacement part for your car, rather than waiting for it to be delivered from inventory held elsewhere. The connection to the circular economy is clear, 3D print technology removes the need to hold large, potentially redundant inventories of spare parts. Maintenance and repair business models become more financially attractive; products are designed to be repaired, upgraded and maintained thus prolonging their lives and eco-effectiveness.

The following paragraphs explain how HP's version of 3D printing addresses some of the current restrictions and technical difficulties of the existing technology.

In additive manufacturing technology – commonly called "3D printing" – objects are built from selective addition of material rather than by molding or by traditional methods of subtractive machining, where material is removed by cutting and grinding. Candidates for 3D printing include the functional and aesthetic components of machines, consumer and industrial products that are produced in short runs – typically less than 1000 units, and, in particular, highly customized and high-value products that may be one-of-a-kind. Because 3D printing builds objects from cross-sections, complex parts – previously requiring multiple elements that were welded or assembled together – can now be built either as a monolithic structure or from fewer subcomponents. For example, some types of 3D printing can produce parts with hollow internal structures and complex 3D internal passages (for air or other fluids) that once required several sections to be fitted together with sealing surfaces between them.

HP's vision for 3D printing is the revolution of part manufacturing (how parts are made) and the part distribution supply chain (where and when parts are made). In the near term, affecting the creative process by making far more useful parts available to a much broader audience. And in the longer term, disrupting supply chains with 3D printing technology. In order for that disruption to occur, there must be significant changes in the economics of 3D printing and in the standards for maintaining quality.

Current 3D printing machines could be categorized in two groups, machines that produce smooth parts with good detail, and machines that produce parts with good strength. Because of the materials that are currently used to produce smooth parts with good detail, this group of machines does not make parts with good strength. In contrast, because of point energy needed to produce parts with good strength, this group of machines does not produce smooth parts with detail. Further, many existing processes fuse or cure the materials together at a focused point, for example using a focused laser beam to fuse, or using a single nozzle to extrude. This point-processing limits the build speed of these technologies. In the end, adoption of current technologies may be limited by imperfect parts, and slow productivity.

As with many 3D printing processes, HP Multi Jet FusionTM technology starts by laying down a thin layer of material in the working area. Next, the carriage containing an HP Thermal Inkjet array passes from left-to-right, printing chemical agents across the full working area. The layering and energy processes are combined in a continuous pass of the second carriage from top-to-bottom. The process continues, layer-by-layer, until a complete part is formed. At each layer, the carriages change direction for optimum productivity.

High productivity can lead to challenges in making quality parts. For parts to work, it's important to ensure that the material has been properly fused and that part edges are smooth and well-defined. To achieve quality at speed, HP invented a proprietary multi-material printing process where the materials are applied by HP Thermal Inkjet arrays. In addition to fusing and detailing agents, HP's technology can employ additional materials to transform properties at each volumetric pixel (or voxel). Color and even different materials can be used in the same print run to produce complex, multi-dimensional parts.

To realize this full potential of 3D printing, HP's vision is to develop a 3D printing platform designed to become an industry standard, and HP is inviting creative collaboration in materials for 3D printing. These breakthroughs in materials and agent-material interactions can power the widespread adoption of 3D design and hardware innovation resulting in a digital transformation of manufacturing as widespread and profound as the way HP's Thermal Inkjet solutions changed traditional printing. Software to manage the design process is equally important, the current shortcomings of the CAD-based format in terms of processing time and object dimensional precision are a barrier for the production of complex, high-precision parts by new technologies such as HP Multi Jet Fusion technology. Furthermore, this format only allows geometric representation, so it does not allow voxel-by-voxel (volumetric pixel) information to be carried from the CAD software to the printer. To realize the full potential of 3D printing, the roadmaps of 3D printers and 3D CAD software must be aligned, and the roadmaps must be accompanied by a change to a more information-rich file format.

Comparison to commercially available 3D printing technologies has demonstrated clear advantages to HP's technology and its material set to define new levels of part quality, high part functionality, at 10 times the build speed and at much improved economics. A key feature of the technology is the potential to modify material properties to produce controlled variations of the mechanical and physical characteristics within a part, i.e. parts can have different materials built-in during the manufacturing process, instead of being welded or connected later. This can enable many new possibilities in the design and performance of parts built by 3D printing.

7 Conclusion

Moving from a linear economy to a circular one will require disruptive innovation. For more than 30 years, HP technologies have disrupted and led printing technologies in a wide range of markets. This chapter has described how HP is designing products and services which meet and enable circular economy applications. HP has a long history of resource efficiency – of doing more with less – this is accepted good business sense. HP also knows that there are good business advantages in extending into the circular economy.

- Saving resources = lower virgin material spend (greater profitability)
- New, convenient business models = happy customers (and repeat business)

The extension of effort does not come without its challenges:

- High grade plastics: consistency – quantity – quality
- Closing our own material loops (from service models)
- Expanding service offerings to more customers

HP is working to overcome internal and industry perceptions of recycled content materials, business model profitability and the need to market new business values to customers. There is further work to align incentives throughout HP's supply chain; from customers (users) through channel partners (resellers) to manufacturing design teams. Managing across geographies is key to multinational businesses, being able to move products (both new and for refurbishment) across the world is vital to establishing the economies of scale which make circular business models financially viable.

It is clear that the new style of IT enables many future and current circular economy initiatives from car sharing, community garden/power tool sharing and developing further connections between people and manufactured goods. The 'internet of things' has huge potential to retain and grow control over dispersed resources. The interconnection of uniquely identifiable embedded computing devices (e.g. sensors) within the existing internet infrastructure is expected to offer advanced connectivity of devices, systems and services that goes beyond machine-to-machine communications (M2M) and covers a variety of protocols, domains and applications. M2M applications will allow both wireless and wired systems to communicate with other devices, in situations such as industrial automation, logistics, Smart Grids, Smart Cities, health and defence for monitoring and control purposes.

Through collaborative technologies and partnerships, and by engaging the innovation potential of others, HP looks to lead the proliferation of full system solutions that can allow inventors to design and build assemblies that have form and function surpassing what can be imagined and manufactured today.

About HP Hewlett Packard (HP) delivers innovation in printing, personal computing, software, services and IT infrastructure. HP offers the industry's broadest portfolio, most expansive scope and deepest industry expertise to deliver value and improved outcomes for customers in almost every country in the world. The company is at the forefront of technological innovations that advance the way society lives and works, enabling it to play a vital role in enabling sustainable growth.

Operating in more than 170 countries, Hewlett Packard has long been a leader in global citizenship – one of its seven corporate objectives since 1957. With more than seven billion people seeking greater prosperity worldwide, balancing economic growth with environmental sustainability calls for innovation and leadership. Working towards a more sustainable world, HP responds to this challenge by improving the efficiency of products and solutions, supply chain and operations.

By combining the expertise of HP's 308,000 people, its innovative technology portfolio and collaborative partnerships, the company is working to develop and share solutions that streamline and replace resource-intensive processes. Building on its size and scale, HP believes it is uniquely positioned to advance solutions that improve lives and make the world a better place. The business will move forward by reducing HP's own environmental footprint, and that of its customers, while helping people prosper and companies thrive.

References

Allenby, B. (2006). The ontologies of industrial ecology? *Progress in Industrial Ecology An International Journal, 3*, 28–40.

Boulding, K. (1966). The economics of the coming spaceship earth. In H. Jarrett (Ed.), *Environmental quality in a growing economy* (pp. 3–14). Washington, DC: Resources for the Future.

Braungart, M., McDonough, W., & Bollinger, A. (2007). Cradle to cradle design: Creating healthy emissions – A strategy for eco-effective product and system design. *Journal of Cleaner Production, 15*(13–14), 1337–1348.

Buckingham, F. (2014, December 24). Top 10 sustainability campaigns 2014. *The Guardian.* Retrieved February 19, 2015, from http://theguardian.com/sustainable-business/2014/dec/24/top-10-sustainability-campaigns-2014

Ellen MacArthur Foundation and McKinsey Co. (2014). *Towards the circular economy: Aaccelerating the scale up across global supply chains.* Geneva: World Economic Forum.

Ellen MacArthur Foundation and McKinsey Co. (2015). *Project MainStream – A global collaboration to accelerate the transition towards the circular economy.* Geneva: World Economic Forum.

Global Footprint Network. (2014). *Global footprint network at a glance.* Retrieved February 19, 2015, from http://footprintnetwork.org/en/index.php/GFN/page/at_a_glance

Hewlett Packard. (2014a). *Living progress report 2013.* Retrieved February 24, 2015, from http://www8.hp.com/us/en/hp-information/global-citizenship/reporting.html

Hewlett Packard. (2014b). *HP multi jet fusion TM technology: A disruptive 3D printing technology for a new era of manufacturing.* Retrieved February 19, 2015, from http://h10124.www1.hp.com/campaigns/ga/3dprinting/4AA5-5472ENW.pdf

Hewlett Packard. (2015). *Internal data from commodity procurement specialists.* Hewlett Packard.

IPCC. (2014). *Climate change 2014: Impacts, adaptation, and vulnerability.* Geneva: International Panel on Climate Change.

Masuda, S. (2014). *What does the concept circular economy mean to a large corporation: Case of Hewlett Packard and the IT/high-tech industry.* MSc thesis, IIIEE Lund University.

McKinsey Global Institute. (2013a). *Resource revolution: Tracking global commodity markets trends survey 2013.* Retrieved February 19, 2015, from www.mckinsey.com/insights/energy_resources_materials/resource_revolution_tracking_global_commodity_markets

McKinsey Global Institute. (2013b). *Disruptive technologies: Advances that will transform life, business, and the global economy.* Retrieved February 24, 2015, from www.mckinsey.com/insights/business_technology/disruptive_technologies

Reichheld, F. (2003, December). One number you need to grow. *Harvard Business Review.* Retrieved February 19, 2015, from https://hbr.org/2003/12/the-one-number-you-need-to-grow/ar/1

Ross, D. (2014, March 26). *Millennials don't care about owning cars, and car makers can't figure out why.* Fast Company, Co.Exist. Retrieved February 19, 2015, from http://fastcoexist.com/3027876

Schmidheiny, S. (1992). *Changing Course: A global business perspective on development and the environment.* WBSCD, 1992 Rio Earth Summit. Cambridge: Massachusetts Institute of Technology.

Chapter 18
The Industrial Ecology of the Automobile

Roland Geyer

Abstract For the last 100 years, virtually every automobile was an internal combustion vehicle (ICV) powered by either gasoline or diesel and mostly made from steel. Even as the ICV was identified as a source of serious environmental impact, it continued to outcompete others, arguably more environmentally benign, transportation modes. Banning lead from gasoline, requiring catalytic converters, and increasing powertrain efficiency allowed the ICV to respond to environmental criticism and continue its dominance over other transportation technologies. Today, well over one billion ICVs are in use worldwide.

Since the turn of the last century, however, this dominance is beginning to be contested, not so much from other transportation modes but from alternative automotive designs and fuels, such as biofuels, lightweight materials, and fuel cell, hybrid, and battery electric powertrains. All of these alternatives are meant to decrease the environmental impacts of cars, but in all cases there is concern about trade-offs, unintended consequences, and regrettable substitutions. This chapter discusses history and recent developments of automobiles from an industrial ecology perspective. Such a perspective is necessary to determine the extent to which the emerging automotive technologies can genuinely reduce rather than simply shift the environmental impacts of automobiles.

Keywords Industrial ecology of automobiles • Environmental sustainability of cars • Biofuels • Advanced powertrains • Lightweight automotive materials

1 Introduction

Since time immemorial people and goods had been transported by horse-drawn carriages. This changed in the late nineteenth century, when self-propelled carriages started to appear in Europe and the United States. In 1897 the New York Times predicted that "the mechanical wagon with the awful name automobile […] has

R. Geyer (✉)
Bren School of Environmental Science and Management, UCSB,
3426 Bren Hall, Santa Barbara, CA 93106-5131, USA
e-mail: geyer@bren.ucsb.edu

© The Author(s) 2016
R. Clift, A. Druckman (eds.), *Taking Stock of Industrial Ecology*,
DOI 10.1007/978-3-319-20571-7_18

come to stay." The newspaper went on to say that "man loves the horse, and he is not likely ever to love the automobile" (Cohn 2009). We all know which one of those two predictions was wrong.

After a century of undisputed domination, the gasoline- or diesel-powered internal combustion vehicle (ICV) finally has to contend with some serious competition. The staple automotive material, steel, has also come under considerable competitive pressure. Interestingly, all contenders, be they biofuels, hydrogen, hybrid or pure electric powertrains, aluminum, or fiber-reinforced polymers, are all marketed as ways to reduce the environmental impacts of cars. For this reason, the demand for industrial ecology expertise, especially life cycle assessment, has increased significantly in the automotive world. While all these developments are relatively recent, the history of environmental concerns caused by cars is almost as old as the history of the car itself.

The modern automobile, or car, was first created in Europe in the late nineteenth century by inventors and entrepreneurs such as Karl Benz, Gottlieb Daimler, and Wilhelm Maybach. It is based on four-stroke gasoline or diesel engines, invented, among others, by Nikolaus Otto and Rudolf Diesel, even though cars using steam engines and electric motors were also developed at that time. While electric vehicles (EVs) enjoyed considerable success in the early twentieth century, continuous improvement of ICV design and performance together with a steady decline in ICV prices and increasing availability of gasoline lead to an eventual demise of the EV industry by 1920. After the turn of the century, supply and demand of ICVs started to increase rapidly, both in Europe and in the United States. While France was initially the largest producer of vehicles, it was soon overtaken by the United States, which introduced and perfected mass production of vehicles. No car epitomizes the affordable, mass-produced automobile more than the Model T, introduced by Henry Ford in 1908. Over 15 million models were produced worldwide by the time Ford ceased production of the Model T in 1927.

The first environmental drama began to unfold in 1921 when Thomas Midgley, who was working for Charles Kettering at the General Motors Research Corporation, discovered tetraethyl lead's (TEL) excellent antiknock properties and patented it (Kitman 2000). Both were aware of viable antiknock alternatives to TEL which couldn't be patented, such as ethanol. The toxicity of lead had been known for several thousand years, and the proposal to use TEL as gasoline additive almost instantly sparked public health controversies. Acute lead poisoning was common in the early TEL production plants. In fall 1924, 5 of 49 TEL workers in Standard Oil's Bayway Refinery in New Jersey died of acute lead poisoning, and 32 had to be hospitalized. As a result, New York City, New Jersey, and Philadelphia banned leaded gasoline, jeopardizing GM, DuPont, and Standard Oil's plan to make TEL the leading antiknock additive. To address and preempt growing public health concerns about TEL, General Motors commissioned research from the US Bureau of Mines in 1923 and asked the Surgeon General Hugh Cumming to hold public hearings in 1925. The hearings were inconclusive and charged an expert committee to further investigate the public safety of TEL. While some experts mentioned the risk of chronic exposure, all official reports and statements focused on the risk of acute lead poisoning and eventually declared TEL's use as gasoline additive safe. The existence of less toxic alternatives, such as ethanol, was ignored by both the industry and relevant public health officials.

By the early 1960s, TEL was in virtually all US gasoline and was quickly expanding in the rest of the world. Around the same time cars were identified as a major source of photochemical smog in highly motorized areas such as Los Angeles. That ICVs powered by (leaded) gasoline cause significant environmental problems finally became undeniable when scientists started to notice dangerous and rising levels of lead in the environment and human blood, and the smog caused by cars went from bad to worse. In the early 1970s, US car makers decided to use catalytic converters to meet the emerging tailpipe emission standards. This was bad news for TEL, which poisons catalytic converters. At the same time the recently founded US EPA started to consider phasing out leaded gasoline to reduce chronic lead exposure. TEL was eventually banned in California in 1992 and in the rest of the United States in 1996. In the EU, catalytic converters became mandatory in 1990, and lead was finally banned in 2000.

By then, the use of lead in gasoline had caused catastrophic levels of lead pollution. While lead levels in human blood decrease quickly in regions where leaded gasoline is banned, TEL is still used in many developing economies, and elevated levels of lead can be found in virtually every corner of the earth. Banning TEL required the use of an alternative antiknock. The United States and other countries decided to use methyl tertiary butyl ether (MTBE) to replace lead, a typical material substitution approach to pollution prevention. Unfortunately, MTBE is highly water soluble, and even small fuel spills can contaminate large amounts of groundwater. MTBE may also be a carcinogen. This is an example of the environmental trade-offs that are frequently involved in substitution approaches. As a result, the use of MTBE has been phased out in the United States, which now uses ethanol as antiknock and oxygenate, the same substance that was ignored in the 1920s. There seems to be a certain amount of reinventing the wheel in environmental problem solving. The rediscovery of reusable bags, containers, and packaging come to mind here.

Three-way catalytic converters are classic end-of-pipe technology designed to control pollution. They are extremely successful in reducing CO, NO_X, and hydrocarbon emissions from vehicles but require platinum and slightly reduce powertrain performance. More importantly, it could be argued that they have enabled staggering levels of ICV ownership and use. This means that photochemical smog is still a major problem in areas like Los Angeles, only now caused by vast numbers of low or ultralow-emission vehicles as opposed to the fewer cars with high emissions in the 1960s. Also, catalytic converters do nothing to CO_2, so the enormous proliferation of ICVs, partially enabled by this end-of-pipe technology, leads to an equal increase in automotive greenhouse gas (GHG) emissions, which has finally come under scrutiny. In 2006, the UNFCCC reported rising GHG emission trends and noted that "in particular, transport remains a sector where emission reductions are urgently required but seem to be especially difficult to achieve."

Initially, environmental automotive regulation focused on the air pollutants CO, VOC, NO_X, and PM. After the oil crisis in 1973, the United States also added fuel economy standards. Today, over 70 % of the global new vehicle market is subject to GHG and/or fuel economy standards (Miller and Façanha 2014). This worldwide commitment to automotive emission reductions has led car manufacturers to rethink the automobile. The prevalent car design, the steel-based ICV powered by gasoline or diesel, is being challenged by alternative fuels, powertrains, and structural materials. The following sections will discuss these developments from an industrial ecology perspective. Such a perspective is necessary to determine whether these alternatives offer overall environmental impact reductions or instead shift burdens to other life cycle stages or other environmental concerns.

2   Biofuels

Biofuels are not an invention of the modern environmental movement but were commonplace until coal began to fuel the industrial revolution in the second half of the eighteenth century. The diesel engine at the World Fair in Paris in 1900 ran on peanut oil, and Rudolf Diesel himself believed that vegetable oil would become an important fuel. An early version of Otto's engine ran on ethanol. The Model T was designed to run on gasoline or ethanol, and Henry Ford thought that ethanol was the fuel of the future. In the 1930s gasoline blended with ethanol from corn was proposed in the United States to support its ailing agriculture. High oil prices and oil shortages during World War II and the oil crises in the 1970s briefly renewed US interest in corn ethanol. These phases were short-lived, however, and gasoline and diesel from petroleum became and remained the exclusive fuels for the growing fleet of ICVs in the United States.

The same is true for the rest of the world, with the exception of Brazil, where ethanol from sugarcane has been used to fuel cars since the 1920s. Brazilian ethanol production increased steadily until cheap oil became consistently available after World War II. However, prompted by the oil crises in the 1970s, Brazil launched a National Ethanol Program in 1975 (Garten Rothkopf 2007). Among other things, this program included ethanol subsidies and mandated that all gasoline be blended with ethanol at certain ratios and that ethanol be sold at lower prices than gasoline. As a result, Brazil became the world's largest fuel ethanol producer and consumer by far. In the 1980s oil prices tumbled to historic lows, where they stayed until the end of the millennium. This eroded the economic case for ethanol, and Brazilian production was relatively flat during that period at around 11–15 billion liters per year (EIA 2015).

Between 1981 and 2001, annual corn ethanol production in the United States increased at a slow but steady pace from 0.3 to 6.7 billion liters, which was mainly fostered by subsidies. After 2000, progressive replacement of MTBE with ethanol further helped to increase US production. However, the big boost for US ethanol came with the creation of the Renewable Fuel Standard (RFS) program in the Energy Policy Act of 2005 and its expansion in the Energy Independence and Security Act of 2007. In 2007, the United States produced 18.5 billion liters of

ethanol and overtook Brazil as the world's largest producer. The surge in corn ethanol production in the United States was accompanied by an increasingly heated
debate about its energy and GHG benefits. A growing number of so-called fuel
cycle or well-to-wheel studies became available with a wide range of contradictory
findings. Fuel cycle or well-to-wheel analyses are essentially life cycle assessments
(LCAs) of fuels, even though many of the early studies were from researchers outside of the LCA community and without reference to existing LCA standards.

Studies by Patzek and Pimentel received particular media attention as they found
that, over its life cycle, corn ethanol requires more fossil energy inputs than it has
calorific value and emits more GHGs than gasoline. Studies from other research
groups, however, concluded that cumulative fossil energy demand and life cycle
GHG emissions of corn ethanol are substantially lower than those of gasoline. A
meta-analysis intent on settling the controversy was probably one of the first LCAs
published in the journal *Science*, even though it never mentions the term LCA
(Farrell et al. 2006). Unsurprisingly, the study found that the wide range in results
was due to differences in inventory data, system boundaries, and coproduct allocation. It concluded that the GHG savings of corn ethanol are moderate but those of
cellulosic ethanol substantial. Unfortunately, producing cellulosic ethanol, also
called second-generation biofuel, is much more difficult than starch- and sugar-
based ethanol, since it is very hard to break down the lignocellulosic feedstock in an
economically viable way. So hard, in fact that the US Environmental Protection
Agency (EPA) retroactively reduced the 2013 RFS target volume for cellulosic ethanol from 1 billion gallons to 810,185 gallons (EPA 2014a).

The environmental reputation of biofuels received its next challenge in 2008,
when two studies in the same issue of *Science* reported their findings on the GHG
implications of land use change (LUC) (Fargione et al. 2008; Searchinger et al.
2008). Fargione et al. found that clearing land for fuel crop production creates a
significant "carbon debt" and that biofuels require 17 to 420 years to generate GHG
savings of the same size. Searchinger et al. argued that using feedstock from existing fields does not avoid this issue since it induces indirect land use change (iLUC)
by removing the crop from its prior market. For example, corn used for ethanol is
now missing as animal feed, which causes land conversion for new corn production
elsewhere. Searchinger et al. conclude that corn and cellulosic ethanol have higher
GHG emissions than gasoline when iLUC is included. Naturally, these strong findings were contested by many, including biofuel associations and the US Department
of Energy. California's Low Carbon Fuel Standard (LCFS) and the new RFS include
GHG emissions from iLUC but with conflicting results. The controversy about
LUC and iLUC continues. Both effects are prime examples of consequential LCA
and thus question the usefulness of attributional LCA for environmental decision
making (Plevin et al. 2014). It is interesting to note that none of the original LUC
and iLUC researchers came from the industrial ecology or LCA communities.

The next twist in the biofuel saga came the following year with two more *Science*
publications. The first pointed out that turning fuel crops into electricity for battery
electric vehicles (BEVs) rather than biofuels for ICVs would roughly double crop-
to-wheel conversion efficiency (Ohlrogge et al. 2009). The second showed how this

translates into substantially larger life cycle energy and GHG benefits, even if you consider that BEVs have significantly larger cradle-to-gate production energy inputs and GHG emissions than equivalent ICVs (Campbell et al. 2009). However, one major drawback of any sun-to-wheels transportation pathway based on biomass is that the energy conversion efficiency of photosynthesis is typically below 1 % (Blankenship et al. 2011). This means that vast areas of land are needed to harvest significant amounts of solar energy (McDonald et al. 2009). A much more efficient alternative would be direct photovoltaic conversion into electricity. Such a PV-BEV system is orders of magnitude more land use efficient than even the most optimistic biomass scenarios and has equal or higher energy and GHG benefits (Geyer et al. 2013). PV-powered BEVs are conceptually appealing but have some technical and operational challenges, one of which is the timing of PV power supply and EV charging demand.

3 Powertrains

Electric vehicles had all but vanished by 1920, apart from some niche applications such as the iconic British milk float. The modern era of the EV began when General Motors (GM) unveiled a BEV prototype called Impact at the 1990 Los Angeles Auto Show. This was encouraging news for the California Air Resources Board (CARB), which had been working on a low-emission vehicle (LEV) program to help areas such as Los Angeles meet federal air quality standards (Collantes and Sperling 2008). CARB had come to the conclusion that improvements in conventional powertrains alone would not achieve the required emission reductions. As a result, CARB added a so-called zero-emission vehicle (ZEV) mandate to the LEV program of 1990. The mandate specifies that car sales of the major manufacturers had to be composed of at least 2 % ZEVs by 1998, 5 % by 2001, and 10 % by 2003. A ZEV is defined as having no tailpipe emissions of air criteria pollutants. CARB clearly had BEVs in mind, but since its regulation has to be technology neutral, it pointed out that fuel cell vehicles (FCVs) would also meet the definition. The ZEV mandate is arguably the single biggest driver behind the emergence of alternative powertrains. It is interesting to note that it emerged from concerns over air quality and not oil resources or climate change. In the United States, fuel economy can only be regulated at the federal level. After a White House proposal to increase fuel economy standards failed in congress in 1992, the Clinton administration started the Partnership for a New Generation of Vehicles (PNGV) with the goal to develop dramatically more fuel-efficient powertrains (Malakoff 1999). The research collaborative, which was cancelled in 2001 by the Bush administration, focused on diesel-electric hybrids and FCVs rather than BEVs.

In late 1997 Toyota's Prius, the first mass-produced hybrid-electric vehicle (HEV), went on sale in Japan. A few years later, Honda and Toyota started selling HEVs in the United States. In contrast, only a number of concept vehicles were created under the PNGV program. Measured in ZEV sales, California's ZEV mandate was also not a success. Between 1996 and 2003 just over 4,400 BEVs, such as GM's EV1 and Toyota's electric RAV4, were leased or sold (Bedsworth and Taylor 2007). The ZEV mandate had to be amended many times to make it achievable. First, the 1998 and 2001 ZEV sales requirements were dropped. Next, new vehicle categories and alternative compliance pathways were created. It became possible to substitute BEV sales with larger sales of HEVs and smaller sales of FCVs. If this sounds all very complicated that's because it is. Thanks to the LEV program and its ZEV mandate, California has now a veritable zoo of vehicle categories. Ten years after GM introduced the Impact, BEVs were all but forgotten again. HEV sales climbed steadily, though, and more and more car manufacturers offered hybrid-electric versions of their models. At the same time FCVs were increasingly seen as the automotive endgame, with car companies and governments making bold announcements about the impending rollout of hydrogen cars and infrastructure. While mass-produced FCVs always appeared to be another 5 years away, BEVs returned with a roar in the form of the Tesla Roadster in 2008. Since then many BEV and plug-in hybrid-electric (PHEV) models have entered the market, the most successful of which are the Nissan Leaf, the Chevy Volt, and the plug-in Prius. And just when people started to wonder whether hydrogen cars were a pipe dream after all, Toyota revealed the Mirai at the 2014 Los Angeles Auto Show, the first commercially available FCV.

All four challengers of the incumbent ICV involve an electric motor and a traction battery. This allows all of them to recover and store the car's kinetic energy through regenerative braking. However, motors and batteries differ in size and the way they are used. In parallel HEVs, the electric motor is combined with an internal combustion engine (ICE), and both provide torque to the wheels. With typical values between 1 and 2 KWh, HEVs have the smallest traction batteries and thus the smallest all-electric driving range. HEVs still use liquid fuels, typically gasoline or diesel, as their exclusive energy source. In PHEVs the traction battery can be charged directly from an external electric power source. With typical values between 5 and 10 KWh, it is larger than in HEVs, which increases all-electric driving range. BEVs and FCVs use only electric motors for traction and typically don't contain any internal combustion engines. An interesting exception is the Chevy Volt which has a gasoline engine but uses it only to charge the battery. FCVs have batteries for intermediate energy storage but use hydrogen tanks for main energy storage. The fuel cell converts the hydrogen into electricity. Hydrogen is an energy carrier, not an energy source, and needs to be produced first. Earlier plans for on-board hydrogen production, e.g., through hydrocarbon reforming, are no longer being pursued. In BEVs the only energy storage device is the battery, and the only traction device is the motor. BEVs therefore have the simplest powertrains but also require the largest batteries.

Battery technology, in particular cost and energy density, has improved substantially over the years and is a key determinant in alternative powertrain choice and design. GM's EV1 used lead-acid batteries. Toyota's HEVs use nickel metal hydride (NiMH) batteries with roughly double the energy density. All BEVs use lithium ion (Li-ion) chemistries with roughly four times the energy density of lead-acid batteries. It is thus the energy density of Li-ion batteries that enabled the latest reemergence of the BEV, even though they still have smaller driving ranges and longer charging times than ICVs. It has also been pointed out that EVs are only as clean as the electricity they use, a somewhat obvious observation for industrial ecologists (Moyer 2010).

Relative to the incumbent ICV, alternative/advanced powertrains have higher tank-to-wheel energy efficiency but also higher cradle-to-gate production impacts, due to the nature of their components, such as batteries, fuel cells, and electric motors (Demirdöven and Deutch 2004; ANL 2014). In the case of HEVs, it is relatively simple to show that the fuel savings far outweigh the additional production impacts. Life cycle comparisons of the other alternative powertrains are complicated by the fact that they use electricity and hydrogen as fuel, which can be produced in many different ways (Samaras and Meisterling 2008; Notter et al. 2010; Hawkins et al. 2012). Currently, most hydrogen is produced through steam reforming of hydrocarbon fuels. To eliminate the need for fossil fuels, it is frequently stated that the hydrogen for FCVs should ideally come from electrolysis of water powered by renewable electricity. However, it would be considerably more energy efficient to use renewable electricity directly in BEVs rather than convert it into hydrogen through electrolysis and then back into electricity in a fuel cell. The detour via hydrogen has the advantage, though, that hydrogen is easier to store than electricity.

4 Lightweight Materials

In addition to more efficient powertrains, the PNGV also researched lightweight materials for vehicle mass reduction. Such a mass reduction increases the fuel economy of the vehicle without reducing its size. The use of lightweight materials is usually also seen as necessary to compensate for the higher mass of advanced powertrains. A material is regarded as lightweight if it achieves significant mass reduction relative to mild steel without compromising other design parameters, but there is no precise definition. The considered materials are typically aluminum and magnesium alloys, fiber-reinforced polymers, and advanced high-strength steels (AHSS) (DOE 2014). With the exception of AHSS, the primary production of lightweight materials has significantly higher environmental impacts than mild steel production. In fact mass reduction potential appears to be correlated to production impacts (Geyer 2013). Again, LCA is required to quantify the trade-off between the increase in material production emissions and the decrease in vehicle use phase emissions. The trade-off needs to be studied on a case-by-case basis, but different studies of similar cases frequently yield conflicting results. There is significant debate about the amount of mass reduction lightweight materials can achieve in practice, since this is not directly observable and has to be either modeled or derived from analysis of proxy data sets. The same is true of the relationship between vehicle mass reduction and fuel economy improvement. Initial use of simplistic rules of thumb is slowly being replaced by physics-based powertrain models (Koffler and Rohde-Brandenburger 2010). It turns out, for example, that the regenerative braking and the higher efficiency of advanced powertrains significantly reduce the impact of vehicle mass reduction on fuel economy. This challenges the gospel that advanced powertrains require lightweight materials. Other sources of uncertainty are the assumed total mileage of the vehicle and, as always, the inventory data of the involved processes, such as material and fuel production.

By far the most contentious issue, however, is the question of how recycled content and end-of-life recycling impacts the net environmental benefits of lightweight automotive materials (Geyer 2008). The controversy over how to account for material recycling is generic to LCA and not specific to vehicle mass reduction. There is a plethora of literature explaining, comparing, and reviewing the various existing recycling methodologies. In the case of lightweight automotive materials, changing recycling methodology can change the rank-ordering of the results, which is highly unsatisfactory. Consequential system expansion is the only way to determine the actual effects of material recycling. Environmental studies of lightweight materials, just like those of biofuels, therefore call into question the usefulness of attributional LCA for public policy making. Attempts at consequential LCA, on the other hand, highlight the large uncertainties intrinsic to consequential analysis. Car manufacturers all know and use LCA and are well aware of its ambiguities in particular with regard to recycling. Policy makers are currently reluctant to change automotive emission regulations from tailpipe to life cycle, regardless of the fact that the latter perspective is superior in principle.

As a result, all public policy on automotive GHG emissions focuses of fuel economy or tailpipe CO_2 (Miller and Façanha 2014). None use a full life cycle perspective; in particular vehicle production impacts are ignored by all of them. Many car manufacturers therefore see lightweight materials as an important way to meet these standards. So far, Ford made the boldest move and decided to make the body structure of the 2015 model of its most successful vehicle, the F150 pickup truck, entirely aluminum. Ford states that this enabled mass reductions of up to 700 pounds (318 kg) and fuel economy improvements of up to 20 % relative to 2014 model. While it is clear that such a dramatic change to America's best-selling vehicle is an enormous economic gamble, it is unclear what the net climate change impacts of this move are. Rather than trying to predict the consequences of such a change, say through consequential LCA, we are now running the experiment. Luckily, this experiment is bound to have a less dramatic outcome than the one of adding lead to gasoline.

5   Conclusions

The use of automobiles experienced phenomenal growth ever since cars started being mass-produced just over 100 years ago. Today, well over one billion vehicles are in use worldwide (OICA 2015). In 2013 alone, over 65 million cars and almost

22 million commercial vehicles were added. Thanks to rapidly developing econo-mies like China and India, there is no end of this growth in sight.

Serious efforts to reduce the environmental impacts of this ever-growing vehicle fleet are relatively recent. In the EU, catalytic converters became mandatory only 25 years ago, and lead was banned only 15 years ago. The United States moved earlier to reduce air pollutants from cars but is lagging in terms of fuel efficiency. In fact, the fuel economy of new light-duty vehicles in the United States declined between 1987 and 2004 (EPA 2014b). This trend was driven by increases in vehicle weight, power, and acceleration and also the growing share of so-called sports utility vehi-cles (SUVs), wiping out all advances in engine and powertrain efficiency. These trends are currently flat or at least increasing more slowly.

It is unlikely, though, that this is enough to reduce the environmental impacts from a huge and growing global car fleet to acceptable levels, which is why more and more decision makers are looking for a new automotive paradigm. It is cur-rently unclear what will be the future fuel, powertrain, or even material of the car. It is clear, however, that the tools and concepts of industrial ecology could and should play a vital role in evaluating environmental trade-offs and avoiding unintended consequences. Humans have a substantial track record of causing large environmen-tal problems, the conventional ICV being one of them. Yet humans are also starting to build a track record of solving environmental problems. Let's hope that with the enlightened use of industrial ecology, the future automobile will be one such solution.

References

ANL. (2014). GREET 2 2014. Argonne National Laboratory (ANL). Retrieved January 30, 2015, from https://greet.es.anl.gov

Bedsworth, L. W., & Taylor, M. R. (2007). Learning from California's zero-emissions vehicle pro-gram. *California Economic Policy, 3*(4), 1–19. Public Policy Institute of California.

Blankenship, R. E., et al. (2011). Comparing photosynthetic and photovoltaic efficiencies and rec-ognizing the potential for improvement. *Science, 332*, 805–809.

Campbell, J. E., Lobell, D. B., & Field, C. B. (2009). Greater transportation energy and GHG offsets from bioelectricity than ethanol. *Science, 324*, 1055–1057.

Cohn, S. (2009). *It happened in Chicago.* Guildford: The Globe Pequot Press.

Collantes, G., & Sperling, D. (2008). The origin of California's zero emissions vehicle mandate. *Transportation Research Part A, 42*, 1302–1313.

Demirdöven, N., & Deutch, J. (2004). Hybrid cars now, fuel cell cars later. *Science, 305*, 974–976.

DOE. (2014). *Lightweight materials R&D program.* DOE/EE-1039. United States Department of Energy (DOE). Retrieved January 30, 2015, from http://energy.gov/eere/vehicles/downloads/vehicle-technologies-office-2013-lightweight-materials-rd-annual-progress

EIA. (2015). *Fuel ethanol production.* Energy Information Agency (EIA). Retrieved January 30, 2015, from http://www.eia.gov/cfapps/ipdbproject/IEDIndex3.cfm

EPA. (2014a). *EPA issues direct final rule for 2013 cellulosic standard.* EPA-420-F-14-018. United States Environmental Protection Agency (EPA). Retrieved January 30, 2015, from http://www. epa.gov/otaq/fuels/renewablefuels/documents/420f14018.pdf

EPA. (2014b). *Light-duty automotive technology, carbon dioxide emissions, and fuel economy trends: 1975 through 2014.* EPA-420-S-14-001. United States Environmental Protection Agency (EPA). Retrieved January 30, 2015, from http://www.epa.gov/otaq/fetrends.htm

Fargione, J., Hill, J., Tilman, D., Polasky, S., & Hawthorne, P. (2008). Land clearing and the biofuel carbon debt. *Science, 319*, 1235–1238.

Farrell, A. E., Plevin, R. P., Turner, B. T., Jones, A. D., O'Hare, M., & Kammen, D. M. (2006). Ethanol can contribute to energy and environmental goals. *Science, 311*, 506–508.

Garten Rothkopf. (2007). *A blueprint for green energy in the Americas, Chapter IV.* Brazil. Inter-American Development Bank. Retrieved January 30, 2015, from http://www.gartenrothkopf. com/research-and-analysis/custom-research-publications.html

Geyer, R. (2008). Parametric assessment of climate change impacts of automotive material substitution. *Environmental Science & Technology, 42*(18), 6973–6979.

Geyer, R. (2013). *UCSB auto materials GHG model, Version 4.* Retrieved January 30, 2014, from http://www.worldautosteel.org/life-cycle-thinking/greenhouse-gas-materials-comparison-model

Geyer, R., Stoms, D., & Kallaos, J. (2013). Spatially-explicit life cycle assessment of sun-to-wheels transportation pathways in the U.S. *Environmental Science & Technology, 47*(2), 1170–1176.

Hawkins, T. R., Singh, B., Majeau-Bettez, G., & Hammer Strømman, A. (2012). Comparative environmental life cycle assessment of conventional and electric vehicles. *Journal of Industrial Ecology, 17*(1), 53–64.

Kitman, J. L. (2000). The secret history of lead. *The Nation*, March 2, 2000. Retrieved January 10, 2015, from http://www.thenation.com/article/secret-history-lead

Koffler, C., & Rohde-Brandenburger, K. (2010). On the calculation of fuel savings through lightweight design in automotive life cycle assessments. *The International Journal of Life Cycle Assessment, 15*, 128–135.

Malakoff, D. (1999). U.S. supercars: Around the corner, or running on empty? *Science, 285*, 680–682.

McDonald, R. I., Fargione, J., Kiesecker, J., Miller, W. M., & Powell, J. (2009). Energy sprawl or energy efficiency: Climate policy impacts on natural habitat for the United States of America. *PLoS ONE, 4*, e6802.

Miller, J. D., & Façanha, C. (2014). *The state of clean transport policy.* Washington, DC: International Council on Clean Transportation. Retrieved January 30, 2015, from http://www. theicct.org

Moyer, M. (2010). The dirty truth about plug-in hybrids. *Scientific American, 303*(1), 54–55.

Notter, D. A., Gauch, M., Widmer, R., Wäger, P., Stamp, A., Zah, R., & Althaus, H.-J. (2010). Contribution of Li-Ion batteries to the environmental impact of electric vehicles. *Environmental Science & Technology, 44*(17), 6550–6556.

Ohlrogge, J., Allen, D., Berguson, B., DellaPenna, D., Shachar-Hill, Y., & Stymne, S. (2009). Driving on biomass. *Science, 324*, 1019–1020.

OICA. (2015). *Production statistics and vehicles in use.* Organisation Internationale des Constructeurs d'Automobiles (OICA). Retrieved January 30, 2015, from http://www.oica.net

Plevin, R. J., Delucchi, M. A., & Creutzig, F. (2014). Using attributional life cycle assessment to estimate climate-change mitigation benefits misleads policy makers. *Journal of Industrial Ecology, 18*(1), 73–83.

Samaras, C., & Meisterling, K. (2008). Life cycle assessment of greenhouse gas emissions from plug-in hybrid vehicles: Implications for policy. *Environmental Science & Technology, 42*, 3170–3176.

Searchinger, T., et al. (2008). Use of U.S. croplands for biofuels increases greenhouse gases through emissions from land-use change. *Science, 319*, 1238–1240.

Chapter 19
Quantifying the Potential of Industrial Symbiosis: The LOCIMAP Project, with Applications in the Humber Region

Malcolm Bailey and Andrew Gadd

Abstract The Humber region, in North East England, is a major hub of industrial activity and trade. It has seen applications of industrial symbiosis for many years, initially centred on 'top-down' infrastructure projects with large capital investment but subsequently following a 'bottom-up' approach engaging industries in the area. Reductions in GHG emissions and waste generation have already been impressive. The possibilities for further savings, recognising the European Union's aspirations for deep GHG cuts and the objectives of the A.SPIRE partnership involving 114 stakeholders from the process industries in Europe, have been explored in the LOCIMAP (low-carbon industrial manufacturing parks) project, which involved partners from across Europe. Industrial symbiosis has been central in the plans for LOCIMAP from the outset. Studies conducted for LOCIMAP have revealed that more substantial savings require industrial symbiosis to be designed in, rather than developed once facilities exist. Major further savings depend on co-location of activities in eco-industrial parks to enable systematic process integration, but following this approach raises further questions, including:

- How can such systems be engineered without compromising safety?
- What are the implications for system resilience?
- How does close integration affect operations such as maintenance?

The project has shown that we have the engineering ability to achieve deep reductions in energy use and GHG emissions provided industries can be located in eco-industrial parks with interactions designed according to thermodynamic principles. Barriers to realising this concept, to achieve a new industrial revolution, include an economic and fiscal system which means that design for optimal economic performance leads to different outcomes from designing for optimal environmental performance.

Keywords Eco-industrial parks • Energy integration • Humberside • Industrial symbiosis • Resource innovation

M. Bailey (✉) • A. Gadd
Link2Energy Ltd, 1-3 Bigby Street, DN20 8EJ Brigg, North Lincolnshire
e-mail: malcolm@link2energy.co.uk; andrew@link2energy.co.uk

© The Author(s) 2016
R. Clift, A. Druckman (eds.), *Taking Stock of Industrial Ecology*,
DOI 10.1007/978-3-319-20571-7_19

343

1   Introduction: Brief History of Industrial Symbiosis in the Humber Region

The Humber is the largest river system in the UK and is responsible for draining 20 % of the UK's land mass. It captures drainage waters as far south as Birmingham through the Trent River system and north into Yorkshire through the Ouse. The river itself is the boundary between Yorkshire and Northern Lincolnshire. It is a major navigation channel and its banks are home to significant industrial clusters. The natural geography and geology of the Humber bring significant advantages for the location of industry. The deep water channel swings first to the south bank and then to the north. The ports of Immingham, Grimsby and Hull are located at these touching points and in combination make up the largest tonnage port complex in the UK. Twenty-seven per cent of UK's refining capacity is located here; 25 % of UK's rail freight originates here, much of which finds home in the steel and power industries; just north of the Humber estuary, 20 % of UK's gas is landed at Easington, fed by the Langeled pipeline, the world's longest underwater pipeline, via the Sleipner gas processing platform (Fig. 19.1).

In addition to ports and logistics, the chemical industry is a highly significant cluster. With a turnover of £6 billion, it provides employment for 10,000 people within 120 companies that include global brands such as BP and Croda Chemicals. The food sector, based historically on the fishing industry, has been dominant in the

Fig. 19.1 Arial overview of Immingham/South Killingholme (industrial areas in *orange*)

towns of Grimsby and Hull. Grimsby, known as Europe's Food Town, is home to a high concentration of food producers, supported by cold storage, logistics, engineering and packaging services. The town is synonymous with seafood, including brands such as Findus and Young's, and represents 70 % of the entire UK seafood processing capacity. Steel processing at Scunthorpe has its origins in the iron deposits of the Jurassic limestone escarpment that is found at either side of the estuary.

Seventeen per cent of UK's power-generating stations including Drax, Eggborough, Ferrybridge, Cottam, Radcliffe, Humber Power and Centrica lie inland from the Humber region. With only 8 % of UK's electricity demand, the Yorkshire and Humber region is a net exporter of power to the rest of the country. The heavy industrial base of the wider Yorkshire and Humber economy is the source of 23 % of the CO_2 emissions of England and Wales (89 Mte) (Environment Agency) and over six million tonnes of commercial and industrial waste arisings (Defra 2011, 2014) (Fig. 19.2).

Fig. 19.2 Strategic location of the Humber within the North Sea basin

Immingham CHP is one of the largest combined heat and power (cogeneration) plants in Europe. The 1,220 MWe facility provides steam and electricity to Phillips 66's Humber Refinery, steam to the neighbouring Lindsey refinery and merchant power into the UK market. With more recent regional investment in wind power and the bioethanol plant at Saltend that add to the existing CCGT and CHP power plants, the Humber has been positioned as the 'Energy Estuary'.

This rich industrial activity of the Humber and the wider Yorkshire and Humber region have made it a natural home for applying industrial symbiosis principles (see Chertow 2007 and Chap. 5). The early work on industrial symbiosis on Humberside around the turn of the millennium was centred on the large capital projects of the region. These included the Immingham CHP plant, studies on a 'Humber bundle' considered to link the chemical and gas generation plants of the north and south bank through a strategic pipelines crossing beneath the river, together with a further major study of material streams within the Humber's chemical industry. This 'top-down' approach, considering major infrastructure initiatives, was later replaced with a 'bottom-up' widespread engagement of industrial partners taken by the NISP-Humber from 2003 onwards (Mirata 2004). This was one of the three pilot regions at the start of the UK National Industrial Symbiosis Programme (NISP); it became the NISP-Yorkshire and Humber region in 2005. Over the subsequent 5 years alone, this programme engaged with 700 companies in Yorkshire and Humber and documented CO_2 reductions of 780,000 tonnes per annum for its clients and a reduction of 1,400,000 tonnes in material being landfilled.

Link2Energy Ltd ran these latter two industrial programmes on behalf of International Synergies Ltd. Initially the funding was from the regional development agency Yorkshire Forward and subsequently from the Department of Food and Rural Affairs (DEFRA). From 2012 Link2Energy Ltd developed its own independent commercial programme *Re:Sourcing UK* with a focus on high-value opportunities and innovation. It is against this background and experience that Link2Energy Ltd was invited to provide industrial symbiosis expertise into the European FP7-funded project LOCIMAP, below.

The strategic importance of the Humber region has also been recognised by a study carried out by the University of Surrey on the Evolution and Resilience of Industrial Ecosystems (ERIE) project. Two further papers draw extensively on the work in the Humber region, 'Habitat' Suitability Index Mapping for Industrial Symbiosis Planning (Jensen et al. 2012) and Quantifying 'Geographic Proximity': Experiences from the United Kingdom's National Industrial Symbiosis Programme (Jensen et al. 2011).

2 The LOCIMAP Project

Europe's 2020 growth strategy (EC 2012) commits to limiting greenhouse gas emissions by 20 % compared to 1990 levels, creating 20 % of Europe's energy needs from renewables and increasing energy efficiency by 20 %. Within this broad strategy, an international non-profit public-private partnership (PPP) has been set up: A Sustainable Process Industry through Resource and Energy Efficiency (A.SPIRE)[1] with the private sector as a partner. A.SPIRE has set further targets for 2030, below. It was launched as part of the Horizon 2020 framework programme, which is in turn the biggest EU research and innovation programme ever with nearly €80 billion of funding available over 7 years (2014 to 2020).

A.SPIRE represents more than 114 industrial and research process industry stakeholders from over a dozen countries spread throughout Europe. It was established through the joint efforts of 8 industry sectors: chemical, steel, engineering, minerals, non-ferrous metals, cement, ceramics and water. The mission of A.SPIRE is to ensure the development of enabling technologies and best practices along all the stages of large-scale existing value chains that will contribute to a resource-efficient process industry. Through purposeful cooperation across all sectors and regions, A.SPIRE has developed a strategic roadmap that addresses research, development and innovation activities as well as policy matters towards the realisation of its 2030 targets. The ultimate goal is to promote the deployment of innovative technologies and solutions required to reach long-term sustainability for Europe and its process industries in terms of global competitiveness, ecology and employment. The 2030 targets are:

1. A reduction in fossil energy intensity of up to 30 % from current levels through a combination of, for example, energy-saving processes (including enhanced use of optimisation techniques, monitoring and modelling via ICT tools), process intensification, energy recovery, sustainable water management, cogeneration (i.e. combined heat and power) and progressive introduction of alternative (renewable) energy sources within the process cycle.
2. A reduction of up to 20 % in non-renewable, primary raw material intensity compared to current levels by increasing chemical and physical transformation yields and/or using secondary or renewable raw materials. This may require more sophisticated and more processed raw materials.

The A.SPIRE PPP objective is to develop the enabling technologies and solutions along the value chain, required to reach long term sustainability for Europe in terms of global competitiveness, ecology and employment.

The low-carbon industrial manufacturing parks (LOCIMAP) project was funded by the European Commission under the Framework 7 programme. It was set up to explore this very conundrum, examining not only technical but also economic and business (and even political) factors as it sought to contribute to the development of a roadmap for a closed-loop economy in a major continent with the longest industrial history and a relatively stable population. The project was managed by North East Process Industry Cluster (NEPIC) and included participation from four industrial parks: the chemical park at BASF Española SL in Spain, Wilton in the UK represented by Sembcorp Utilities (UK) Ltd, Kokkola Industrial Park in Finland and Kalundborg Municipality Denmark.

---

[1] http://www.spire2030.eu

LOCIMAP considered the potential for the integration of a number of high CO2-emitting industries and had industrial partners within its team to bring in a high level of detail on process operations within the key sectors; these were Cemex UK Cement Limited, Papiertechnische Stiftung (PTS) from Germany representing the pulp and paper industry, VDEh-Betriebsforschungsinstitut GmbH (BFI) the iron and steel industry also from Germany and Terreal of France for the ceramic industry. In addition Phillips 66 provided key information on petrochemical operation. Four technical specialist companies participated in the technical and business analyses: Parsons Brinckerhoff Sp. z.o.o. from Poland, Svenska Miljöinstitutet (IVL) from Sweden, Link2Energy Ltd from the UK and the Institut Europeen D'Administration Des Affaires (INSEAD) from France. External communications were managed by the European Chemical Site Promotion Platform (ECSPP) from the Netherlands.[2]

The project concept was to re-examine the structure of industrial parks using best practice benchmarks from across leading parks in the EU and elsewhere, the focus being to substantially improve energy and resource efficiency, reduce CO_2 emissions and improve competitiveness. From the outset, substantial advantages were anticipated through migration towards more completely integrated manufacturing, centred on concentrating activities in new industrial park structures supported by dedicated services optimisation.

At one of the LOCIMAP steering groups, a chance remark was made that what we need to underpin the future of manufacturing in Europe is a new industrial revolution. The context was of course a comment on the magnitude of the change that this project was looking to define in order to make a tangible impact on the competitiveness of European industry. From the outset, the ambition was to much more than the squeezing out of a few percentage points in process efficiency. Rather the goal was to explore the potential for a step change breakthrough.

Conscious of the old adage that 'if you always do what you've always done, then you always get what you've always got', this challenge for an industrial revolution further emboldened the project to be courageous in its thinking. The LOCIMAP project therefore looked to explore a range of parameters that could make such a change and has included not only technical but also economic and business (and even political) factors. At the heart of the analysis was the construction of a 'virtual industrial park' where full optimisation of the industrial processes could be considered with all technical and business constraints removed.

Even before a presentation of the options, the reader may already, and rightly, be thinking that any such technical changes, no matter how good, may be replicated globally; therefore any grounds for optimism may be a respite rather than a long-term solution and still leave us with the original question as to how we secure the competitiveness of the European manufacturing base. The response to this is, in part, to refer to the potential of the closed-loop economy. Europe alone, of all the major continents, has a relatively stable population; if we consider that we have all the materials we need within the economy, then we can indeed close the material

---

[2] http://www.locimap.eu

loops. Coupled with the high-value innovative and entrepreneurial landscape that we need to make this happen, do we indeed have the basis for this new industrial revolution? This question and how we might take steps towards this within a low-carbon economy are addressed in general in Chaps. 6 and 7. The LOCIMAP project sought to address the question specifically within the context of industrial parks.

3 LOCIMAP: Guiding Principles

Industrial symbiosis was recognised as the key principle to achieving this industrial revolution by the LOCIMAP project. It represents the cornerstone of not only a low-carbon industrial manufacturing park but a low-carbon economy. This business principle, if not the name, has been in operation as long as industry has existed; it is greatly underexploited.

A number of sectors have prepared roadmaps to a low-carbon industry, e.g. Cement Technology Roadmap 2009 (OECD et al. 2009), Steel's Contribution to a Low-Carbon Europe 2050 (Wörtler et al. 2013), Paving the Way to 2050: The Ceramic Industry Roadmap (Cerame-Unie 2012), Unfold the Future: The Forest Fibre Industry 2050 (CEPI 2011) and Roadmap to a low-carbon bio-economy: An Aluminium Roadmap to a low carbon Europe: Lightening the Load (EAA 2012). These are highly professional documents, but (with the possible exception of the cement industry and to some extent the paper industry) they are in many cases limited in their vision, principally restricting the analysis to their own sector. To be truly effective, industrial symbiosis needs to be cross-sectoral. More than that, it should go beyond the optimisation of resources in a given location to the selection of the location at the point of business investment. Ultimately, the driver for the uptake is simply that of best practice business delivering both competitivity and environmental benefits.

The requirement is to deploy the principles of industrial symbiosis in a bold and strategic way. For example, why shouldn't the blast furnace off-gas from the iron and steel industry be used as syngas within the chemical industry? To illustrate the point, many factories within the chemical or petrochemical industry have as the unavoidable result of established processes – and despite the efforts of generations of engineers – significant quantities of low-grade heat. Similarly, many food factories have a demand for cooling. The potential for the chemical industry supplying refrigeration capacity through absorption chilling opens up a solution that is not available within single sectors. This is a far-ranging comment since it implies a breadth of imagination at not only the business and engineering level but also that of strategic planning. Though it may not be always practicable for two sectors to be cheek by jowl geographically, this does not prevent them from having a thermodynamic umbilical link. Again, these examples draw in the requirement to be strategic and to include local government, planners and politicians alongside business and engineering.

For industrial symbiosis to deliver its full commercial potential in terms of engineering efficiency, it needs to move beyond the first-generation approach of material transfer. This is certainly not to dismiss the benefit of 'long-radius' synergies involving material transfers from further afield; the positive carbon credit for use of material by-products against the extraction and processing of virgin materials is eminently quantifiable. However, real progress is only achievable by including the integration of heat and power in 'short-radius' synergies with associated heat and power supply networks in dedicated industrial eco-parks. It will clearly be an economic impossibility to optimise utility systems that rely on close proximity between partner organisations unless they really are co-located. The park concept is key to this. Furthermore, within a CO_2 minimisation agenda, supply chain integration is likely to be subservient to the industrial symbiosis objective. This is not the case now, where logistics represent the primary threads holding the supply chain together. This shift will require reconfiguration or even a redefinition of the supply chain, but this will need an effective powerful driver either of long-term policy or of carbon price.

Residual 'low-grade' heat presents a major opportunity for further improving the already good CO_2 performance of our industrial parks. However, we suggest that a change in mindset is needed so that its use is considered from the outset, as part of the overall process design. Whilst district heating systems for residential areas are good, they suffer from high seasonality of demand; what is required is a constant demand of industrial proportions. The example already cited, regarding the development of district cooling systems powered by residual heat through absorption chilling systems, requires bold planning moves such as co-location in the food industry, with its typically high demand for refrigeration, cold stores, data centres, etc., alongside the sources of residual heat, particularly the chemical, petrochemical and power industries.

As a consequence of this study, Link2Energy Ltd has proposed such a system for the South Humber Bank; waste heat from the petrochemical plants is being considered to provide cooling for a cluster of food companies 10 km away with the two sites potentially linked through a utility corridor within the coastal industrial strip.

However, not all residual heat is low grade. By way of example, the energy flows through an integrated steel complex dwarf those of most other industries, and the ability to recover waste heat from the cooling of products and from the slags, though inevitably difficult and a technological challenge (that is being grasped), is a significant source of potential high-grade heat and hence of reduced emissions.

The recovery of heat is clearly only of value if there is a home for it. Industrial clustering on parks is paramount to capitalise on this potential. The challenge is clear: how can we better integrate our industries and deliver collective CO_2 emission reductions rather than leave industries isolated and subject to carbon leakage pressures?

Furthermore, each future park may benefit from having its own technology centre for evaluating the optimum utility configurations of the resident industries in a dynamic setting, for assessing new opportunities for the valorisation of by-products and for the deployment of new technology. There are certainly examples of this across Europe (e.g. CPI at Wilton, Chemelot Campus, etc.)

Most of the specific sector roadmaps cite *technology advances* within their plan for 2020 and 2050 targets. In most instances, the ways suggested to bring about a reduction in CO_2 footprint are sector specific. An example is the ULCOS project.[3] ULCOS stands for ultra-low carbon dioxide (CO_2) steelmaking. It is a consortium of 48 European companies and organisations from 15 European countries that have launched a cooperative research and development initiative to enable drastic reduction in carbon dioxide (CO_2) emissions from steel production. The consortium consists of all major EU steel companies, of energy and engineering partners, research institutes and universities and is supported by the European commission. The aim of the ULCOS programme is to reduce the carbon dioxide (CO_2) emissions of today's best routes by at least 50 %. Other initiatives include the deployment of high-efficiency kilns as examples of material changes such as the development of artificial pozzolans within the cement industry (OECD et al. 2009). However, if the challenge is to deliver a low-carbon economy, then the ultimate requirement is to base the design of that (industrial) economy on thermodynamic principles (Bakshi et al. 2011).

LOCIMAP brought together an array of existing and emerging technologies that individually offer an ability to recover and to transfer heat from one process to another and which are key to achieving the ambition for integration. Further, the management of such integrated systems will demand an overarching control philosophy and deployment of ICT (see LOCIMAP White Paper 3).

Link2Energy Ltd is active in developing opportunity within both the Humber and Tees river basins for the deployment of flameless oxy-combustion FPO systems. Suitable for a wide variety of wet waste materials including renewables, the technology is capable of very high operating efficiency and low emissions due to the absence of nitrogen. It has also pioneered the development of hydrothermal carbonisation for the treatment of organic materials including poultry litter and is applicable in several industries within the Humber hinterland.

Process techniques such as *pinch technology* make it possible to define minimum utility consumption for individual processes and also the optimum energy target for integration of quite disparate processes. This powerful technique has been used extensively within the process industries in particular. LOCIMAP extended the technique to the application across a virtual industrial park. The end result is a powerful blueprint for integrating processes from the chemical, petrochemical, pulp and paper, fine chemical and biofuel sectors together with some from the iron and steel and non-ferrous metal industries.

The LOCIMAP project has sought to minimise the release of carbon through intelligent synergies between industries. Processes are under development elsewhere to use CO_2 as a feedstock. Some products may be manufactured from flue-gas CO_2, e.g. cyclic carbonates, but the market for these products is small in comparison with present emissions. Carbon capture and storage is the ultimate backstop, and many of the individual sector roadmaps forecast the importance of this technology within their 2050 view on CO_2 reductions. It may become increas-

---

[3] http://www.ulcos.org

ingly important, and indeed an asset, for future industrial parks to be geographically located on a carbon capture and transportation highway umbilically linked to a carbon storage facility. Carbon management and CO_2 *sequestration* will be at the heart of a future low-carbon industrial park and become part of the utility system.

In the Yorkshire and Humber region, National Grid is helping to develop solutions to reduce the carbon dioxide (CO_2) emissions from power stations and industrial plants. A solution being explored is carbon capture, transportation and storage (CCS) technology – capturing carbon dioxide emissions and transporting them to be stored permanently beneath the seabed in natural porous rock formations or depleted oil and gas fields. If approved, the Yorkshire and Humber CCS Cross-Country Pipeline project[4] will involve the construction of a cross-country pipeline and a subsea pipeline to transport carbon dioxide from fossil fuel power stations and industrial plants in the region to a permanent geological storage site beneath the North Sea. The onshore pipeline would be 75 km long and would use the same sort of technology as the national high-pressure gas pipeline network, owned and operated by National Grid. It would be up to 24″ (about 600 mm) in diameter and buried at least 1.2 m below ground. The carbon dioxide would be transported in liquid form at a pressure of 150 barg. The subsea pipeline would be the same size and on the seabed. Offshore, the carbon dioxide would be transported at a pressure of up to 200 barg to a geological storage site beneath the North Sea. The pipeline would have the capacity to transport up to 17 million tonnes of carbon dioxide every year. The long-term aspiration is for the pipeline to form the foundation of a regional CCS network, potentially capturing tens of millions of tonnes of carbon dioxide every year.

As noted above, the major themes within LOCIMAP, the optimisation of steam and power systems, cannot be realised within *supply chain integration* unless the manufacturing units are co-located. It will clearly be an economic impossibility to integrate utility systems that rely on close proximity between partner organisations unless that is the case. Within the CO_2 agenda, supply chain integration is subservient to the industrial symbiosis question. This may even lead to a reconfiguration or even a redefinition of the supply chain.

The *waste industry* will play an increasing role within the industrial landscape of such symbiotic parks through the provision of feedstock. Whilst industrial symbiosis and the exchange of industrial by-products as feedstocks are vital in future industrial parks, the importance of post-consumer waste as a feedstock will also grow. For some elements, e.g. copper, it is recorded that there is more material in the technosphere rather than the geosphere, and there is much concern over the availability of a range of other 'critical raw materials'. In some cases the concentration of these materials is greater in post-consumer and industrial wastes than in the virgin ore; some process are natural concentrators of the ore, e.g. the levels of germanium and gallium in coal ashes are inevitably almost 100 times that in the coal. Despite many critical raw materials being nonindigenous to Europe, recycling and recovery rates of such elements and compounds are still amazingly low.

---

[4] http://www.ccshumber.co.uk

These changes in feedstock supply will have a bearing on energy demands and CO_2 emissions. The aluminium industry is a good and current case of an industry that has already shifted focus towards a recycled feedstock and shows the improved economics and environmental performance through recycling of aluminium (e.g. cans) compared with the life cycle implications of producing virgin metal (see Chap. 6).

Resource innovation, the recovery of component parts of 'waste streams', whether that is critical raw materials from mineral based industries, proteins and flavonoids from industrial food wastes or phosphates from water discharges, represents an improvement in utilisation of finite resources which will need to shape future policy and approach. Avoiding the closure of material loops is not a future option; this is not a matter of principle, but ultimately one of economics and the emerging industries will be integral to a changing mix on the future industrial park.

Again, to lean on the thermodynamic argument for materials as well as for our utility studies, the more dissipated our resources, the higher the entropy and the more the energy required to recover them. It is surely better to sprout a new industry on a collocated industrial park to recover critical material from fly ash before we provide it as an ingredient to make cement.

Link2Energy has a legacy of successful material exchanges that minimise material being sent to landfill. However more recent examples of resource innovation in the Humber area relate to the development of a marine biorefinery for the extraction of phospholipids from fish and particular salmon skins. The project, which was funded by the Innovate for Growth competition run by the Technology Strategy Board, engaged a Grimsby-based food factory, academia and a local company specialising in extraction technology. It has replicated this collaborative approach with academia and industry for a number of other high-value resource innovation projects. These include the extraction of proteins and peptides from reject potatoes, flavonoids from waste citrus fruit and rare earth metals from by-product residues from the mineral industry. The company is also engaged with valorisation of alkaline leachates from the steel industry as part of a 3-year study funded by the Natural Environment Research Council (NERC).

To realise the benefits identified in LOCIMAP, it will be necessary to challenge existing *approaches to business*. These challenges are twofold. The intra-park challenge goes beyond utility platform sharing and into process integration. The extra-park collaboration goes beyond intercompany exchange into public-private sector partnerships.

The key advantage of the low-carbon park is that it provides the location where minimum energy and lowest cost can exist together. The model for that business may take many forms and will be determined by culture and public sector policy and support. Enlightened self-interest may prove sufficiently strong to engineer some of these changes. LOCIMAP has excellent exemplars from within its own partners as to what can be achieved already through industrial and industrial-municipal collaborations; examples are provided by the parks at Tarragona, Kokkola, Wilton and Kalundborg.

But, as outlined in LOCIMAP White Paper 4, the low-carbon future requires developments of new approaches and public engagement that can be effective in delivering business solutions at the park level. The project view is that the establishment of 'Synergy Management Services' organisations is probably the best way to go. These need to be led by the park operator or by the industrial cluster with support from the local public sector with an interest in sustainability themselves.

One such example is the Saltend Chemicals Park, a cluster of world-class chemicals and renewable energy business at the heart of Humberside, established by BP Chemicals Ltd in 2009. Today a number of leading organisations operate on the 370-acre site, sharing an established infrastructure and extensive provision of services, feedstocks and utilities, enabling them to drive down costs, increase efficiency and boost profitability. The site has seen £500 million of investment in recent years and its products range from clothing to paints, pharmaceuticals and packaging.

4 Prospects

If the size of the prize from this level of integration were small, there would be little drive for pursuing the concepts outlined. However, the conclusion of the LOCIMAP work suggests that the prize is very high. In fact, it is so high that such integrated systems could deliver not only the EU 2020 targets but those set by the SPIRE project also. In contrast with the ever-increasing squeeze on existing assets through traditional resource efficiency, the benefit achieved by the coordinated and symbiotic co-location promises major benefits at both thermodynamic and business levels (Fig. 19.3).

Such a strategic shift raises big questions. How can we engineer such systems without compromising safety? What are the implications for system resilience? How can such a closely integrated arrangement manage operational necessities such as maintenance schedules? Such questions would need to be answered in time – for the present, it is sufficient to recognise the scale of the prize on offer through such an approach.

What happens if we do not take this bold approach? The project highlighted the concerns by its partners regarding *carbon leakage*, and this is seen as a hidden barrier. Each manufacturer, and each park, provided information on the legislative and taxation structures within which they operate. They commented on their driving forces in terms of the costs involved in related investments for improvement and in buying carbon credits. Beyond that, they were invited to outline any company or sector initiatives at the 2050 time horizon. In essence the feedback to the project confirmed that carbon trading has the potential to increase economic pressures on the more energy-intensive sectors to leave the European economy. This will disproportionately affect parks.

The project further examined present and future potential by-product linkages between the sectors represented within LOCIMAP. Whilst the loss of an industry

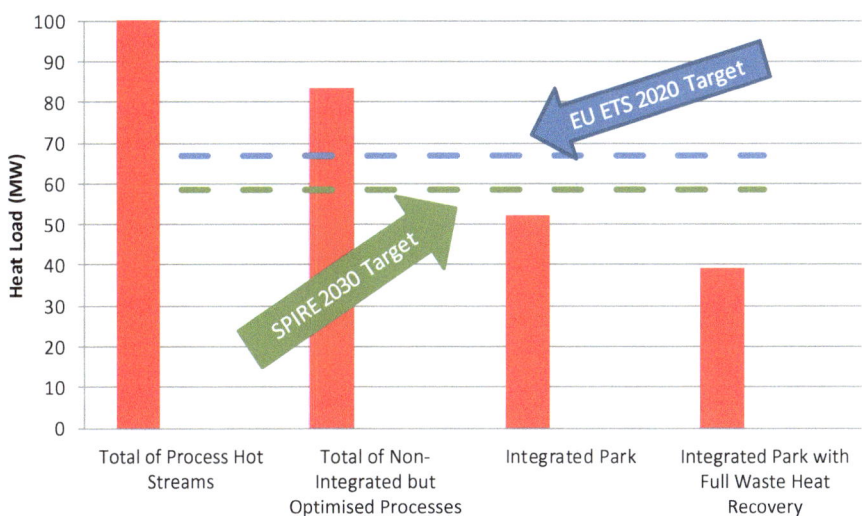

Fig. 19.3 Impact of process integration within a virtual 'LOCIMAP' park

sector from the European industrial mix has negative impact enough on the social and economic perspective, the LOCIMAP studies have further highlighted the loss that an industry makes on the overall potential integration and optimisation which is far wider that when considering the industry in isolation. The loss in potential not only for heat and power integration but for material by-product integration and the closing of material loops has a quantifiable and negative impact on environmental performance of the whole. LOCIMAP makes the case for the evaluation of the impact of CO_2 emissions not only 'up the stack', which the EU ETS addresses, but also enabled by that industry's presence; in essence, this is an argument for a consequential approach to evaluation (see Chaps. 2 and 3). The case is made for the retention and development of those industries which are world leading in terms of CO_2. Diversity of industry types is seen as the cornerstone of successful and resilient industrial symbiosis exemplified by a low-carbon park; the positive impact on resilience is undervalued by the omission of by-product synergy quantification.

5   Conclusion

We conclude that the application of these principles, through the appropriate design of industrial parks and the development of opportunities for process integration, has the potential to exceed the targets set within EU ETS under which the majority of our participating industries operate and also the targets within SPIRE. The question is: what are the barriers to achieving this? The project sees three major challenges to existing business models:

- This low-carbon future requires developments of new approaches and public engagement that can be effective in delivering business solutions at the park level. Some of the beneficial opportunities are already available and could be implemented now within existing industrial clusters. What is needed is a culture change to think and work cross-sectorally. This process is not easy and will need stimulus but is essentially one of knowledge transfer. The mechanism is one of allowing normal business drivers to effect change once opportunities are visualised. The establishment of 'Synergy Management Services' organisations is probably the best way to go. These need to be led by the park operator or by the industrial cluster with support from the local public sector with an interest in sustainability themselves.
- The second barrier lies beyond industrialists, engineers and business owners and relates to local and national government and planning. A new investment is not best served by offering location at a number of greenfield sites, rather by the optimum integration with existing facilities. The answer to a low-carbon economy is a thermodynamic one, and planning needs to allow for this, even to the extent of (physically) linking, for example, the chemical and food industries. Policies need to be created which encourage such developments.
- The third barrier is that some of the opportunities are not economic under the existing landscape. This becomes a societal and a political issue. What is the price of carbon? If we really want a low-carbon economy, how do we introduce measures that deliver the low-carbon benefits *without* disadvantaging European manufacturing within a global playing field? The fundamental issue is that, given a brief to design the most economic system, the outcome will be different from designing the lowest carbon system. The work shows that we have the engineering capability to deliver low-carbon systems; the question is a societal one as to whether we have the will and understanding to change the rules to align economic and environmental performance.

Can we go so far as to say we have the ingredients for a new industrial revolution? Does delivering low-carbon manufacturing within the landscape of 'closing loops' contribute to this? What is the future role of the waste industry? The LOCIMAP project sought to inform opinion on all these points. It goes beyond the bilateral synergies rightly documented as exemplars to a much more strategic approach to energy and resource efficiency that demands that the requirement for a low-carbon industry inform primary planning and design specifications.

So the question is: what do we need to do to deliver a low-carbon manufacturing park? The answer is not 'green', or '42'; it's thermodynamic! The task is to communicate this to policymakers and to translate these principles back to the source of their inspiration: concrete experience from the contributing industrial parks of Tarragona, Wilton, Kokkola and Kalundborg and of course to the Humber.

Acknowledgements The work presented draws extensively but not exclusively on White Paper 5 of project 296010 LOCIMAP. This coordination and support action was funded by the European Commission under the Framework 7 programme; their support is appreciated. The partners in the project are NEPIC, BASF, Link2Energy, ECSPP, INSEAD, Sembcorp, PTS, IVL Svenska, Kokkola Industrial Park, Kalundborg Kommune, Parsons Brinckerhoff, Cemex, Terreal and BFI. This report acknowledges the collaborative effort of all the above organisations.

References

A full list of White Papers from the project is available from www.locimap.eu: White Paper 1: The challenges facing the European Industrial Parks; White Paper 2: Industrial Symbioses; White Paper 3: Smart Future Industrial Parks; White Paper 4: New Operational & Organisational Structures; White Paper 5: 10 Principles for a Low Carbon Future.

Bakshi, B. R., Gutowski, T. G., & Sekulic, D. P. (Eds.). (2011). *Thermodynamics and the destruction of resources*. New York: Cambridge University Press.

CEPI. (2011). *Unfold the future: The Forest Fibre Industry, 2050 roadmap to a low carbon bioeconomy*. Brussels: Confederation of European Paper Industries. Retrieved April 20, 2015, from http://www.unfoldthefuture.eu/uploads/CEPI-2050-Roadmap-to-a-low-carbon-bio-economy.pdf

Cerame-Unie. (2012). *Paving the way to 2050. The ceramic industry roadmap*. Brussels: European Ceramic Industry Association. Retrieved April 20, 2015, from http://www.ceramfed.co.uk/uploads/popular_downloads/04ed1d019530eec2cfe5fd2f4e174a19bbd363ae.pdf

Chertow, M. R. (2007). "Uncovering" industrial symbiosis. *Journal of Industrial Ecology, 11*(1), 11–30.

Defra. (2011). *Commercial and industrial waste survey 2009*. London: Department of the Environment, Food & Rural Affairs.

Defra. (2014). *New methodology to estimate waste generation by the commercial and industrial sector in England*. London: Department of the Environment, Food & Rural Affairs.

EAA. (2012). *An Aluminium roadmap to a low carbon Europe: Lightening the load*. Brussels: European Aluminium Association. Retrieved April 20, 2015, from http://www.alueurope.eu/wp-content/uploads/2012/03/03_An-aluminium-2050-roadmap-to-a-low-carbon-Europe.pdf

EC. (2012). *Europe 2012: Europe's growth strategy*. Brussels: European Commission.

Jensen, P. D., Basson, L., Hellawell, E. E., Bailey, M. R., & Leach, M. (2011). Quantifying 'geographic proximity': Experiences from the United Kingdom's National Industrial Symbiosis Programme. *Resources, Conservation and Recycling*. doi:10.1016/j.resconrec.2011.02.003.

Jensen, P. D., Basson, L., Hellawell, E. E., & Leach, M. (2012). 'Habitat' suitability index mapping for industrial symbiosis planning. *Journal of Industrial Ecology, 16*(1), 38–50.

Mirata, M. (2004). Experiences from early stages of a national industrial symbiosis programme in the UK: Determinants and coordination challenges. *Journal of Cleaner Production, 12*(8–10), 967–983.

OECD, IEA, & World Business Council for Sustainable Development. (2009). *Cement technology roadmap 2009: Carbon emissions reductions up to 2050*. Retrieved April 20, 2015, from http://www.wbcsdcement.org/pdf/technology/WBCSD-IEA_Cement%20Roadmap.pdf

Wörtler, M., Schuler, F., Voigt, N., Schmidt, T., Dahlmann, P., Lüngen, H. B., & Ghenda, J.-T. (2013). *Steel's contribution to a low-carbon Europe 2050: Technical and economic analysis of the sector's CO2 abatement potential*. London: BCG. Retrieved April 20, 2015, from http://www.bcg.de/documents/file154633.pdf

Index

© The Author(s) 2016
R. Clift, A. Druckman (eds.), *Taking Stock of Industrial Ecology*,
DOI 10.1007/978-3-319-20571-7